AN INTRODUCTION TO FREE RADICALS

AN INTRODUCTION TO FREE RADICALS

JOHN E. LEFFLER
Florida State University

A Wiley-Interscience Publication
JOHN WILEY & SONS, INC.
New York · Chichester · Brisbane · Toronto · Singapore

This text is printed on acid-free paper.

Library of Congress Cataloging in Publication Data:

Leffler, John E.
An introduction to free radicals / by John E. Leffler.
p. cm.
"A Wiley-Interscience publication."
Includes index.
ISBN 0-471-59406-7 (alk. paper)
1. Free radicals (Chemistry) I. Title.
QD471.L5 1993
541.2′24--dc20 92-35288

Printed in the United States of America

10 9 8 7 6 5 4 3 2 1

To Nell

PREFACE

Upon this gifted age, in its dark hour,
Rains from the sky a meteoric shower
Of facts . . . they lie unquestioned,
uncombined.
Wisdom enough to leech us of our ill
Is daily spun; but there exists no loom
. . . to weave it into fabric.

Edna St. Vincent Millay

Some years ago I wrote a book called *The Reactive Intermediates of Organic Chemistry*. Merely one small volume, it covered carbocations and carbanions as well as free radicals. Readers and reviewers liked the book and seemed to think that the subject was dealt with quite adequately.

Today, treatments of just one type of intermediate tend to consist of several or many volumes representing the work of many authors, each responsible for one or two chapters.

My object in this book is to extract from the vast literature on free radicals a relatively modest selection of the most fundamental and generally useful material. Hence it is an *introduction* to free radicals, a foundation on which readers may build to meet their more specialized needs.

J.E.L.

Tallahassee, Florida
January 1993

CONTENTS

AN INTRODUCTION TO FREE RADICALS

1

BASIC CONCEPTS: ELECTRONIC AND NUCLEAR SPIN ANGULAR MOMENTUM

In any modern approach to the chemistry of free radicals certain ideas are so extremely useful that it is desirable to introduce them at the outset. Primary among such ideas is that of electronic spin angular momentum [1–4], which is important for understanding such things as covalent bonding, the nature of singlet and triplet states, the distinction between triplets and diradicals, electron spin resonance spectroscopy, and chemically induced dynamic nuclear spin polarization (CIDNP).

1.1 THE QUANTIZATION OF SPIN ANGULAR MOMENTUM

The classical nature of the easily perceived parts of our environment makes the use of classical models in visualizing quantum mechanical objects inevitable. Such models can be of real value, but only if certain reservations, or mental footnotes, are firmly attached to the classical picture and carried along with it [2]. In the case of "spinning" electrons or nuclei and their associated angular momenta and magnetic moments, the indispensable footnotes are the rules for the quantization of angular momentum and those for determining which quantities are quantum mechanical *observables*. The latter are often properties of the system as a whole rather than of a single particle. We will begin our discussion simply, with a system that consists of only one particle.

1.2 RULES FOR A SINGLE PARTICLE

Whether a particle is an electron or a nucleus, the uncertainty principle permits us to know only certain properties of its spin angular momentum all at one time. That is to say, a knowledge of one set of properties precludes any knowledge of certain other properties. One set of properties that can be known all at one time consists of the square of the spin angular momentum vector (i.e., its length or magnitude) together with the length and direction of *just one* component of the angular momentum vector. This component is parallel to the magnetic field used in the measurement. By convention it is called the z component.

1.2.1 Eigenvalues

All of the quantum mechanical observables are eigenvalues associated with a corresponding eigenoperator and an eigenfunction. A familiar example of an eigenvalue is the energy in the Schrödinger equation. The reader may recall that for each wave function $\mathbf{\Psi}_i$ (the eigenfunction) there is a value of the energy E_i (the eigenenergy). The result of operating on $\mathbf{\Psi}_i$ with the energy eigenoperator **H** (the Hamiltonian operator) is equal to $\mathbf{\Psi}_i$ multiplied by E_i.

$$\mathbf{H}(\mathbf{\Psi}_i) = E_i(\mathbf{\Psi}_i) \tag{1.1}$$

Like the energy in Eq. 1.1, each observable property (eigenvalue) having to do with the angular momentum has an operator (eigenoperator) that goes with it. Operating on the spin wave function has the same effect as multiplying the spin wave function by an eigenvalue of the angular momentum property in question. Equation 1.2, for example, says that operating on the electron spin wave function $<\alpha|$ with the $\mathbf{S}^2$, which corresponds to the square of the spin angular momentum, has the effect of multiplying it by the square of the magnitude of the electron spin angular momentum, $(\frac{3}{4})(h/2\pi)^2$.

$$\mathbf{S}^2 <\alpha| = \left(\frac{3}{4}\right)\left(\frac{h}{2\pi}\right)^2 <\alpha| \tag{1.2}$$

1.2.2 Spin Eigenfunctions

A single electron or proton is allowed to exist in just two spin states. These are represented by the Greek letters α and β, called spin wave "functions." Unlike the more familiar symbols for spatial wave functions such as 1s or 2p, the spin wave functions are not shorthand for some more informative mathematical expression. The spatial wave function 1s, for example, can be replaced with an expression that provides a map of the square root of the electron density. In contrast to this, the functions α and β do not represent any more informative description of the spin state. Whatever physical reality underlies the concept of electron spin, it does not include any fiduciary mark on the side of the electron by means of which rotation might be observed directly rather than inferred from the existence of an angular momentum. In short, the "spinning" of an electron is not observable. The only observable quantities are those associated with the spin angular momentum.

1.2.3 Spin Quantum Numbers

The quantum number j, which will be encountered in the eigenequation for the square of the magnitude, or length, of the spin angular momentum vector, has the value 1 for the deuteron and the ^{14}N nucleus, $\frac{1}{2}$ for the electron, the proton, and the ^{13}C nucleus, and zero for the ^{12}C nucleus. The units are $\hbar$, Planck's constant h divided by 2π.

For nuclei, j is zero if both the mass and atomic numbers are even, half-integral if the mass number is odd, and integral if the mass number is even and the atomic number odd.

A related quantum number m_j will be encountered in the eigenequation for the *z component* of the spin angular momentum. The allowed values of m_j range from j to $-j$ in steps of one unit. Thus for the electron, for which the "spin" quantum number j is $\frac{1}{2}$, m_j can be $\frac{1}{2}$ and $-\frac{1}{2}$. For the deuteron (spin $j = 1$), $m_j = 1, 0$, or -1. For ^{12}C, m_j can only be zero.

A general way of indicating the spin state is to attach the quantum numbers j and m_j to the wave function as subscripts, as in $\psi_{j,mj}$, for example, However, the spin states of the spin $\frac{1}{2}$ particles such as the electron and the proton are referred to so frequently that it has proved convenient to use the Greek letters α and β instead of their equivalents, $\psi_{1/2,1/2}$ and $\psi_{1/2,-1/2}$.

1.2.4 The Angular Momentum Magnitude Equation

The eigenequation for the magnitude of the angular momentum is Eq. 1.3. The eigenoperator is $\mathbf{J}^2$. The eigenvalue, the length of the angular momentum vector squared, is $j(j + 1)(h/2\pi)^2$. The spin eigenfunction is $\psi_{j,mj}$. The equation is a general one, and can be used for the magnitude of the spin angular momentum vector of any particle in any spin state.

$$\mathbf{J}^2(\psi_{j,mj}) = j(j + 1)\left(\frac{h}{2\pi}\right)^2 \psi_{j,mj} \tag{1.3}$$

Because it is often necessary to treat nuclei and electrons at the same time, and we need to keep from getting them confused, a further convention is to use **I** for the eigenoperator when referring to nuclei and **S** when referring to electrons, rather than the **J** of the general equation (Eq. 1.3). For example, Eq. 1.3a and 1.3b, which deal with electrons, use $\mathbf{S}^2$ for the angular momentum magnitude operator:

$$\mathbf{S}^2 <\alpha| = \left(\frac{3}{4}\right)\left(\frac{h}{2\pi}\right)^2 <\alpha| \tag{1.3a}$$

$$\mathbf{S}^2 <\beta| = \left(\frac{3}{4}\right)\left(\frac{h}{2\pi}\right)^2 <\beta| \tag{1.3b}$$

Since the particle in Eqs. 1.3a and 1.3b is an electron, $j(j + 1) = (\frac{1}{2})(\frac{3}{2}) = \frac{3}{4}$, and the length of the angular momentum vector is $(\frac{3}{4})^{1/2}(h/2\pi)$.

The half brackets $<|$ enclosing the spin functions α and β in Eqs. 1.3a and 1.3b mean that these spin functions are to be treated as vectors. This notation comes from Dirac's dual vector space approach to quantum mechanics. The symbols $<|$ and $|>$ are called *bras* and *kets*.

The magnitude of the spin angular momentum is shown pictorially as the length of the vectors lying on the conical surfaces in Fig. 1.1.

1.2.5 The *z* Component of Angular Momentum

Equation 1.4 is the general eigenequation for the projection of the spin angular momentum on the *z* axis. The length of that component is the eigenvalue $m_j(h/2\pi)$ shown in the equation.

$$\mathbf{J}_z(\psi_{j,mj}) = m_j \left(\frac{h}{2\pi}\right) (\psi_{j,mj}) \tag{1.4}$$

Since the quantum number m_j is limited to values ranging from j to $-j$ in steps of one unit, there are only two m_j numbers for the electron, $+\frac{1}{2}$ and $-\frac{1}{2}$. These correspond to the spin states α and β. The eigenequations are Eqs. 1.5a and 1.5b.

$$\mathbf{S}_z <\alpha| = +\left(\frac{1}{2}\right)\left(\frac{h}{2\pi}\right) <\alpha| \tag{1.5a}$$

$$\mathbf{S}_z <\beta| = -\left(\frac{1}{2}\right)\left(\frac{h}{2\pi}\right) <\beta| \tag{1.5b}$$

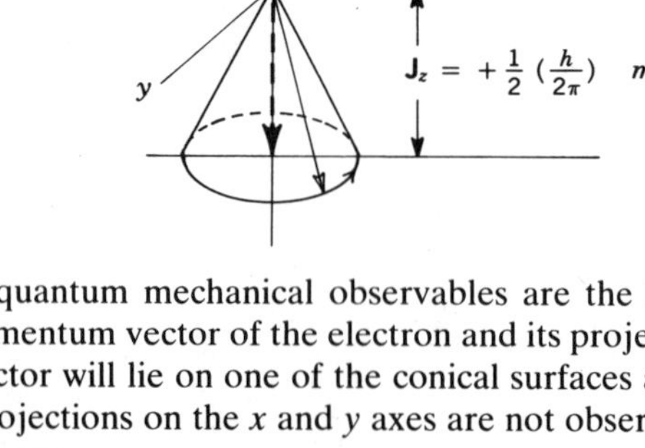

FIG. 1.1 The quantum mechanical observables are the length (magnitude) of the spin angular momentum vector of the electron and its projection on the magnetic field (*z*) axis. The vector will lie on one of the conical surfaces and precess about the field axis. Since its projections on the *x* and *y* axes are not observable, neither is its precise position on the cone.

The projections are shown in Fig. 1.1 labeled with their magnitudes $+(\frac{1}{2})(h/2\pi)$ and $-(\frac{1}{2})(h/2\pi)$.

For particles of spin $j = 1$, such as the deuteron or the ^{14}N nucleus, the rule that m_j can be anything from j to $-j$ in steps of one unit means that there are three possible spin states and three eigenequations (Eqs. 1.6a–1.6c).

$$\mathbf{I}_z(\psi_{1,1}) = +\left(\frac{h}{2\pi}\right)(\psi_{1,1}) \quad (1.6a)$$

$$\mathbf{I}_z(\psi_{1,0}) = 0 \quad (1.6b)$$

$$\mathbf{I}_z(\psi_{1,-1}) = -\left(\frac{h}{2\pi}\right)(\psi_{1,-1}) \quad (1.6c)$$

The eigenvalues of Eqs. 1.6a–1.6c are illustrated in Fig. 1.2. Vectors precessing about the z axis and lying on the surfaces of the two cones correspond to spin states with m_j equal to 1 and -1. Vectors precessing about the z axis, and lying on the surface of the disc correspond to m_j equal to zero, have a zero z component.

Usage of the Word "Spin." This word is used in two different ways depending on the context:

In one context it refers to the value of j: the deuteron is a particle of spin (j) 1, the electron is a particle of spin (j) $\frac{1}{2}$.

In other contexts it refers to the m_j value: an electron can have a spin (m_j) $\frac{1}{2}$ or $-\frac{1}{2}$, or spin α or spin β.

Still more loosely, an electron can be said to have spin up or spin down, although only the z component is oriented directly parallel (up) or antiparallel to the z axis.

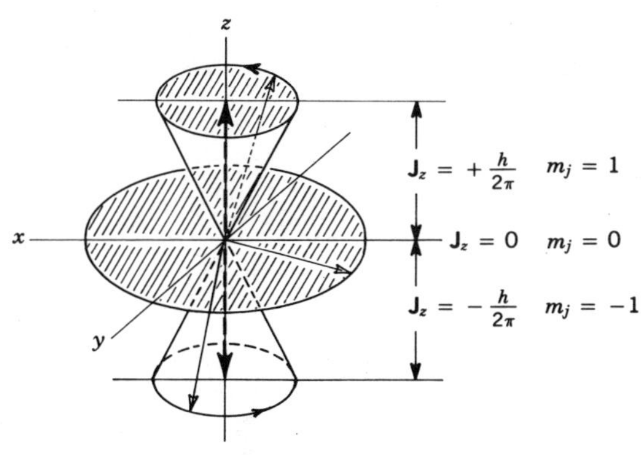

FIG. 1.2 The quantization of angular momentum for a nucleus of spin (j) equal to 1.

1.3 RULES FOR A SET OF PARTICLES

When a system has several electrons or several nuclei close enough together to interact magnetically, it is no longer possible to know both the energy of the molecule and at the same time the individual spin angular momentum of each particle. Another way of saying this is that the individual spin operators $\mathbf{S}^2$ and $\mathbf{S}_z$ no longer commute with the Hamiltonian energy operator of the molecule.

It is possible, nevertheless, to invent a "user friendly" set of *total* spin angular momentum operators which *do* commute with the Hamiltonian. The spin angular momentum properties that go with these operators can be known at the same time as the energy of the molecule.

To simplify the discussion we will begin with a set consisting of only two particles, but the generalization to larger ensembles, and the treatment of nuclei as well as electrons, is straightforward.

It will be found that the equations for sets of interacting particles look very much like those for a single particle. The definitions of the operators and quantum numbers are simple modifications of the previous ones.

1.3.1 A Two-Electron Spin System

The new total spin operators are defined in terms of single electron operators $\mathbf{S}_1$ (for electron 1) and $\mathbf{S}_2$ (for electron 2), each of which operates only on its own electron. The total spin operator $\mathbf{S}$ is given by Eq. 1.7.

$$\mathbf{S} = \mathbf{S}_1 + \mathbf{S}_2 \tag{1.7}$$

Since these operators act like vectors, Eq. 1.7 represents vector addition. The magnitude of the resulting vector $\mathbf{S}$ is given by Eq. 1.8a.

$$\mathbf{S}^2 = \mathbf{S}_1^2 + \mathbf{S}_2^2 + 2\mathbf{S}_1\mathbf{S}_2 \tag{1.8a}$$

Replacing the individual vectors in Eq. 1.8a with their x, y, and z components gives Eq. 1.8b.

$$\mathbf{S}^2 = \mathbf{S}_1^2 + \mathbf{S}_2^2 + 2\mathbf{S}_{1x}\mathbf{S}_{2x} + 2\mathbf{S}_{1y}\mathbf{S}_{2y} + 2\mathbf{S}_{1z}\mathbf{S}_{2z} \tag{1.8b}$$

1.3.2 The $\mathbf{S}^2$ Eigenequation

The equation for the magnitude of the total spin angular momentum vector is identical in form to Eq. 1.2 for a single particle.

$$\mathbf{S}^2(\psi_{s,ms}) = s(s + 1)\left(\frac{h}{2\pi}\right)^2 (\psi_{s,ms}) \tag{1.9}$$

The quantum number s in Eq. 1.9 is allowed to take on the values Σj_i, $\Sigma j_i - 1, \ldots 0$. The j_i are the spin quantum numbers of the single particles i and the sum is taken over all the particles.

In our example, where the particles are electrons, and there are just two of them, $j_1 = j_2 = \frac{1}{2}$, and so Σj_i is 1. The allowed values for s are therefore 1 and 0.

The two eigenvalues are then $2(h/2\pi)^2$ and 0.

$$\mathbf{S}^2(\psi_{1,ms}) = 2\left(\frac{h}{2\pi}\right)^2(\psi_{1,ms}) \tag{1.10a}$$

$$\mathbf{S}^2(\psi_{0,0}) = 0 \tag{1.10b}$$

1.3.3 The $\mathbf{S_z}$ Eigenequation

The operator $\mathbf{S_z}$ is simply $S_{1z} + S_{2z}$. As in the analogous equation for a single particle, the quantum number m_s is allowed to take on the values s, $s - 1, \ldots 0 \ldots -s$. The general form of the eigenequation for the z component of the total electron spin angular momentum of the two-electron system is Eq. 1.11. Putting in the allowed values of s and m_s gives four spin states in all. These are described by Eqs. 1.12a–1.12d.

$$\mathbf{S_z}(\psi_{s,ms}) = m_s\left(\frac{h}{2\pi}\right)(\psi_{s,ms}) \tag{1.11}$$

$$\mathbf{S_z}(\psi_{1,1}) = \left(\frac{h}{2\pi}\right)(\psi_{1,1}) \tag{1.12a}$$

$$\mathbf{S_z}(\psi_{1,0}) = 0 \tag{1.12b}$$

$$\mathbf{S_z}(\psi_{1,-1}) = -\left(\frac{h}{2\pi}\right)(\psi_{1,-1}) \tag{1.12c}$$

$$\mathbf{S_z}(\psi_{0,0}) = 0 \tag{1.12d}$$

1.3.4 Identification of the Spin Functions

At this point the reader may be wondering what became of the functions α and β that were supposed to replace the more cumbersome subscripted $\boldsymbol{\psi}$ notation. In fact, each of the spin functions $\psi_{s,ms}$ can be replaced by some linear combination of the one-electron functions α and β, provided that certain rules are followed:

First, the expressions must not imply that it is possible to distinguish between the two electrons. That is, an expression like $\alpha(1)\beta(2)$, which suggests that electron 1 definitely has spin α and electron 2 definitely has spin β, is not allowed. More precisely, the allowed expression must be either *symmetric* or *antisymmetric* with respect to interchanging the positions of the two electrons in the expression. A symmetric function is one that is not changed at all by the interchange of electrons, and an antisymmetric function is one that merely changes sign.

The function $\alpha(1)\alpha(2)$ remains the same when the indices 1 and 2 are interchanged, so this function is symmetric and passes the symmetry test. The function $\alpha(1)\beta(2)$ changes to $\alpha(2)\beta(1)$ and is therefore not allowed. The function $\alpha(1)\beta(2) - \alpha(2)\beta(1)$ merely changes sign and is therefore antisymmetric and allowed, as far as the first test is concerned.

The second requirement is that the expression must obey one of the four eigenequations Eqs. (1.12a–1.12d).

To test the symmetric function $\alpha(1)\alpha(2)$, for example, we note that

$$\mathbf{S}_z[\alpha(1)\alpha(2)] = (\mathbf{S}_{z1} + \mathbf{S}_{z2})[\alpha(1)\alpha(2)] \tag{1.13a}$$

$$= \mathbf{S}_{z1}[\alpha(1)\alpha(2)] + \mathbf{S}_{z2}[\alpha(1)\alpha(2)] \tag{1.13b}$$

$$= \alpha(2)\mathbf{S}_{z1}[\alpha(1)] + \alpha(1)\mathbf{S}_{z2}[\alpha(2)] \tag{1.13c}$$

$$= \left(\frac{h}{2\pi}\right)[\alpha(1)\alpha(2)] \tag{1.13d}$$

Hence $\mathbf{S}_z$ operating on the function $[\alpha(1)\alpha(2)]$ has the effect of multiplying it by an eigenvalue $(h/2\pi)$. The function $\alpha(1)\alpha(2)$ must therefore correspond to $\psi_{1,1}$ (Eq. 1.12a).

A similar test of the function $\alpha(1) + \alpha(2)$ shows that this function, although symmetric, must be rejected because it does not obey any of the eigenequations (Eqs. 1.12a–1.12d).

Four functions that do fit both criteria are shown in Table 1.1.

Comments on the Four Spin Functions (Table 1.1). Neither of the individual terms $\alpha(1)\beta(2)$ and $\alpha(2)\beta(1)$ in the second and fourth functions of the table has any physical meaning taken by itself, but merely serves to make the spin function symmetric or antisymmetric by balancing the other term.

The center triplet function $\psi_{1,0}$ and the singlet function $\psi_{0,0}$ resemble each other in that both have a zero component in the z direction, but they also *differ* in an important way. The singlet function can not have anything but a zero component on *any* axis because its magnitude is zero. The center triplet

TABLE 1.1 Four Functions $\psi_{s,ms}$

$\psi_{s,ms}$	Function of α and β	
$\psi_{1,1}$	$\alpha(1)\alpha(2)$	(Triplet sublevel)
$\psi_{1,0}$	$(\frac{1}{2})^{1/2}[\alpha(1)\beta(2) + \beta(1)\alpha(2)]$	(Triplet sublevel)
$\psi_{1,-1}$	$\beta(1)\beta(2)$	(Triplet sublevel)
$\psi_{0,0}$	$(\frac{1}{2})^{1/2}[\alpha(1)\beta(2) - \beta(1)\alpha(2)]$	(Singlet)

function has a nonzero magnitude but has a zero z component, because it is orthogonal to that axis.

Since the concept of individual electron spin angular momenta, as opposed to total spin angular momentum of the molecule, does violence to physical reality, it is important not to infer too much from descriptions of the spin state that seem to be couched in terms of individual electron spins. For example, a system in the single state $\Psi_{0,0}$ is often described as having antiparallel electron spins, indicated by the symbol $\uparrow\downarrow$. Similarly, the triplet state is often described as having parallel electron spins ($\uparrow\uparrow$). The three subspecies of the triplet functions are sometimes represented by two parallel arrows pointing up (1, 1), down (1, −1), or sideways (1, 0). Of course, only the z component of the total spin is oriented up or down relative to the z axis.

1.3.5 Doublets, Triplets, and Double Doublets

A free radical in its electronically unexcited state is called a doublet because it is really a mixture of radicals in two states, $\psi_{1/2,-1/2}$ and $\psi_{1,2,1/2}$, of nearly the same energy. A double doublet system has two electrons far enough apart so that the overlap of their spatial wave functions is negligible. Such a system might be a molecule with two electrons on atoms that are far apart and not connected by any conjugated structure, or it might consist of a pair of free radicals too far apart to interact. The term *double doublet* refers to the fact that each of these independent single-electron systems has two allowed spin states. Each of these electrons could be in its $m_s + \frac{1}{2}$ or its $m_s - \frac{1}{2}$ spin state. In particular, it is possible for both of them to be in the $+\frac{1}{2}$ state. Although both this system and a triplet state spin 1 subspecies might conceivably be represented by the symbol $\uparrow\uparrow$, it would be misleading to do so.

In the double doublet, both electrons have a z component of angular momentum equal to $+(\frac{1}{2})(h/2\pi)$, but the two angular momentum vectors are precessing about the z axis in a completely uncorrelated manner: their phase relationship is random. In the triplet subspecies, the angular momentum vectors of the two electrons precess about the z axis *in phase*. As a consequence their total spin angular momentum can be represented by a single vector precessing about the z axis with a constant z component $h/2\pi$, 0, or $-h/2\pi$.

When a pair of radicals, initially close enough together for their electron spins to interact, diffuse apart, they undergo a transition from their initial singlet or triplet subspecies state to the double doublet state. The spins are said to become dephased in this process.

1.4 THE TOTAL WAVE FUNCTION

A total wave function, or spinorbital, consists of a spatial wave function multiplied by a spin wave function. The total wave functions for sets of particles with half-integral spins, such as the electron or the proton, must be antisymmetric with respect to interchange of the particles. For example, the symmetric ground state wave function $\psi_a(1)\psi_a(2)$ must be multiplied by the antisymmetric singlet spin wave function, as in Eq. 1.14, to make the total wave function antisymmetric.

$$\Psi = \psi_a(1)\psi_a(2) \times (\tfrac{1}{2})^{1/2}[\alpha(1)\beta(2) - \beta(1)\alpha(2)] \tag{1.14}$$

An excited singlet in which one electron is in a lower energy orbital ψ_a and the other in a higher energy orbital ψ_b would be represented by Eq. 1.15.

$${}^1\Psi^* = (\tfrac{1}{2})^{1/2}[\psi_a(1)\psi_b(2) + \psi_a(2)\psi_b(1)] \times (\tfrac{1}{2})^{1/2}[\alpha(1)\beta(2) - \beta(1)\alpha(2)] \tag{1.15}$$

A triplet involving ψ_a and ψ_b has an antisymmetric spatial function, and has to be multiplied by any one of the three symmetric triplet spin functions.

$${}^3\Psi^{*1,0} = (\tfrac{1}{2})^{1/2}[\psi_a(1)\psi_b(2) - \psi_a(2)\psi_b(1)] \times (\tfrac{1}{2})^{1/2}[\alpha(1)\beta(2) + \beta(1)\alpha(2)] \tag{1.16}$$

$${}^3\Psi^{*1,-1} = (\tfrac{1}{2})^{1/2}[\psi_a(1)\psi_b(2) - \psi_a(2)\psi_b(1)] \times \beta(1)\beta(2) \tag{1.17}$$

$${}^3\Psi^{*1,1} = (\tfrac{1}{2})^{1/2}[\psi_a(1)\psi_b(2) - \psi_a(2)\psi_b(1)] \times \alpha(1)\alpha(2) \tag{1.18}$$

REFERENCES

1. Carrington, A.; McLachlan, A. D. *Introduction to Magnetic Resonance,* Harper and Row: New York, 1967.
2. La Paglia, S. R. *Introductory Quantum Chemistry,* Harper and Row: New York, 1971.
3. Pauli, W. *Phys. Rev.* **1940,** *58,* 716.
4. Uhlenbeck, G. E.; Goudsmit, S. A. *Nature* **1926,** *117,* 264.

2

ELECTRON SPIN RESONANCE

In the absence of a magnetic field the two electron spin isomers that comprise the free radical doublet state, $\psi_{1/2,-1/2}$ and $\psi_{1/2,1/2}$, are degenerate and have the same energy. In a magnetic field, however, the energy of the spin β species ($\psi_{1/2,-1/2}$) decreases and that of the spin α species ($\psi_{1/2,1/2}$) increases. The difference in energy, known as the electronic Zeeman or fine splitting, increases in proportion to the magnetic field strength as shown in Fig. 2.1.

2.1 THE ESR TRANSITION

For free radicals in magnetic fields of a few thousand gauss the electronic Zeeman splitting corresponds to the energy of a photon in the microwave range of frequencies. The transitions responsible for the electron spin resonance (ESR) absorption correspond to changes in the spin wave function of the unpaired electron from the $m_j - \frac{1}{2}$ or β electron spin state to the higher energy $m_j + \frac{1}{2}$, or α state. A free radical in the 3400-G magnetic field of a typical ESR spectrometer will absorb at about 9400 MHz.

The energy levels corresponding to the electronic spin β and spin α states are further split because of interactions of the electron spin with the spins of any magnetic nuclei that may be present. This splitting, which is much smaller than the Zeeman splitting, is called the hyperfine splitting. Because of the hyperfine splitting, a typical ESR spectrum consists of several lines, each line corresponding to one of the possible nuclear spin states. To take the simplest possible example, the ESR spectrum of the hydrogen atom has two lines, one contributed by the hydrogen atoms whose nuclei happen to have spin α and

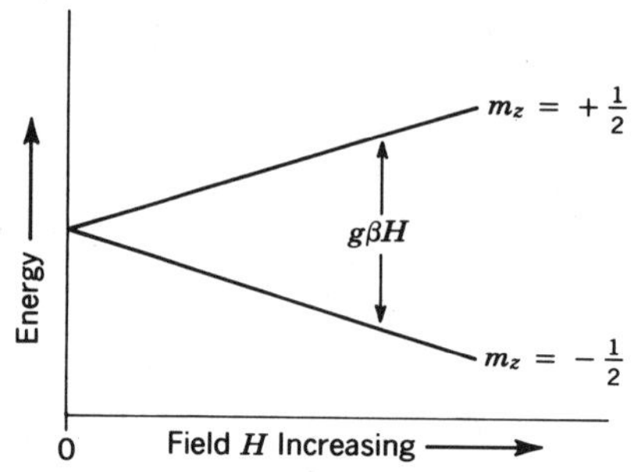

FIG. 2.1 The effect of a magnetic field on the energies of the spin α ($\Psi_{1/2,1/2}$) and spin β ($\Psi_{1/2,-1/2}$) species.

the other to the atoms whose nuclei happen to have spin β. Because these two populations are nearly equal, the two ESR lines are of equal intensity.

For practical reasons, most instruments use a fixed microwave frequency and detect the ESR lines by scanning across a small range of magnetic fields until a field strength is found that causes absorption of the fixed microwave frequency selected. In terms inherited from the early military use of microwaves for radar, an X-band instrument is one that uses wavelengths from 2.85 to 6.09 cm, and a K-band instrument one that uses wavelengths from 1.33 to 2.66 cm. As a result of the decision to scan the magnetic field strength rather than the microwave frequency, it is customary to express differences in the position of lines in an ESR spectrum in gauss rather than in frequency units. For example, the hyperfine splitting of the H· spectrum is about 500 G. This is an unusually large splitting and results from the fact that the electron has an unusually large density at the nucleus because it is in the hydrogen 1*s* orbital. This widely separated pair of lines is often seen in photochemical experiments, because irradiation of a glass or quartz container produces hydrogen atoms.

The spectrum of di-*t*-butylnitroxyl in Fig. 2.2 is another example. This radical is a mixture of nuclear spin isomers in which the spin 1 ^{14}N nucleus is in the -1, 0, or $+1$ nuclear spin state. These subspecies are present in very nearly equal concentrations and give rise to three equally intense lines in the ESR spectrum. The hyperfine splitting is 15.45 G. Actually, each of

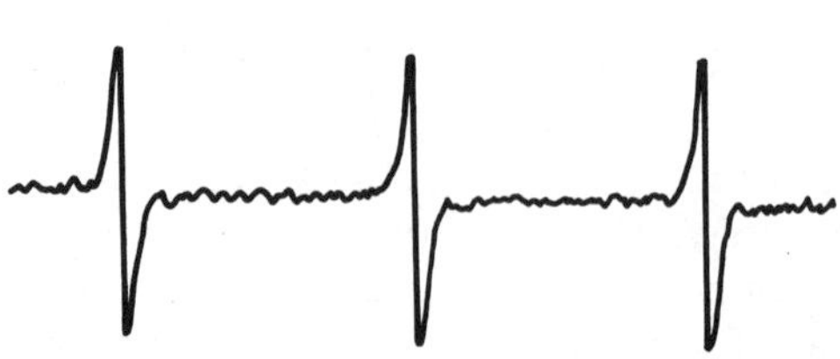

FIG. 2.2 First derivative spectrum of the di-*t*-butylnitroxyl radical. The distance between the lines is 15.45 G [30].

the three ^{14}N nuclear spin isomers is itself a mixture of isomers corresponding to different values of the proton spins, but in this radical the interaction of the unpaired electron with the protons is so small that the proton splitting lines are not resolved in the spectrum. They merely contribute to the widths of the three ^{14}N lines.

The ordinate of Fig. 2.2 is the first derivative of the intensity of the microwave absorption rather than the intensity itself. Thus the maxima and minima correspond to the points of maximum positive and negative *slope* in an ordinary absorption spectrum, whereas the crossings of the abscissa correspond to absorption maxima and minima. The reason most ESR spectra are presented in this form is that they are unresolved, so that a plot of the absorption rather than its first derivative is hard to read, looking like a very broad line with shoulders.

2.2 THE ZEEMAN SPLITTING

The energy difference in Fig. 2.1, called the Zeeman or *fine splitting,* is proportional to the magnetic field strength H, as shown in Eq. 2.1a.

$$\Delta E = \mathbf{g_e}\beta_e H \tag{2.1a}$$

$$h\nu = \Delta E \tag{2.1b}$$

The other symbols in Eq. 2.1a are the Bohr magniton β_e and $\mathbf{g_e}$, which is simply called the g tensor. The value of the Bohr magniton is -9.2732×10^{21} erg G^{-1}.

The value of $\mathbf{g_e}$ depends slightly on the environment of the electron in the radical and hence on the structure of the radical. For example, radicals in which the electron is mainly on carbon have isotropic g_e values near 2.0025; those in which the electron is mainly on oxygen have values near 2.01. To this extent it is analogous to the chemical shift parameter in NMR, but the shifts are small and less useful.

2.3 THE HYPERFINE SPLITTING

Because of interactions between the electron spin angular momentum and the angular momenta of magnetic nuclei in the radical, the Zeeman energy levels are further split, as we have seen in the examples of H· and di-*t*-butylnitroxyl.

Figure 2.3 shows the ESR spectrum of the methyl radical. The four lines, resulting from the interaction of the electron with the three hydrogen nuclei, are spaced 22.9 G apart.

The interaction between the electron and nuclear spins in a radical takes place by two mechanisms. One of these is the interaction of the magnetic

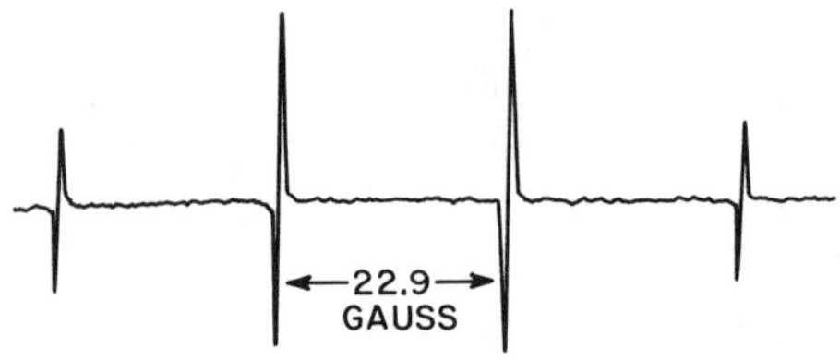

FIG. 2.3 The ESR spectrum of $CH_3\cdot$ at $-138°$. Reprinted with permission from P. J. Krusic and J. K. Kochi, *J. Am. Chem. Soc.*, **1969,** *91,* 3942. Copyright 1969 American Chemical Society.

dipoles through space and is called the dipole–dipole interaction. The other depends on the probability of finding the electron at the nucleus in question, and is called the Fermi contact interaction. The contact interaction requires the electron to be in a spatial orbital that has a nonzero density at the nucleus.

The dipole–dipole interaction complicates the ESR spectrum of radicals that are immobilized in crystals, but averages to zero for radicals in solution because of the tumbling motion. Under these conditions the splitting of the ESR spectrum is due only to the Fermi contact or isotropic hyperfine interaction. The spectrum of a radical immobilized in a crystal depends on the orientation of the crystal with respect to the magnetic field direction.

2.3.1 The Fermi Contact Interaction

Energy Levels for a Single Nucleus. For a single magnetic nucleus of spin quantum number I, the z components of the magnetic moment are allowed the values $m_i = \mathrm{I}, \mathrm{I} - 1, \mathrm{I} - 2, \ldots - \mathrm{I}$, in the usual units of $h/2\pi$. These $2\mathrm{I} + 1$ different nuclear spin subspecies will cause the ESR spectrum to have $2\mathrm{I} + 1$ lines. A single proton or ^{13}C, for example, will split the free electron signal into two lines. The details of this splitting are best appreciated by considering the energy level diagram in Fig. 2.4.

Starting at the left in Fig. 2.4, the energy level E_0 represents the energy of the radical at zero magnetic field. When the field is turned on, the electronic Zeeman splitting adds to or subtracts from this energy according to the spin state of the electron: up by an amount $\mathbf{g_e}\beta_e H/2$ for the radicals in the $+\frac{1}{2}$ spin state and down by a like amount for the radicals in the $-\frac{1}{2}$ electron spin state. The next column of energy levels in the figure takes the nuclear Zeeman splitting into account, subtracting $\mathbf{g_h}\beta_h$ for the radicals whose proton happens to be in the $+\frac{1}{2}$ spin state and adding a like amount for those whose proton is in the nuclear spin $-\frac{1}{2}$ state. The difference in sign of the electronic and nuclear Zeeman splittings is due to the difference in sign of the charges of the electron and the proton.

In the last column of the figure it is assumed that the spatial wave function occupied by the electron has a nonzero density at the nucleus, and a contribution from the Fermi contact interaction is added. This hyperfine interaction

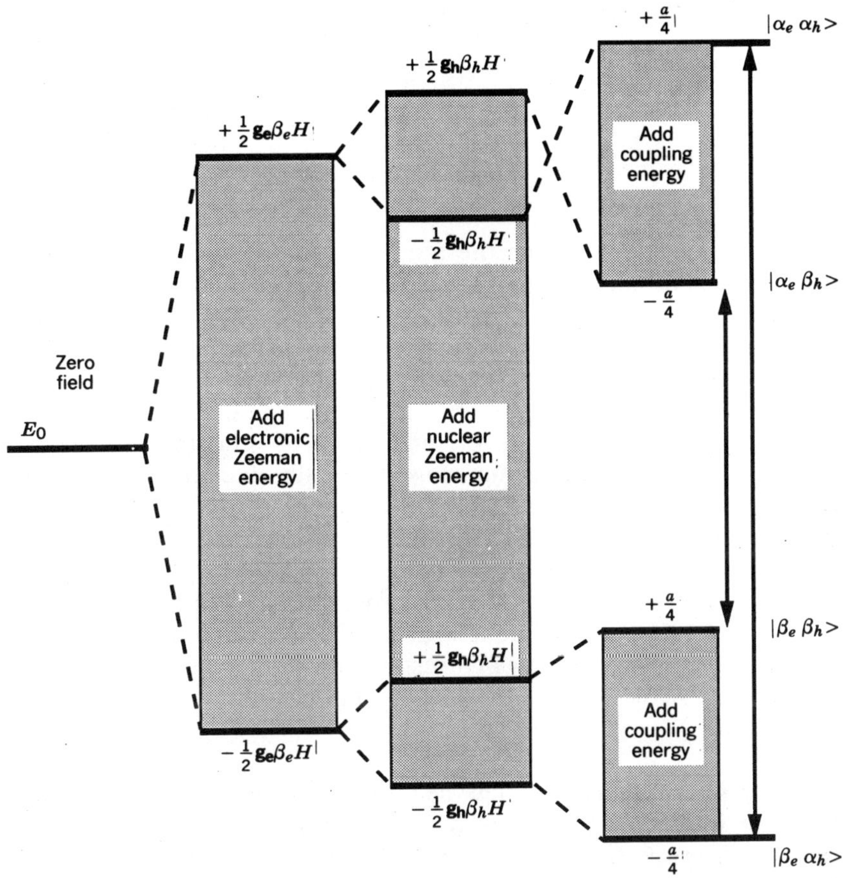

FIG. 2.4 Interaction of an electron spin with one proton spin. Since β_e is 1838 times the nuclear magneton β_h, the nuclear Zeeman energy has been greatly exaggerated for the sake of clarity. Adapted from reference 1.

energy is equal to $m_s m_i a$, in which a is the hyperfine coupling constant. The coupling constant a can be either positive or negative, and can range in magnitude from a small fraction of a gauss to the 507 gauss found for atomic hydrogen. In Fig. 2.4, a has been taken as positive.

Since β_e is 1838 times as large as β_h, electronic Zeeman splittings tend to run about a thousand times as large as the nuclear Zeeman splittings. In Fig. 2.4 the latter have been exaggerated so that they will remain visible.

The total energies of the four spin isomers $|\alpha_e\alpha_h>$, $|\alpha_e\beta_h>$, $|\beta_e\alpha_h>$, and $|\beta_e\beta_h>$ are given by Eq. 2.2.

$$E = E_0 + \underbrace{m_s \mathbf{g_e} \beta_e \mathbf{H}}_{\text{Electronic Zeeman}} + \underbrace{m_i \mathbf{g_h} \beta_h \mathbf{H}}_{\text{Nuclear Zeeman}} + \underbrace{a/4}_{\text{Hyperfine Coupling}} \tag{2.2}$$

TABLE 2.1 Nuclear Spins in the Methyl Radical

I_z =	$\frac{3}{2}$	$\frac{1}{2}$	$-\frac{1}{2}$	$-\frac{3}{2}$
	ααα	ααβ	ββα	βββ
		αβα	βαβ	
		βαα	αββ	

The allowed transitions among the four spin isomeric subspecies are those that change m_s but not m_i. They are indicated by the double-headed arrows of Fig. 2.4

Equation 2.2 is the first-order approximation to the ESR spectrum. The next higher (second order) approximation [1] adds a term of the form $(a/2)^2/(\mathbf{g_e}\beta_e H + \mathbf{g_h}\beta_h H)$ to the energy of the $|\alpha_e\beta_h>$ level and subtracts a like amount from the energy of the $|\beta_e\alpha_h>$ level. Since the field strength is in the denominator, the effect of this correction is usually noticeable only at low fields. A microcomputer program is available for including second-order effects in a simulated spectrum [2].

Splitting by Nuclei of Spin I. The first-order spectrum of a radical in which the odd electron interacts with one nucleus of spin I will have 2I + 1 equally spaced and equally intense lines corresponding to the allowed values of m_i. For example, ^{14}N or D (spin 1) will have three lines, and ^{11}B (spin $\frac{3}{2}$) will have four lines.

Splitting by Several Nuclei. When the odd electron is in a spatial orbital that allows it to interact with several nuclei, each nucleus contributes its own Zeeman term $m_i\mathbf{g_n}\beta_n H$ and its own Fermi contact interaction term $m_s m_i a$ to the energy.

The methyl radical (spectrum in Fig. 2.3) has three equivalent hydrogen nuclei, and can have any of the nuclear spin combinations shown in Table 2.1.

The four columns of Table 2.1 correspond to four values of the total proton I_z, $\frac{3}{2}$, $\frac{1}{2}$, $-\frac{1}{2}$, and $-\frac{3}{2}$. The three nuclear spin states corresponding to the same total $I_z = \frac{1}{2}$ have been represented by single terms like ααβ even though a rigorous representation would avoid the implication that a particular proton has a particular spin. In any case, each term in a given column contributes

TABLE 2.2 Pascal's Triangle

n	Intensities
1	1 1
2	1 2 1
3	1 3 3 1
4	1 4 6 4 1

TABLE 2.3 Number of Lines and Intensities for Nuclei of Spin 1

n	Intensities								
1				1	1	1			
2			1	2	3	2	1		
3		1	3	6	7	6	3	1	
4	1	4	10	16	19	16	10	4	1

equally to the intensity of the corresponding ESR line. This is the reason for the 1:3:3:1 ratio of intensities in Fig. 2.3.

The relative intensities of the lines from any number of equivalent hydrogen nuclei can be found either from the coefficients of the binomial theorem or from Pascal's triangle (Table 2.2).

The number of lines and the relative intensities for a set of n equivalent nuclei of spin 1 are shown in Table 2.3.

Nonequivalent Nuclei. If nonequivalent nuclei are present in the radical, the spectrum can be constructed by introducing the splitting due to each nucleus, one at a time. Thus one hydrogen splits the free electron signal into two lines, a second hydrogen splits each of these into two lines, and a ^{14}N would split each of those into three lines. The order in which the nuclei are selected in this procedure makes no difference. In the example just given the total number of lines will be $2 \times 2 \times 3 = 12$ unless some of the lines happen to coincide.

Applying this system to the methyl radical we would predict $2^3 = 8$ lines, but three of them coincide at the $I_z = \frac{1}{2}$ position and another three coincide at the $-\frac{1}{2}$ position.

If groups of equivalent nuclei are present, the procedure is to split the free electron signal into the pattern predicted for one of the groups, and then split each of those lines into the pattern predicted for the next group, and so on. In the ethyl radical, for example, the protons of the CH_2 group give a 1:2:1 triplet, then each of these lines is split into a 1:3:3:1 quartet by the CH_3 protons. The overlapping of these patterns gives the spectrum shown in Fig. 2.5.

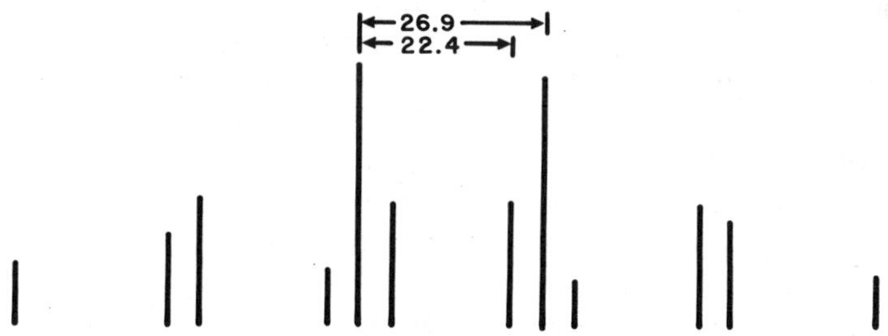

FIG. 2.5 A spectrum for the ethyl radical constructed using an a of 26.9 G for the methyl group and an a of 22.4 G for the methylene group protons.

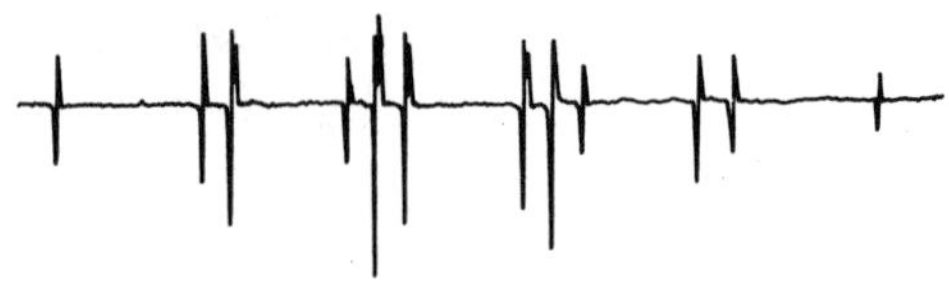

FIG. 2.6 The first derivative ESR spectrum of $CH_3CH_2\cdot$ at $-114°$. Reprinted with permission from P. J. Krusic and J. K. Kochi, *J. Am. Chem. Soc.*, **1969,** *91,* 3942. Copyright 1969 American Chemical Society.

An experimental spectrum for the ethyl radical is shown in Fig. 2.6. The most intense peaks show a slight unresolved splitting due to second-order effects. Some of the peaks expected to coincide exactly in the first-order approximation are found to be at slightly different positions.

2.4 THE ZEEMAN SPLITTING FACTOR $\mathbf{g_e}$

Under conditions that permit the radicals to tumble freely, the Zeeman factor $\mathbf{g_e}$ can be treated as a scalar constant equal to $\frac{1}{3}$ of the trace of the $\mathbf{g_e}$ tensor. We will not be concerned with the components of this tensor, but they can be obtained from ESR spectra taken of radicals at various constant orientations to the magnetic field. The value of the isotropic, or scalar, Zeeman splitting factor depends on the structure of the radical and hence contains useful information.

2.4.1 Values of $\mathbf{g_e}$ (Isotropic)

The value of $\mathbf{g_e}$ can be obtained from a single-line spectrum using Eq. 2.1. In the first-order approximation multiline spectra are symmetrical, and $\mathbf{g_e}$ can be obtained from the field strength at the center. To obtain more accurate g values the observed line positions are first corrected for second-order effects.

Spin-Orbital Coupling. The g value of an unpaired electron in a molecule differs slightly from that of a free electron because the electron spin angular momentum and the orbital angular momentum of a bound electron are not entirely separable quantities. To the extent that spin-orbital coupling occurs, the effective spin angular momentum will depend on the nature of the space wavefunction in which the electron is found.

Because spin-orbital coupling is more important near nuclei of higher nuclear charge, the g factor deviates more from the free electron value (2.002319) when such nuclei are present in the part of the radical in which the unpaired electron spin density is high. The g values for hydrocarbon radicals are within about two hundredths of one percent of the free electron value (Table 2.4). Radicals like cumylperoxy, in which the electron is essentially confined to a

TABLE 2.4 Isotropic Electronic Zeeman Factors

Radical[a]	g_e	Reference
Free electron	2.002319304376	7, 8
Hydrogen atom (1s)	2.00228	9
Methyl	2.00255	10, 11
Ethyl	2.00260	10–12
Alkyl in general	ca. 2.00257	10
Allyl	2.00254	10
Triphenylmethyl	2.00260	5, 13
$HOCH_2$	2.0033	11
HO_2CCH_2	2.0033	11
$HOCH_2CH_2$	2.0027	11
$CH_3CH{\cdot}OEt$	2.0031	11
$H(C{=}O)CH_2$	2.0045	11
Phenyl	2.0020	14
2,4,6-Tri-*t*-butylphenyl	2.00238	15
Vinyl	2.00220	10
Anthracene (+)	2.00257	5
Anthracene (−)	2.00271	5
Naphthalene (−)	2.00276	5
Benzene (−)	2.00285	5
Benzoquinone (−)	2.00468	5
Tetrachlorobenzoquinone (−)	2.00582	5
Cumylperoxy	2.0155	16
Benzoyloxy	2.0123	17
$(CF_3)_2NO{\cdot}$	2.0046	18
C_{60} (−)	1.99757	3, 4

[a]Ion radicals are indicated by (+) and (−).

single oxygen atom, have g values about seven tenths of a percent higher than the free electron value. Oxygen more remote from the odd electron has a smaller effect. Some examples illustrating these points are given in Table 2.4.

The g value for the buckminsterfullerene anion radical, $C_{60}{}^{-}\cdot$, is unusual, falling in the range associated with transition metals [3,4].

Within a group of closely related radicals, such as a series of aromatic anion or cation radicals, or a series of semiquinone radicals, the small differences in the isotropic g value are linearly related to λ, the energy level coefficient of the Hückel molecular orbital containing the unpaired electron ($E = \alpha + \lambda\beta$) [5,6]. The difference, $\Delta g, = g - g_0 = (31.9 - 16.6\lambda) \times 10^{-5}$, where g_0 is the free electron g value.

The presence of more than one radical in a mixture giving a complicated ESR spectrum can be detected by repeating the spectrum using a different microwave frequency. The position of the center of an ESR spectrum depends (Eq. 2.1) on the magnetic field or on the fixed microwave frequency. On the other hand, the hyperfine splitting by the magnetic nuclei, like proton–proton coupling in NMR, is independent of field strength to a good approximation. If the pattern changes very much when the spectrometer frequency is changed, more than one radical is present and the radicals have different g values. The relative positions of lines in the spectrum of a *single* radical species will not change.

2.5 THE HYPERFINE CONSTANT

The hyperfine constant for the Fermi contact interaction of the unpaired electron with a given nucleus depends on the unpaired spin density at that nucleus. This in turn will depend not only on the type and number of intervening bonds but also on the conformation of the radical.

Assuming for the moment that we know the expectation value, or unpaired spin density $|\psi(0)>^2$ at the nucleus, the hyperfine constant, or splitting factor, is given by Eq. 2.3.

$$a = -(8\pi/3)\mathbf{g_e}\beta_e\mathbf{g_N}\beta_N|\psi(0)>^2 \qquad (2.3)$$

The other quantities are the $\mathbf{g_e}$ for the electron, the Bohr magniton β_e, $\mathbf{g_N}$ for the nucleus, and the nuclear magniton β_N. The $\mathbf{g_N}$ for a nucleus is equal to the magnetic moment of the nucleus divided by its spin I. The actual calculation of a is found to require much more accurate wave functions than are needed for most other molecular properties.

2.5.1 Splitting Factors in Planar Pi Radicals

For planar π radicals in which the protons or other magnetic nuclei are in the nodal plane of the orbital containing the unpaired electron, molecular orbital calculations at the simple Hückel level of approximation predict zero for the spin density at the nucleus and hence also for the hyperfine constant. Experiment, however, gives substantial a values for magnetic nuclei in nodal planes. Examples are the protons in methyl radical, the α protons in the ethyl radical, and the protons on the rings of aromatic radical anions and radical cations.

The hyperfine splitting constant for the methyl radical protons is 22.9 G. The value for the α protons of the ethyl radical is 22.4, almost as large as the 26.8 G found for the β protons, even though the latter are not in the nodal plane. A planar structure for methyl is supported both by its optical spectrum [32] and by more elaborate *ab initio* calculations.

The hyperfine splitting constant for the benzene radical anion septet is 3.75 G [28]. Because the electron is distributed over six ring positions, this splitting constant should be multiplied by six to compare it with the value for the methyl radical.

Configuration Interaction. Better estimates of the spin density and hyperfine constants are obtained by including configuration interaction or spin polarization in the calculation of the wave function. A *configuration,* in this context, is usually a total wave function in which only one particular orbital contains an unpaired electron. In simple calculations the only configuration considered is the one in which all the lower energy orbitals are doubly occupied and the odd electron is in the first orbital above that closed shell. An improved total wave function is obtained by mixing the closed shell configuration with several configurations of higher energy. In a higher energy configuration one of the lower energy orbitals may be singly occupied, and a high energy orbital may have two electrons. For example, a higher energy configuration for the methyl radical might have two electrons in the carbon $2p_z$ orbital but only one electron in one of the C–H sigma bonding orbitals.

Using improved total wave functions of this sort, the protons of the methyl radical are predicted to have a spin density of about 0.05 at the H nucleus. Since the free hydrogen atom, for which the spin at the nucleus is about 1, has a splitting constant of 507 G, the constant for the methyl radical protons should be about 0.05×507, or 25 G. This is close to the observed value.

Spin Polarization. An alternative way of accounting for the presence of spin density in the nodal plane of a p or π orbital is to consider the electrons with spins $|\alpha>$ and $|\beta>$ as two separate families occupying two different sets of atomic orbitals [28]. By an extension of Hund's rule, there should be a tendency for the spins of two electrons in any two orbitals on the same atom to be parallel rather than antiparallel. Applying this idea to the methyl radical, if the electron in the carbon $2p_z$ orbital has spin β, then the electron in the carbon $2sp^2$ orbital should also tend to have spin β. But this atomic orbital is part of the C–H bond, hence the H 1s electron paired with it should have an enhanced probability of having spin α. The concept is illustrated in Fig. 2.7.

The spin polarization model also predicts the sign of the hyperfine constant. Although the sign is not needed to predict the splitting pattern (to first order), it is needed in connection with CIDNP, which is discussed in another chapter. The sign of a for a methyl radical can be deduced from Fig. 2.6 as follows: note that the spin of the electron in the carbon $2p_z$ is predominantly $|\beta>$ if the radical is in its lower spin state. The spin of the electron in the H 1s orbital is then $|\alpha>$, or opposite in sign. Hence the sign of the hyperfine constant is negative.

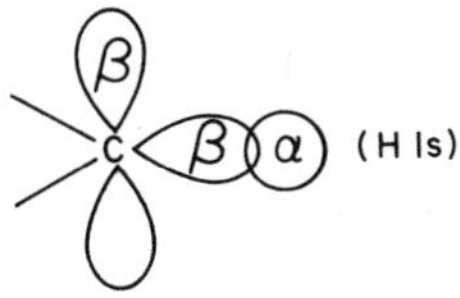

FIG. 2.7 Spin polarization.

For more complicated radicals the theoretical prediction is more complicated and the result may be somewhat uncertain, especially for small splitting factors, since it depends on the balance of terms of opposite sign in a multiterm expression [33].

Signs of the splitting factor can also be obtained experimentally from line shapes [34], from solvent effects on the magnitude of a [34,35], and by electron-nuclear double resonance [36].

2.5.2 Isotope Effects

Values of the isotropic hyperfine constants for some representative radicals are given in Table 2.5. The effect of the nuclear spin on the value of $\mathbf{g_N}$ in

TABLE 2.5 Hyperfine Splittings

Radical	$a_n{}^a$	Reference
Mu· (muonium)	1593^{be}	19
H·	506.6^f	10, 20
D·	78^e	21
CH_3·		
H	-23.0^g	10
^{13}C	+38.3	22
CD_3·	3.58^h	10, 22
Allyl·		
H at C1 and C3, exo	−14.81	23
H at C1 and C3, endo	−13.90	
H at C2	+4.06	
Benzyl·		
α	−16.35	24
ortho	−5.14	
meta	+1.75	
para	−6.14	
Triphenylmethyl·		
α^{13}C	+23.5	25
ortho	+2.53	13, 26
meta	−1.11	
para	+2.75	
Phenoxy·		
ortho	+6.60	27
meta	+1.96	
para	−10.40	
Benzene^{-}·		
H	-3.75^i	28
^{13}C	+2.80	28
Naphthalene^{-}·		
α	−5.3	28
β	−0.9	

TABLE 2.5 (*Continued*)

Radical	$a_n{}^a$	Reference
Phenyl·		
ortho	+17.4	14
meta	+5.9	
para	+1.9	
2,4,6-Tri-*t*-butylphenyl·		15
H meta	7.31	
H of ortho *t*-butyl	0.298	
^{13}C α	122.5	
^{13}C ortho	6.16	
^{13}C methyl of ortho	2.02	
$CH_2{=}CH\cdot$		
α	+13.4	10
β, trans	+68.5	
β, cis	+34.2	
$CH_2{=}C\cdot{-}CH_3$		
CH_3	+19.5	10
β, trans	+57.9	
β, cis	+32.9	
$CH_3CH_2\cdot$		
α	−22.38	10
β	+26.87	
$CH_2MuCH_2\cdot$		
βMu	130	19
$CH_3CH_2CH_2\cdot$		
α	−22.08	10
β	+32.2	
γ	+0.38	
$(CH_3)_3C\cdot$		
β	+22.7[10]	10
$\alpha^{13}C$	45[c,d]	29
Di-*t*-butylnitroxyl·		
^{14}N	15.45	30
$(C_6H_5)_2CO$ anion radical		
ortho	2.49	31
meta	0.83	
para	3.32	

[a]For H unless otherwise indicated.
[b]Muonium can be considered to be an isotope of hydrogen with mass 0.1134 and spin 1. Since its magnetic moment exceeds that of hydrogen by a factor of 3.1833, its hyperfine coupling constants tend to be about three times as large as that for hydrogen.
[c]Temperature dependent.
[d]The minimum value, found at 220 K.
[e]1:1:1.
[f]1:1.
[g]1:3:3:1.
[h]1:3:6:7:6:3:1.
[i]1:6:15:20:15:6:1.

Eq. 2.3 is illustrated by the data for H· versus D· and CH_3· versus CD_3·. Since the electronic wave function is not very much affected by the nuclear substitution, the isotope effect on a_H/a_D is approximated by the ratio $\mathbf{g_H}/\mathbf{g_D}$. This is equal to $(\mu_H/S_H)/(\mu_D/S_D)$ or $(2.793/[\frac{1}{2}])/(0.857/1)$, which is a factor of about 6.5.

2.5.3 Conformational Effects

Although H/D isotope effects on the value of a are close to those predicted from the magnetic moments and spins, the agreement is not exact because isotopic substitution can cause small changes in conformation. The methyl radical, as already noted, is believed to be quite accurately planar, and this is confirmed by the ^{13}C $a = 38$ of $^{13}CH_3$·. Deviations from planarity should cause rehybridization of the orbital containing the odd electron from p toward sp^3. For a pure sp^3 the ^{13}C splitting factor should be about 300 gauss [37].

The substitution of deuterium for hydrogen in $^{13}CH_3$ has two effects of interest. First, the ratio a_H/a_D is about 1% lower than the ratio predicted from the magnetic moments of H and D. Second, a comparison of the values of a_{13C} in $^{13}CH_3$· and in $^{13}CD_3$· suggests either that the bonds themselves bend during the vibration of the radical, a concept known as incomplete orbital following, or that the anharmonicity of the vibration needs to be taken into account [38].

Unlike the methyl radical, $^{13}CF_3$· has a large 271-G a_{13C} and is therefore distinctly nonplanar. The a_{13C} of the central carbon of the t-butyl radical, which is temperature dependent owing to the rapid pyramidal inversion, passes through a minimum value of 45 G at 220 K. The average deviation from planarity at this temperature is about 11.5° [39].

Sigma and Pi Radicals. One of the uses of the hyperfine splitting factor in structural investigations is to distinguish between sigma and pi radicals. In the phenyl radical, a typical sigma radical, the odd electron occupies a sigma sp^2 orbital although pi orbitals are present. As expected, the splitting factors for C_6H_5· (Fig. 2.8) are quite different from those of the pi anion radical C_6H_6·- and also quite different from what would be expected for the polarized pi radical structures also given in the figure.

Benzoyl radical (Fig. 2.9) could equally well be a sigma or a pi radical. In the sigma structure the half-occupied orbital is sp^2 and is in the plane of the ring. In the pi structure the carbonyl group is linear and the half-occupied p

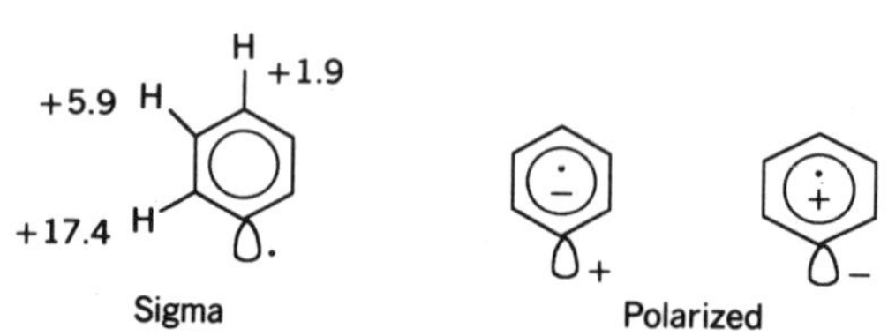

FIG. 2.8 Sigma versus polarized structures for the phenyl radical.

Sigma Pi

FIG. 2.9 Sigma versus pi structures for the benzoyl radical.

orbital is orthogonal to the ring and parallel to the benzene pi orbitals. Using the benzyl radical as a model, the pi structure would be expected to have large positive a_H values at the ortho and para positions and smaller, negative, values at the meta positions. In fact, the only observable proton splitting is that due to the meta protons, for which a_H 1.16 G. The splitting by meta protons is an example of the *trans*-W-plan effect discussed in another section. The sigma and relatively undelocalized nature of the radical is also confirmed by the large 128.2-G carbonyl ^{13}C splitting and by the low 2.0014 g value [40].

Remote Nuclei and Geometric Considerations. As can be seen from the example of the n-propyl radical in Table 2.5, the spin density and the hyperfine factor drop sharply after the β position. Still more remote nuclei will not ordinarily give resolvable splittings, although they may contribute to the width of the lines of the ESR spectrum. However, splitting constants depend not just on the number of bonds connecting a magnetic nucleus with the site of the unpaired electron, but also on the orientation of those bonds [21,41].

An example of the dependence of the hyperfine splitting constant on geometric considerations is the effect of changing the angle (θ) between the carbon $2p$ orbital on the alpha carbon and the plane defined by the H–C–C bonds. The relationship is given in Eq. 2.4, in which ρ is the spin density at the α carbon. The spin density is ordinarily taken as unity unless the unpaired electron is known to be delocalized.

$$a_{\beta H} = [A + B\cos^2(\theta)]\rho \tag{2.4}$$

The splitting factor is least when the angle is 90°, which puts the proton in the nodal plane of the p orbital. For alkyl radicals A is about zero and B about 53.7 G. The splitting factors for β hydrogen in radicals like ethyl, in which fast internal rotations about the C–C bond make the three hydrogens equivalent, are close to that expected if $\cos^2(\theta)$ is replaced by its average value.

Equation 2.4 can also be used to distinguish between conformers of β-substituted ethyl radicals [42].

Splittings due to hydrogens more remote than the β carbon atom of a radical can sometimes be resolved when a special configuration brings the topologically remote hydrogen into a position geometrically close to the localized orbital of the odd electron. The 2.7-gauss doublet in the ESR spectrum of the radical derived from fluorenone oxime (**I** below) indicates that the oxygen and the odd electron are close to just one of the two topologically equivalent ortho protons [43].

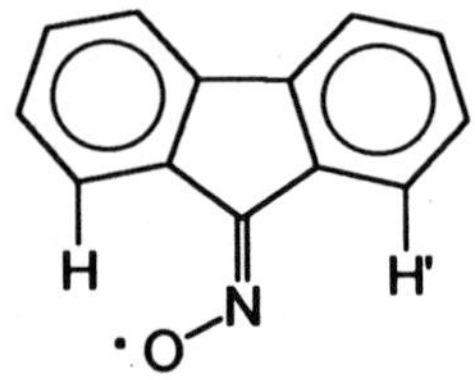

I

There is also a tendency for the long-range coupling between an electron and a proton to be greater if the intervening bonds form a trans planar W-plan zigzag [44–46]. In the semidione anion radical (II), the four equivalent protons indicated give a 1:4:6:4:1 quintet, $a = 2.09$ G. The topologically equivalent hydrogens endo to the carbonyl system do not give a resolvable splitting even though they are geometrically closer to the semidione group than are the four exo hydrogens. A similar enhancement of the coupling is seen in NMR when two protons are connected by bonds in a trans planar pattern.

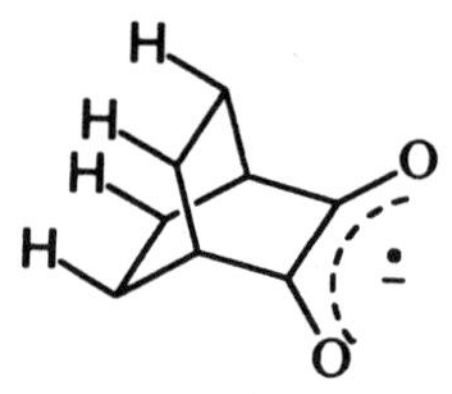

II

The splitting factors of the bridgehead H's of the bicyclic radicals III–V have been analyzed in terms of Eq. 2.5 [47],

$$a_{\mathrm{H}} = (A - n_B B + n_C C)^2 \tag{2.5}$$

in which A, B, and C are due respectively to through-space, through-three-bond, and through-two-bond interactions. The coefficients n_B and n_C are the number of paths. The through-three-bond interaction has a negative effect, counteracting the through-space and through-two-bond interactions because of the phase of the molecular orbitals involved.

III

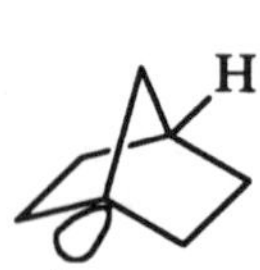

IV

V

2.5.4 Empirical Correlations of Hyperfine Constants

Hyperfine splitting factors are correlated, not only with calculated spin densities but also to some extent with such measures of chemical reactivity as Hammett substituent constants.

The McConnell Equation. In aromatic anion or cation π radicals the proton hyperfine splitting is due to nonzero spin density in the H 1*s* orbital. The difficult calculation of this spin density can be avoided because it is a linear function of the spin density on the carbon to which the proton is bonded [48]. As a result, the proton splitting can be correlated with the easily obtained spin density on carbon. The McConnell equation (Eq. 2.6) [48] contains a calculated π electron spin density at the aromatic carbon to which the proton is bonded (ρ_i^π) and an empirical factor $Q_{\mathrm{CH}}^{\mathrm{H}}$.

$$a_{\mathrm{H}}^{i} = Q_{\mathrm{CH}}^{\mathrm{H}} \rho_i^\pi \tag{2.6}$$

The notation $Q_{\mathrm{BC}}^{\mathrm{A}}$ represents the splitting associated with nucleus A, as a result of polarization of the BC sigma bond, caused by π electron spin density on B [33].

The value of ρ_i^π will depend on the method of calculation, except in the benzene anion radical, in which it is necessarily $\frac{1}{6}$. Inserting this value into the McConnell equation, and using -3.8 G for a_{H}, gives -23 G for $Q_{\mathrm{CH}}^{\mathrm{H}}$. It should be noted however, that the value of a_{H} varies with the solvent and temperature [10]. It has also been found that the value of $Q_{\mathrm{CH}}^{\mathrm{H}}$ needed to predict a_{H} for aromatic cation radicals is significantly greater than the value for anion radicals, this despite the pairing theorem of quantum mechanics which holds that ρ_i^π should be the same for both types of ion radical [49].

For aromatic nitrogen anion radicals the ^{14}N splitting factor is correlated by Eq. 2.7 [50,51]. This equation relates the ^{14}N splitting factor to the spin density ρ_{N} on the nitrogen, but adds a term for the spin density on each adjacent aromatic carbon atom.

$$a_{\mathrm{N}} = Q_{\mathrm{NN}}^{\mathrm{N}} \rho_{\mathrm{N}} + Q_{\mathrm{CC}}^{\mathrm{N}} \rho_i^\pi \tag{2.7}$$

For aromatic nitrile anion radicals $Q_{\mathrm{NN}}^{\mathrm{N}}$ is about 21.57 and $Q_{\mathrm{CC}}^{\mathrm{N}}$ about -13.65. For nitrogen heterocycles the best values appear to be $Q_{\mathrm{NN}}^{\mathrm{N}}$ 19.1 and $Q_{\mathrm{CC}}^{\mathrm{N}}$ 9.1. As might be expected, the splitting factors for ^{13}C in nuclear positions of aromatic ion radicals also depend on spin densities calculated for adjacent carbons [52].

Correlations with Substituent Parameters or Rates. The values of the hyperfine splitting constants of magnetic nuclei at various positions of a radical contain information about the delocalization of the unpaired electron and about the energy of the frontier orbital. Substituents in a radical change not only the electron spin distribution, but also the rates of reactions involving the destruction or formation of the radical. However, the contribution of electron spin delocalization to substituent effects on rates or equilibria is usually obscured by important steric or polar factors.

In many cases radical rates or selectivities are correlated quite well by Hammett or other polar substituent parameters without any consideration of electron spin delocalization. Since delocalization of the electron spin is *itself* dependent on both polar and steric effects, a separation of this factor from substituent effects in general is not easy.

The values of a_H of the methyl protons in the series of substituted anion radicals **VI** (or **VII**) are an example of a correlation with polar substituent constants. A series of sixteen "well behaved" meta or para substituents obeyed Eq. 2.8 with a Hammett ρ value of +1.2 [53].

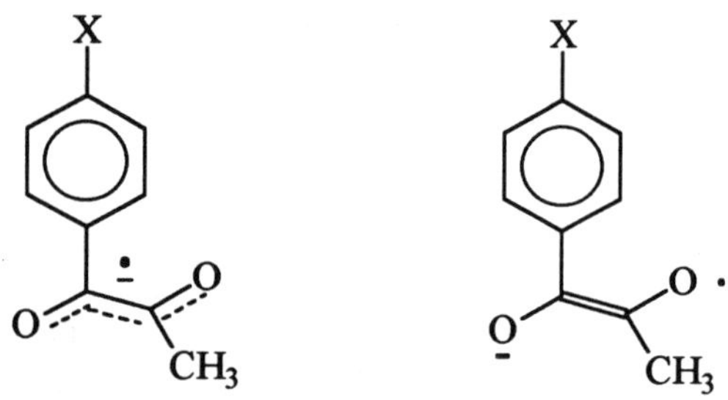

VI **VII**

$$a_H = 3.43 + 1.2\sigma \tag{2.8}$$

Electron-withdrawing substituents, which presumably increase the proportion of the negative charge on the carbonyl adjacent to the ring, evidently decrease the spin density on the carbonyl adjacent to the methyl group. This is in accord with an enhancement of the canonical structure **VII** by electron-withdrawing substituents. Four substituents were not "well behaved" and deviated considerably from the line. These included *p*-CN and *p*-phenyl, for which the a_H values were much smaller than predicted by the line, and for which a further localization of the spin on the carbonyl adjacent to the ring rather than the one adjacent to the methyl group is plausible.

It seems reasonable that a set of special σ· substituent constants could be defined for use with the logarithms of the rate constants of radical reactions, or with other energy quantities associated with radicals, in much the same way as special σ^+ and σ^- have been defined for reactions in which the delocalization of positive or negative charges is particularly important. Ideally, it should be defined by a process or property in which factors more important for intermediates of other types are not involved. In practice, most of the σ· scales that have been proposed have had to correct the defining substituent effect for a polar component, a procedure that is not free from ambiguities [54,55].

A σ_α· scale, defined in a more straightforward way than most, depends on the a_H of the alpha hydrogens of meta or para benzyl radicals **VIII** (Eq. 2.9) [54,55].

$$X—C_6H_4—CH_2\cdot$$

VIII

$$\sigma_{\alpha x}\cdot = 1 - a_{x\text{-benzyl radical}}/a_{\text{benzyl}} \tag{2.9}$$

The scale has been fixed in such a way as to make $\sigma_{\alpha H}\cdot$ zero and the $\sigma_{\alpha x}\cdot$ values for substituents that decrease the spin at the benzylic position positive. The $\sigma_\alpha\cdot$ values are completely uncorrelated with the Hammett σ and σ^+, which indicates a fairly good degree of separation of spin from polar properties. Applied to most radical reactions, a plot of $\sigma_\alpha\cdot$ against log k will not have a good correlation coefficient unless an additional parameter representing the polar effects is included. The best fit *without* the use of a correction for polar effects is for the rates of the rearrangement of methylenearylcyclopropanes **IX** [56] via the diradical, but the correlation coefficient is still only 0.96. Inclusion of a polar substituent term improves the correlation somewhat. The rate constants of reaction 2.10 have themselves been used to define a set of $\sigma\cdot$ parameters [57].

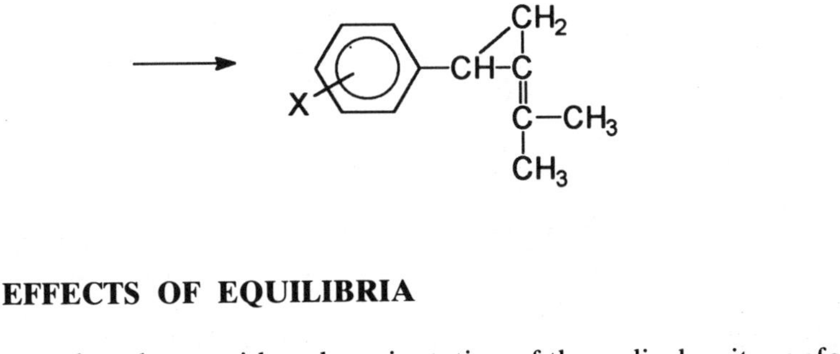

(2.10)

2.6 EFFECTS OF EQUILIBRIA

Processes that change either the orientation of the radical or its conformation can change its electron spin resonance spectrum.

For example, the Zeeman splitting factor $\mathbf{g_e}$ of Eq. 2.1a is a tensor rather than a scalar quantity. The fact that $\mathbf{g_e}$ is a tensor means that it will have different values for different orientations of the radical with respect to the external field. Radicals with different orientations will therefore absorb the microwave photon at different magnetic field strengths. However, if the radicals are rotating rapidly, $\mathbf{g_e}$ is seen by the ESR experiment only as an average value equal to $\frac{1}{3}$ the trace of the tensor. In ordinary solvents at room temperature the rotation of most radicals is fast enough to give isotropic spectra. As the rate at which the radical is tumbling decreases, the isotropic line first broadens and then splits into separate lines. In the extreme case of a radical whose orientation with respect to the field is fixed by the lattice of a host crystal, the effects of different orientations can be studied by reorienting the crystal. For example, isobutyramide radicals (**X**) in a crystal of isobutyramide have the following *g* factors as the magnetic field is oriented along the three crystallographic axes: 2.0033, 2.0035, and 2.0028 [58].

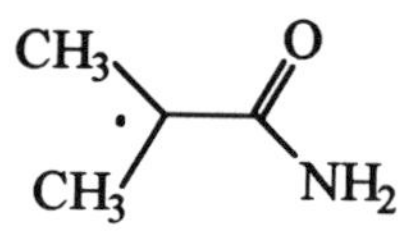

X

As we have seen in connection with conformational effects on the hyperfine constant, (e.g., Eq. 2.4), rigid radicals of different conformations will have different hyperfine splitting factors. It is now time to consider what happens if the radical is not rigid, so that the conformation is constantly changing. Each static conformation will have its own splitting pattern. If the frequency of the conformational equilibration is low, the spectrum will look like that of a mixture of the static conformations. If the frequency of the conformational equilibration is large, as in the case of the rotation of a methyl group about its bond axis, the splitting factors will be the average of all the conformations and the nuclei will be seen as equivalent.

For simplicity, consider the case of a single hydrogen nucleus that can be in either of two conformations, with splitting factors a_1 and a_2. If the exchange between the two conformations is fast, the ESR spectrum will be a doublet separated by the average splitting, $(a_1 + a_2)/2$. With very slow exchange, the spectrum is that of a mixture of the two isomeric radicals; there will be four lines, consisting of an a_1 and an a_2 doublet.

The ESR experiment will see the sample as a single radical rather than as a mixture when the frequency of the change in conformation is large compared to the difference in the frequencies of the two hyperfine interactions. As the conformational change is made slower, the two sharp lines of the fast-exchange doublet first broaden, then gradually resolve into four broad lines, and eventually into the four sharp lines characteristic of the static mixture. In another way of putting it, the rate at which the conformational change interchanges the positions of the protons has to be less than the rate at which the electron spin angular momentum vector precesses about the nuclear spin angular momentum vector: $2.8\ a_H \times 10^6$ hertz, where a is in gauss.

In suitable decalin derivatives, not only is the rate of conformational isomerism very near zero, but one conformer is strongly preferred. The stable conformer of *trans*-Δ^2-decalin semidione anion radical (**XI**) is essentially a double chair in which the 1 and 4 positions each have one quasi-equatorial and one quasi-axial proton. The quasi-axial hydrogens have a = 13.18 G and the quasi-equatorial hydrogens a = 6.59 G [59].

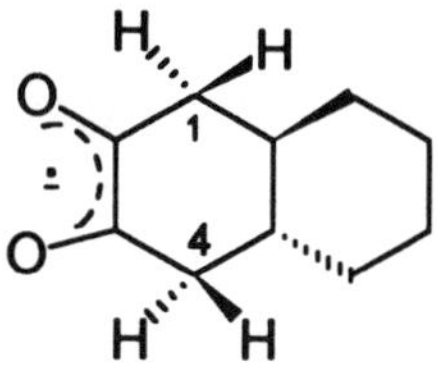

XI

Exchange of a proton with the solvent can also have an effect on the ESR spectrum. An example is the effect of changes in pH on the spectrum of $HOCH_2\cdot$ caused by exchange of the OH proton with that of hydronium ion [60].

REFERENCES

1. Carrington, A.; McLachlan, A. D. *Introduction to Magnetic Resonance,* Harper and Row: New York, 1967.
2. Joela, H. *Comput. Chem.* **1988,** *12,* 189.
3. Allemand, P.-M.; Srdanov, G.; Koch, A.; Khemani, K.; Wudl, F.; Rubin, Y.; Diederich, F.; Alvarez, M. M.; Anz, S. J.; Whetten, R. L. *J. Am. Chem. Soc.* **1991,** *113,* 2780.
4. Krusic, P. J.; Wasserman, E.; Parkinson, B. A.; Malone, B.; Holler, E. R. Jr.; Keizer, P. N.; Morton, J. R.; Preston, K. F. *J. Am. Chem. Soc.* **1991,** *113,* 6274.
5. Segal, B. G.; Kaplan, M.; Fraenkel, G. K. *J. Chem. Phys.* **1965,** *43,* 4191.
6. Stone, E. W.; Maki, A. H. *J. Chem. Phys.* **1963,** *38,* 1999.
7. Dehmelt, H. *Science* **1990,** *247,* 539.
8. Van Dyck, R. S. J.; Schwinberg, P.; Dehmelt, H. *Phys. Rev. Lett.* **1987,** *37,* 747.
9. Bersohn, M.; Baird, J. C. *An Introduction to ESR,* Benjamin: New York, 1966.
10. Fessenden, R. W.; Schuler, R. H. *J. Chem. Phys.* **1963,** *39,* 2147.
11. Norman, R. O. C.; Pritchett, R. J. *Chem. Ind.* **1965,** *2040.*
12. Fessenden, R. W. *J.Chem. Phys.* **1962,** *37,* 747.
13. Kulkarni, S. V.; Trapp, C. *J. Am. Chem. Soc.* **1970,** *92,* 4801.
14. Kasai, P. H.; Clark, P. A.; Whipple, E. B. *J. Am. Chem. Soc.* **1970,** *92,* 2640.
15. Barclay, L. R. C.; Griller, D.; Ingold, K. U. *J. Am. Chem. Soc.* **1974,** *96,* 3011.
16. Bersohn, M.; Thomas, J. R. *J. Am. Chem. Soc.* **1964,** *86,* 959.

17. Yamauchi, S.; Hirota, N.; Takahara, S.; Sakuragi, H.; Tokumaru, K. *J. Am. Chem. Soc.* **1985,** *107,* 5021.

18. Blackley, W. D.; Reinhard, R. R. *J. Am. Chem. Soc.* **1965,** *87,* 802.

19. Webster, W. *Chem. Brit.* **1983,** 110.

20. Wittke, J. P.; Dicke, R. H. *Phys. Rev.* **1956,** *103,* 620.

21. Heller, C.; McConnell, H. M. *J. Chem. Phys.* **1960,** *32,* 1535.

22. Fessenden, R. W. *J. Phys. Chem.* **1967,** *71,* 74.

23. Kochi, J. K.; Krusic, P. J. *J. Am. Chem. Soc.* **1968,** *90,* 7157.

24. Carrington, A.; Smith, I. C. P. *Mol. Phys.* **1965,** *9,* 137.

25. Schreiner, K.; Berndt, A.; Baer, F. *Mol. Phys.* **1973,** *26,* 929.

26. Chesnut, D. B.; Sloan, G. J. *J. Chem. Phys.* **1960,** *33,* 637.

27. Stone, T. J.; Waters, W. A. *Proc. Chem. Soc.* **1962,** 253.

28. Pople, J. A.; Beveridge, D. L.; Dobosh, P. A. *J. Am. Chem. Soc.* **1968,** *90,* 4201.

29. Griller, D.; Ingold, K. U.; Krusic, P. J.; Fischer, H. *J. Am. Chem. Soc.* **1978,** *100,* 6750.

30. Hoffman, A. K.; Hodgson, W. G.; Jura, W. *J. Am. Chem. Soc.* **1961,** *83,* 4671.

31. Ayscough, P. B.; Wilson, R. *J. Chem. Soc.* **1963,** 5412.

32. Herzberg, G. *Proc. Royal Soc.* **1961,** *A262,* 291.

33. Karplus, M.; Fraenkel, G. K. *J. Chem. Phys.* **1961,** *35,* 1312.

34. Fraenkel, G. K. *J. Phys. Chem.* **1967,** *71,* 139.

35. Gendell, J.; Freed, J. H.; Fraenkel, G. K. *J. Chem. Phys.* **1962,** *37,* 2832.

36. deBoer, E.; MacLean, C. *J. Phys. Chem.* **1967,** *71,* 21.

37. Cole, T.; Pritchard, H. O.; Davidson, N. R.; McConnell, H. M. *Mol. Phys.* **1958,** *1,* 406.

38. Schrader, D. M.; Karplus, M. *J. Chem. Phys.* **1964,** *40,* 1593.

39. Fessenden, R. W.; Schuler, R. H. *J. Chem. Phys.* **1965,** *43,* 2704.

40. Krusic, P. J.; Rettig, T. A. *J. Am. Chem. Soc.* **1970,** *92,* 722.

41. Stone, T. J.; Maki, A. H. *J. Chem. Phys.* **1960,** *37,* 1326.

42. Griller, D.; Ingold, K. U. *J. Am. Chem. Soc.* **1974,** *96,* 6715.

43. Gilbert, B. C.; Norman, R. O. C. *J. Chem. Soc. (B)* **1966,** 86.

44. Russell, G. A.; Chang, K. Y. *J. Am. Chem. Soc.* **1965,** *87,* 4381.

45. Kosman, D.; Stock, L. M. *J. Am. Chem. Soc.* **1966,** *88,* 843.

46. Kira, M.; Akiyama, M.; Ichinose, M.; Sakurai, H. *J. Am.Chem. Soc.* **1989,** *111,* 8256.

47. Kawamura, T.; Matsunaga, M.; Yonezawa, T. *J. Am. Chem. Soc.* **1978,** *100,* 92.

48. McConnell, H. M. *J. Chem. Phys.* **1956,** *24,* 764.

49. Bolton, J. R. *J. Chem. Phys.* **1965,** *43,* 309.

50. Henning, J. C. M. *J. Chem. Phys.* **1966,** *44,* 2139.

51. Bloor, J. E.; Gilson, B. R.; Shillady, D. D. *J. Phys. Chem.* **1967,** *71,* 1238.

52. Strauss, H. L.; Fraenkel, G. K. *J. Chem. Phys.* **1961,** *35,* 1738.

53. Strom, E. T. *J. Am. Chem. Soc.* **1966,** *88,* 2065.

54. Dust, J. M.; Arnold, D. R. *J. Am. Chem. Soc.* **1983,** *105,* 6531.

55. Dust, J. M.; Arnold, D. R. *J. Am. Chem. Soc.* **1983,** *105,* 1221.
56. Creary, X. *J. Org. Chem.* **1980,** *45,* 280.
57. Leigh, W. J.; Arnold, D. R.; Humphries, R. W.; Wong, P. C. *Can. J. Chem.* **1980,** *58,* 2537.
58. Hamrick, P. J. J.; Shields, H. W.; Parkey, S. H. *J. Am. Chem. Soc.* **1968,** *90,* 5371.
59. Talaty, E. R.; Russell, G. A. *J. Am. Chem. Soc.* **1965,** *87,* 4867.
60. Fischer, H. *Mol. Phys.* **1965,** *9,* 149.

3

OVERVIEW OF DIRADICALS AND TRIPLET STATES

3.1 SPINORBITALS

In diradicals two frontier orbitals have a special importance and also are convenient for classifying the species. Assuming that each of these orbitals is occupied by one electron, the species may be either an electronic singlet (antiparallel spins) or a triplet (parallel spins). The complete electronic wave function, the "spinorbital," must of course account for the spatial distribution of the electrons as well as their spin angular momentum. For our purposes, it is a good enough approximation to neglect fine points such as configuration interaction, and also to assume that the spinorbital can be represented as the product of two parts, one describing only the distribution of the two frontier electrons in space, and the other describing only their spins. For the spin parts, as discussed in Chapter 1, there are four possibilities:

$$(\tfrac{1}{2})^{1/2}[\alpha(1)\beta(2) - \beta(1)\alpha(2)] \qquad (0,0 \text{ singlet})$$
$$\alpha(1)\alpha(2) \qquad (1,1 \text{ of triplet})$$
$$(\tfrac{1}{2})^{1/2}[\alpha(1)\beta(2) + \beta(1)\alpha(2)] \qquad (1,0 \text{ of triplet})$$
$$\beta(1)\beta(2) \qquad (1,-1 \text{ of triplet})$$

The spin function to go with each of the spatial orbitals has to be chosen in such a way (Pauli principle) as to make the product of the two functions, the spinorbital, antisymmetric. An antisymmetric orbital changes sign when electrons 1 and 2 trade positions in the wave function.

For example, since the singlet spatial wave function is symmetric, the spin function has to be antisymmetric (i.e., change sign), so that the spinorbital will be antisymmetric. In Eq. 3.1 for the singlet spinorbital ψ_{occ} is an underlying filled shell of orbitals of no interest in the present context, and ψ and ψ' are the frontier spatial orbitals.

$$^1\psi_0 = (\tfrac{1}{2})\psi_{occ}[\psi(1)\psi'(2) + \psi(2)\psi'(1)][\alpha(1)\beta(2) - \beta(1)\alpha(2)] \quad (3.1)$$

The three triplet orbitals use the antisymmetric spatial function and symmetrical spin wave functions as shown in Eqs. 3.2–3.4.

$$^3\psi_1 = (\tfrac{1}{2})^{1/2}\psi_{occ}[\psi(1)\psi'(2) - \psi(2)\psi'(1)][\alpha(1)\alpha(2)] \quad (3.2)$$

$$^3\psi_0 = (\tfrac{1}{2})\psi_{occ}[\psi(1)\psi'(2) - \psi(2)\psi'(1)][\alpha(1)\beta(2) + \beta(1)\alpha(2)] \quad (3.3)$$

$$^3\psi_{-1} = (\tfrac{1}{2})^{1/2}\psi_{occ}[\psi(1)\psi'(2) - \psi(2)\psi'(1)][\beta(1)\beta(2)] \quad (3.4)$$

In case ψ is lower in energy than ψ', Eq. 3.1 represents an excited singlet state. The ground singlet, in which both electrons occupy ψ, and ψ' is empty, is represented by Eq. 3.5. It is not a diradical since there are no singly occupied spatial orbitals. It can be obtained from the excited diradical singlet by moving the electron in the higher energy orbital to the lower energy orbital.

$$^1\psi_0 = (\tfrac{1}{2})^{1/2}\psi_{occ}[\psi(1)\psi(2)][\alpha(1)\beta(2) - \beta(1)\alpha(2)] \quad (3.5)$$

3.2 CLASSIFICATION OF DIRADICALS

3.2.1 Double Doublets

In a *double doublet* diradical (see Chapter 1), each orbital is occupied by one electron, the orbitals are remote from each other, and they are connected only by structures that do not permit conjugation. As a result, overlap is negligible, and the ESR spectrum is like that of ordinary independent radicals.

3.2.2 Singlet Diradicals

In a *singlet* diradical, each orbital is occupied by one electron, and the electron spins are antiparallel. The spinorbital is given by Eq. 3.1. The distinction from the double doublet is that although the two orbitals do not overlap enough to form a strong covalent bond, they are close enough for the precession of the electron spin of one electron to preserve its phase relationship with respect to the precession of the other electron, that is, the spins stay antiparallel. The difference between a singlet diradical and an ordinary molecule is that one "bond" of the diradical is abnormally long and weak (essentially broken), making it a site of high reactivity.

3.2.3 Triplet Diradicals

In a *triplet* diradical the electron spins are parallel and the appropriate spinorbitals are those given in Eqs. 3.2–3.4. The $^3n\pi^*$ triplet state of benzo-

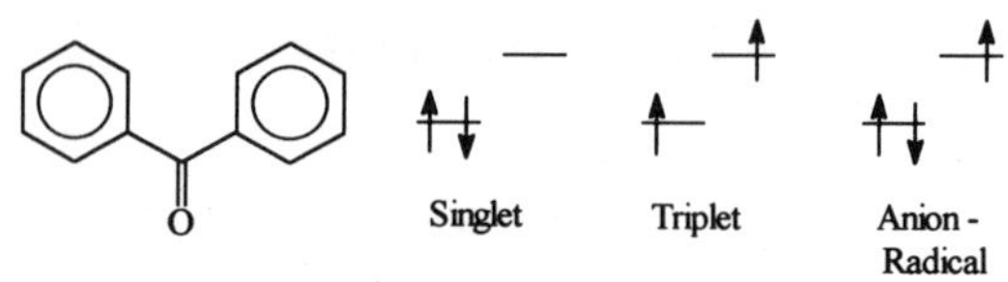

FIG. 3.1 Benzophenone singlet, triplet, and anion-radical.

phenone is a good example. This triplet diradical is made from the singlet ground state by removing one electron from a nonbonding σ_y orbital of the carbonyl oxygen (n) and promoting it to the lowest unoccupied π^* orbital. Because it has a half-occupied π orbital, it is analogous to the π anion–radical benzophenone ketyl (Fig. 3.1). Because it has a half-occupied orbital on oxygen, it is also analogous to ordinary sigma oxygen radicals such as *t*-butoxy.

Since both the benzophenone $^3n\pi^*$ and the *t*-butoxy radical have an unpaired electron localized on oxygen, it is not at all surprising that they have about the same selectivity in the abstraction of a hydrogen atom from pairs of substrates (Fig. 3.2) [1,2].

Another example is the lowest energy triplet of 2-acetonaphthone [3]. This time the triplet is a $^3\pi\pi^*$ state rather than a $^3n\pi^*$ state. Since it has two singly occupied π orbitals and does not have a singly occupied oxygen σ orbital, the triplet no longer resembles an alkoxy radical. The best radical models for this triplet are an aromatic anion radical (extra electron in a π^* orbital) and a cation radical (electron missing from a π orbital).

The triplet ground state of the oxygen molecule shows its radical-like character in various reactions. These include radical–radical combination with R· to give alkylperoxy radicals R–O–O·, and abstraction of H atoms. Triplet oxygen also adds to other diradicals to form cyclic peroxides [4,5]. However, the formation of transannular peroxides from the reaction of oxygen with $^3\pi\pi^*$ states of polynuclear hydrocarbons has a different mechanism: transfer of the excitation energy from the hydrocarbon to oxygen generates $^1\Delta_g$ singlet

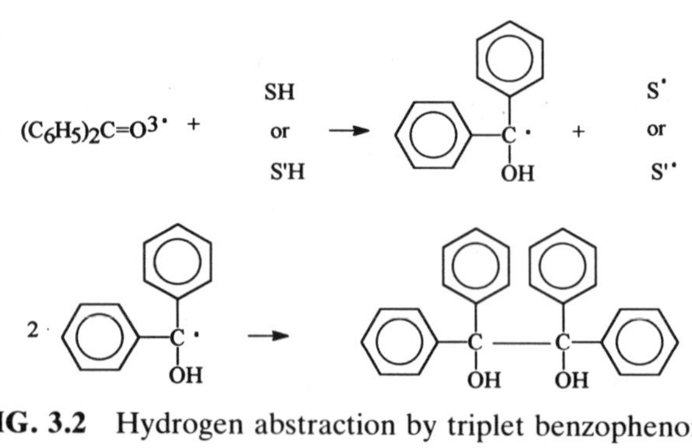

FIG. 3.2 Hydrogen abstraction by triplet benzophenone.

oxygen. The transannular peroxide is formed by the reaction of the singlet oxygen with the singlet hydrocarbon [6].

The thermolysis of photolysis of 1,4-diphenyl-2,3-diazabicyclo[2.2.1]hept-2-ene (Fig. 3.3) gives a highly strained hydrocarbon, and both a singlet diradical and a triplet diradical [7]. The singlet diradical corresponds to the bicyclopentane with a broken bond, but with the spins still coupled. It is converted both into the bicyclopentane and, by intersystem crossing, into the triplet diradical. Both diradicals react with O_2 to give the cyclic peroxide. The bicyclopentane reacts with O_2 only indirectly, by way of the diradicals.

3.2.4 Singlet and Triplet Radical Pairs

Singlet and triplet radical pairs differ from the corresponding diradicals only in that the two moieties are not connected by a covalent bond system, and are therefore free to diffuse relative to each other. As we shall see, the freedom to diffuse has important consequences, but so do the conformational motions of diradicals in which the connecting structure is flexible.

3.2.5 Singlet and Triplet States of Carbenes and Other Reactive Species

In a number of reactive intermediates, including species such as carbenes, nitrenes, electron deficient cationic oxygen compounds, or triplet states of ketones, the two frontier orbitals extend over at least one atom in common. The singlet states of simple carbenes and nitrenes are electronically analogous both to carbenium ions and to carbanions, since they have one vacant atomic orbital and one filled orbital. The reactions of triplets resemble those of radicals, whereas the reactions of the singlet isomers often resemble those of carbenium ions.

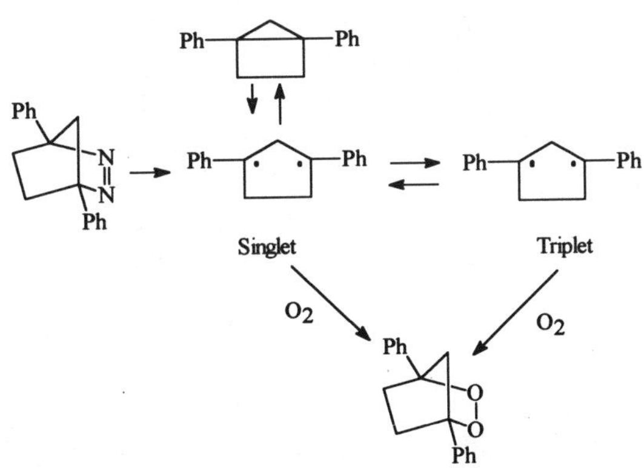

FIG. 3.3 A strained hydrocarbon and its diradicals.

3.3 HYBRID DIRADICALS

Although singlet and triplet radical pairs ordinarily are to a good approximation separate and distinct species rather than singlet–triplet hybrids, this is not always the case. Certain perturbations of the system can make the otherwise forbidden mixing of the singlet and triplet wave functions weakly allowed. Examples of such perturbations are a relativistic effect of nuclei of high charge, which causes coupling of the spin and orbital angular momenta of the electrons, and a perturbation caused by the relative motion of the spin centers as the conformation of the diradical changes. At high magnetic fields the mixing of the singlet is primarily with the $m_s = 0$ level of the triplet. This gives two new wave functions, ψ^+, which is primarily a singlet function, called a quasi-singlet, and ψ^-, a quasi-triplet. A diradical generated by the thermolysis or direct photolysis of a molecular precursor will be a quasi-singlet. Its wave function, in which the mixing coefficient C_1 is very much greater than C_2, is given by Eq. 3.6.

$$\psi^+ = C_1[\text{singlet function}] + C_2[\text{triplet function}] \tag{3.6}$$

A quasi-singlet diradical is more likely to cyclize to a singlet product molecule, whereas a quasi-triplet diradical is more likely to give reactions leading to radical pairs, for example, abstraction of hydrogen from the solvent. Conversion of a triplet diradical to a singlet molecule is a "spin-forbidden" process, and takes place only to the extent that the species are not purely singlet or purely triplet. Quasi-singlet and qausi-triplet radical pairs will be discussed in the next chapter. In diradicals containing very heavy nuclei, such as iodine, the distinction between electronic singlets and triplets is lost entirely.

3.4 THE EXCHANGE INTEGRAL

The properties of a diradical depend a great deal on how strongly the two electrons repel each other and how far apart they are. A very important property for predicting the behavior of a diradical is the exchange integral J_{ab} (Eq. 3.7), which gives the energy of the coulombic repulsion of the two electrons 1 and 2.

$$J_{ab} = \langle \psi_a(2)\psi_b(1)|e^2/r_{12}|\psi_a(1)\psi_b(2)\rangle \tag{3.7}$$

The equation reflects the fact that each of the orbitals ψ_a and ψ_b has one electron, but not specifically electron 1 or electron 2, since electrons cannot be distinguished. It should be noted that J_{ab} is the energy of repulsion of two electrons in different orbitals. The energy of repulsion of electrons in the same orbital is given by a different integral.

The size of J_{ab} depends not only on the distance between the electrons but also on the extent of overlap of the orbitals. If $\psi_a\psi_b$ integrates to zero, J_{ab} is zero. Examples of systems with very small J_{ab} would be orbitals ψ_a and ψ_b that are local wave functions at remote ends of a molecule, or orthogonal wave functions like those in twisted ethylene.

If J_{ab} is zero the electrons do not "see" each other's spins. These diradicals are double-doublets rather than singlets or triplets. Double-doublets have hyperfine splitting patterns and chemical properties like those of the corresponding ordinary radicals or doublets. Both electron spins precess about the magnetic field axis, but their precessions and choice of $m_s + \frac{1}{2}$ or $-\frac{1}{2}$ are completely independent.

When the energy J_{ab} is large compared to hyperfine energies, the system falls into the strong exchange category. Strong exchange causes the molecule to be either an almost pure singlet or an almost pure triplet. In any of the triplet sublevels the electron spins again precess about the magnetic field axis, but the phases are correlated so as to keep the spins parallel. The total spin is therefore equal to 1; the projection on the z axis is 1, 0, or -1.

3.5 DEGENERATE ORBITALS

If two interacting orbitals ψ_a and ψ_b are degenerate, or nearly so, we must replace them with two new orbitals ψ_+ and ψ_-. There will now be two states in which each of the new orbitals has one electron. One of these states, in which the electrons have antiparallel spins, is a singlet. The other, in which the electrons have parallel spins, is a triplet. The spatial parts of the orbitals are given by Eqs. 3.8 and 3.9.

$$\psi_+ = (1/\sqrt{2})[\psi_a(1)\psi_b(2) + \psi_a(2)\psi_b(1)] \quad \text{(singlet)} \tag{3.8}$$

$$\psi_- = (1/\sqrt{2})[\psi_a(1)\psi_b(2) - \psi_a(2)\psi_b(1)] \quad \text{(triplet)} \tag{3.9}$$

The difference in energy between these singlet and triplet states is given by Eq. 3.10.

$$E_+ - E_- = 2J_{\mathrm{ab}} \tag{3.10}$$

The triplet state of such a pair is lower in energy than the singlet because the motions of electrons in the triplet state correlate in such a way as to keep them as far apart as possible. For this reason $2J_{ab}$ is sometimes called the Fermi correlation energy as well as the singlet–triplet splitting. The lower energy expected for the triplet combination of two orbitals is the basis for Hund's rule.

If ψ_a is much lower in energy than ψ_b, there will again be two orbitals, but the lower of the two will be almost purely ψ_a and the upper almost purely ψ_b. The ground state will be a closed shell singlet ${}^1\psi_0$, with wave function equal to $\psi_a(1)\psi_a(2)$, not ψ_-. The energy difference between *this* singlet state and the triplet in which one electron has been promoted to ψ_b is given by Eq. 3.11.

$$E_{\text{triplet}} - E_{\text{singlet}} = E_b - E_a - 2J_{ab} \tag{3.11}$$

If the energies of ψ_a and ψ_b are close and J_{ab} is large, ψ_- can still be the ground state.

Triplets, or systems with large J, usually turn out to be molecules in which the triplet electrons are in orbitals on the same atom, or are in orbitals on adjacent atoms, or are in molecular orbitals whose territories extend over a common nuclear framework.

3.6 THE ESR OF TRIPLETS

The ESR spectrum of a triplet can be quite complicated [8]. One reason for the complicated spectra is that there are several types of transition, allowed $\Delta m_s \pm 1$ transitions between the $m_s + 1, 0, -1$ triplet substates, and a "forbidden" $\Delta m_s = 2$ transition. The latter occurs at about half the magnetic field strength of the others.

The lines due to the $\Delta m_s = \pm 1$ transitions are further split into doublets by the magnetic dipole–dipole interactions of the two electron spins with each other. Because the dipolar interaction operator is a tensor, the splitting due to this interaction will depend on the orientation of the molecule. Triplet spectra are usually studied in glassy or crystalline host media. The energy levels and the ESR spectrum of a triplet dissolved in a glass, and hence randomly oriented, will be different from the energy levels and spectrum of the triplet in a crystalline solid solution. The molecules in a solid solution have orientations imposed by the lattice of the host crystal. The spectrum of the solid solution will change as the host crystal is given different orientations relative to the external magnetic field.

The spin Hamiltonian is often written as Eq. 3.12, in which the first term represents the interaction of the total spin S with the external field, and the second two terms represent the magnetic dipole–dipole interaction. Only two terms are needed for the dipolar part of the Hamiltonian because the trace of the dipolar interaction tensor is zero.

$$H = g\beta BS + D(S_Z^2 - \tfrac{1}{3}S^2) + E(S_X^2 - S_Y^2) \tag{3.12}$$

The magnetic field is represented by B rather than H to avoid confusion with the Hamiltonian. $S = S_1 + S_2$. $S_Z = S_{1Z} + S_{2Z}$. D and E are the zero-field parameters.

3.6.1 The Zero-Field Parameters

The zero-field parameters D and E of Eq. 3.12 contain valuable information about the spatial distribution of the unpaired spins in the molecule.

A large D means that the average distance between the two electrons is small [6,8]. What constitutes a "large" D is best learned by looking at a few examples. In general the motions of the unpaired electrons will be correlated in such a way as to keep them as far apart as possible, subject to constraints by the structure of the molecule and the nature of the two half-occupied orbitals. The D of *t*-butylnitrene [9] is large (1.625 cm^{-1}) because the two electrons are in local orbitals on the same atom. In phenylnitrene one of the orbitals is more delocalized, overlapping with the benzene ring, and D is only 0.99.

In the triplet states of aromatic hydrocarbons the orbitals occupied by the two unpaired electrons extend over the entire molecule. As a consequence, D decreases with the number of aromatic rings. For example, in benzene the value of D is 0.16 cm^{-1}, in naphthalene [10,11] it is about 0.10 cm^{-1}, and in anthracene [8] about 0.072 cm^{-1}. The D value for naphthalene corresponds to an effective distance of about two C–C bond length between the electrons.

The very low D value of 0.0521 cm^{-1} for *p*-phenylenebis(phenylmethylene) [12] (**I**) suggests that the electrons in that molecule tend to be confined in opposite moieties, probably because conjugation is reduced by a lack of coplanarity.

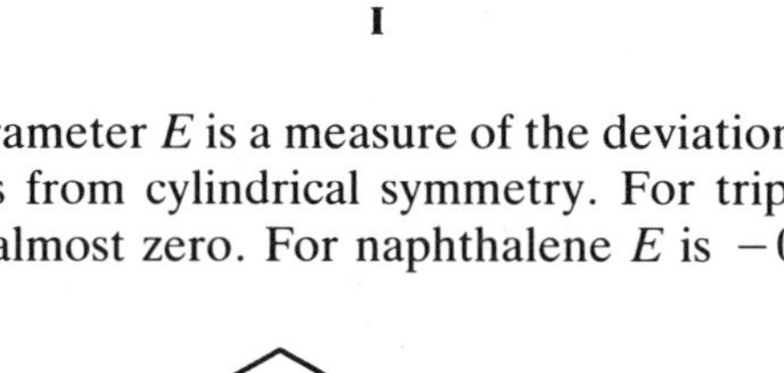

I

The zero-field parameter E is a measure of the deviation of the distribution of the two electrons from cylindrical symmetry. For triplet benzene or triphenylene (**II**) it is almost zero. For naphthalene E is -0.0138 cm^{-1}.

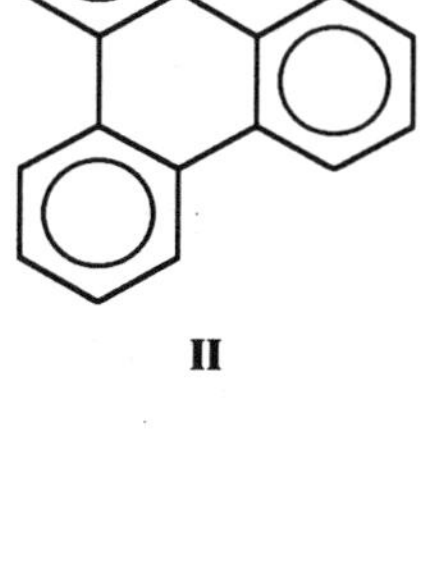

II

In triplet methylenes the absolute value of E/D is proportional to the fraction of s character of the half-filled hybrid orbitals on the central carbon. It can therefore be used to predict the bond angle, except when the angle is constrained by a ring. For an E/D of zero the electrons are in pure p orbitals. In that case the methylene should be linear like the acetylene molecule, with a bond angle of 180°. For E/D equal to unity the bond angle should be 90°.

Diphenylmethylene (**III**) formed by photolysis of the diazo compound in a crystalline host or in a glass has a triplet state with $D = 0.4052\ \text{cm}^{-1}$ and $E = -0.0194\ \text{cm}^{-1}$. The model that best conforms to these values of the zero-field parameters has a bond angle of 150° [13,14]. In **III** one electron occupies an orbital of σ symmetry in the plane of the rings, whereas the other occupies an orbital of π symmetry and interacts with the π orbitals of the rings. The delocalization into the rings is not extensive, however, since D is rather large.

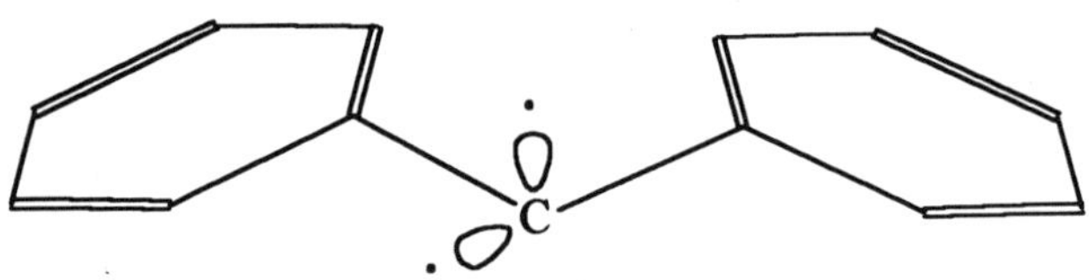

III

Cyclic diazo compounds give cyclic methylenes with D and E values not greatly different from those of the acyclic methylenes. For example the E/D of fluorenylidene suggests an angle greater than 135° [14–16]. Since this is too great for a five-membered ring, the bonds are probably bent, that is, the overlap of the sigma orbitals is not along their long axes.

3.7 HYPERFINE SPLITTING

3.7.1 Triplets

In the strong magnetic field of a typical ESR spectrometer, the hyperfine splitting of a triplet line is derived from a term aS_zI_z in the Hamiltonian, where S_z is the *total* spin $S_{z1} + S_{z2}$ of the two electrons and I_z, as usual, is the nuclear spin. Letting $\psi_a(r)$ and $\psi_b(r)$ be the spatial wave functions that combine to make the spatial wave function of the triplet, the splitting factor a is given by the normalized or weighted average (Eq. 3.13) of the spin density at the nucleus in question.

$$a = A[|\psi_a(0)\rangle^2 + |\psi_b(0)\rangle^2] \tag{3.13}$$

The factor A in Eq. 3.13 depends on the orientation of the triplet relative to the external field. Thus the splitting pattern of a solid sample can change

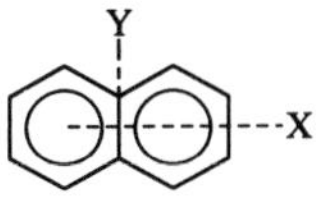

FIG. 3.4 Naphthalene axes.

rather drastically with orientation. For example, if the field is aligned with the x axis of a naphthalene triplet (Fig. 3.4), each of the $\Delta m_s \pm 1$ lines shows a five-line hyperfine structure due to the four protons in the alpha positions of the naphthalene ring, and each of these lines is in turn further split into a quintet by the protons in the beta position. With the field aligned with the y axis of the molecule, on the other hand, all eight protons have the same splitting factor and an unresolved nine-line envelope is seen [10].

The $\Delta m_s = \pm 1$ lines of the methylene from the photolysis of benzene-1,4-diazooxide in p-dichlorobenzene (Fig. 3.5) have a three-line hyperfine structure due to the protons in the 2,6 positions. The value of D for this methylene is 0.3179, due mainly to the spin density of the pi orbital electron at carbon atom 1. The ESR of the related phenoxy radical shows a comparable spin density at that position. Since the spin density at the positions meta to the oxygen in the phenoxy radical is low, it may seem strange that these are the protons whose splitting is seen in the spectrum of the methylene. The reason is that interaction between an odd electron in a pi system and a proton in the nodal plane requires interaction with configurations in which sigma and pi spins are interchanged. The presence of an unpaired electron in the sigma orbital at carbon 1 (Fig. 3.5) causes this orbital to predominate in the sigma–pi spin exchange and makes the protons nearest to this orbital give the largest splitting [17].

Since proton hyperfine structure is often not resolved or not well resolved in the ESR of triplets, electron-nuclear double resonance (ENDOR) is frequently used. For example, preliminary ESR experiments with a central ^{13}C in the fluorenylidene triplet, followed by ENDOR on the protons, allowed all of the matrix elements of the hyperfine tensors of all of the protons to be obtained [18,19].

3.7.2 Exchange Effects

The hyperfine splitting pattern of a diradical depends on the value of J_{ab}. For double-doublets ($J_{ab} \ll a_N$), the weak exchange case, the hyperfine splitting is like that of an ordinary monoradical of similar structure and is characterized

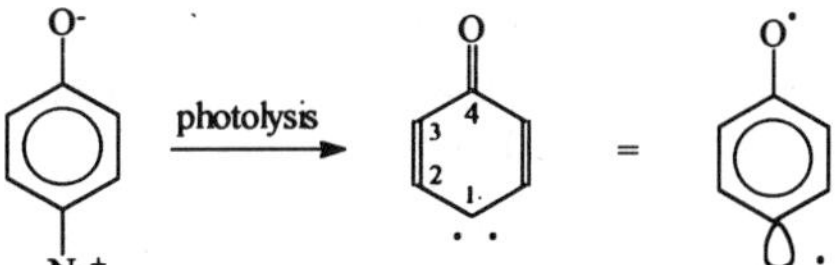

FIG 3.5 Photolysis of benzene-1,4-diazooxide.

by the same splitting constant or constants. This is illustrated by the nitroxide diradical **IV** and the related monoradical **V** [20].

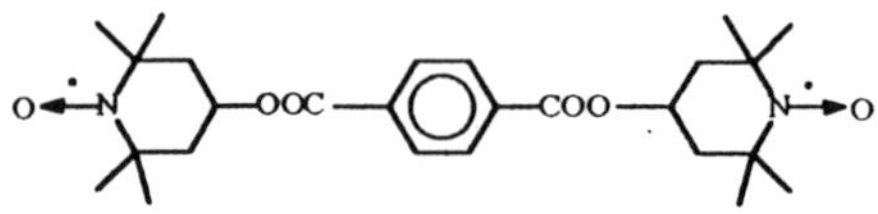

IV

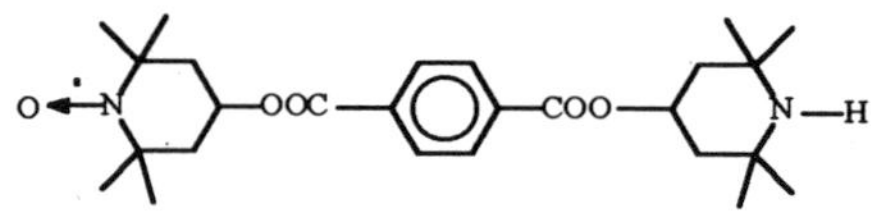

V

The ESR spectra of both **IV** and **V** consist of three lines of equal intensity resulting from the hyperfine splitting by the nitroxide nitrogen. In both spectra a_N is 15.6 G. Each electron "sees" only its own nitrogen nucleus. Actually, the three-line spectrum is the superposition of two identical spectra. There would be more lines if the two halfs of the molecule were not alike.

In the monoradical **VII** the hyperfine splitting again gives a simple three-line spectrum. The splitting factor is 14.80 G. In the nitroxide diradical **VI** the unpaired electrons are closer together than they are in **IV**, making this diradical an example of the strong exchange limit, $J_{ab} \gg a_N$. Each electron sees *both* of the nitroxide nitrogens, and the splitting pattern consists of five lines of relative intensities 1:2:3:2:1. However, since each electron spends only half its time on a given nitroxide site, a_N is only 7.40 G, exactly half that of the related monoradical **VII**.

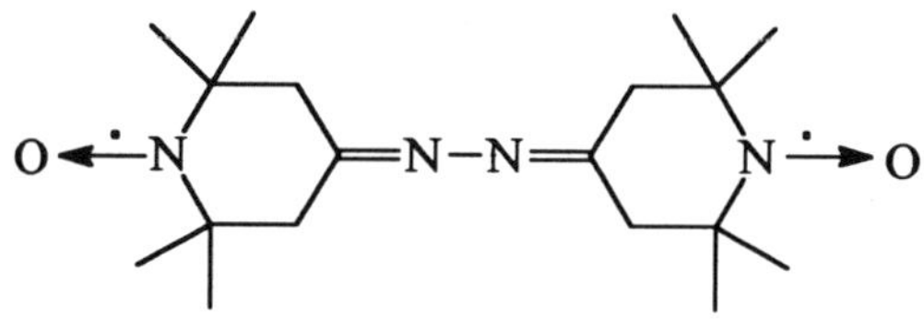

VI

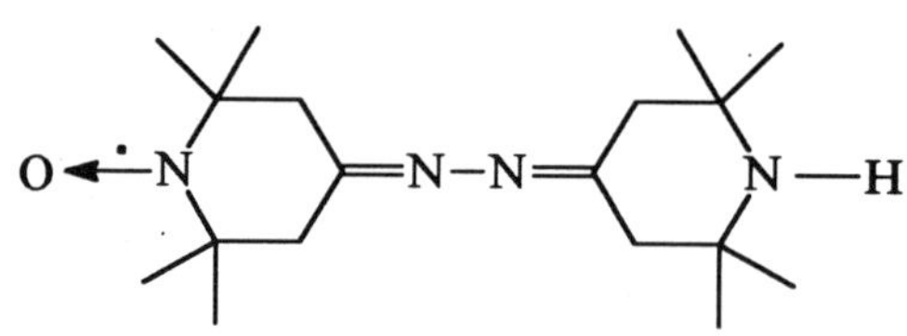

VII

The spectra for galvinoxyl (**VIII**) and a related diradical **IX** [21] illustrate the same principles for a radical with threefold symmetry. Diradical **IX** has a triplet ground state and a seven-line hyperfine structure (from the six meta hydrogens), indicating that $J \gg a_H$ and that the odd electrons are delocalized over all three rings. The splitting factor is $\frac{2}{3}$ that of **VIII**, as expected.

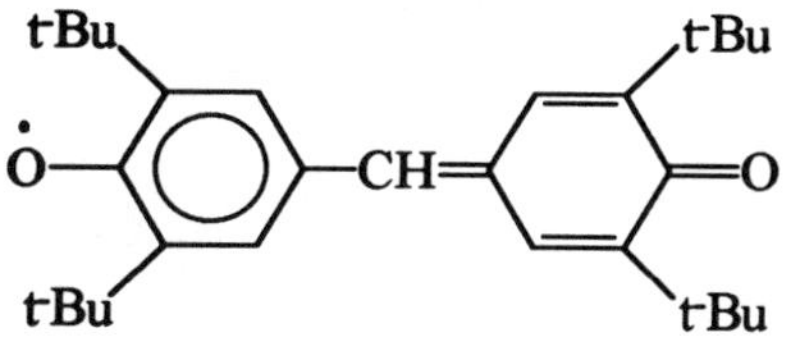

VIII

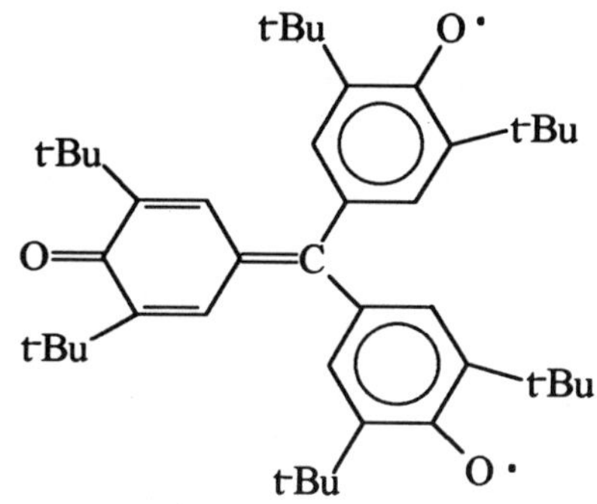

IX

Exchange At Intermediate J Values. The ESR spectra of diradicals **IV** and **VI** illustrated two extremes, weak exchange ($J_{ab} \ll a_N$) and strong exchange ($J_{ab} \gg a_N$). As J_{ab} is progressively increased, the simple splitting pattern of a molecule like **IV** first becomes complicated, with many extra lines, and then becomes simple again as the strong exchange limit is approached. The spectrum is at its most complicated when J_{ab} is the same order of magnitude as the hyperfine splitting constant. The reason for the extra hyperfine lines is that coupling of the electron and nuclear spins causes the $m_s = 0$ singlet energy level to mix with the $m_s = 0$ triplet sublevel. There are still strong $\Delta m_s = \pm 1$ transitions between the triplet $m_s = \pm 1$ and triplet $m_s = 0$ levels, but because of the mixing, new weak transitions appear between the primarily singlet $m_s = 0$ and the triplet $m_s = \pm 1$ levels [22].

Systems in which the singlet and triplet $m_s = 0$ states have been mixed are called quasi-singlets or quasi-triplets, depending on which wave function predominates. As the distance between the odd electrons in a diradical is changed by conformational motions, a quasi-singlet has a small but nonzero chance of evolving into a triplet. Similarly, a quasi-triplet has a small chance of evolving into a singlet. The evolution of quasi states is more important for the chemistry of radical pairs than for diradicals, and is discussed further in another chapter.

Conformational Effects. For the J_{ab} of an unconjugated diradical to be large enough to place the molecule in the strong exchange category, the molecule must be flexible enough to achieve frequent close approaches of the centers bearing the two electrons. In addition, the orientations during the close approaches must allow the orbitals ψ_a and ψ_b to overlap. Since the ESR spectrum will change as the molecule is deformed, radical sites like this anchored on two points of a deformable macromolecule or biological structure can act as a strain gauge.

In conjugated systems in which the electrons can approach each other through appropriate canonical (resonance) structures, the requirement for a large J_{ab} is frequent passage through a nearly coplanar conformation of the double bonds. Analogs of Chichibabin's hydrocarbon **X** show more pronounced radical-like properties when the number of intervening groups (n) is increased, steric hindrance to coplanarity is increased by substituents, or the connections to the central ring are at meta rather than para positions [23].

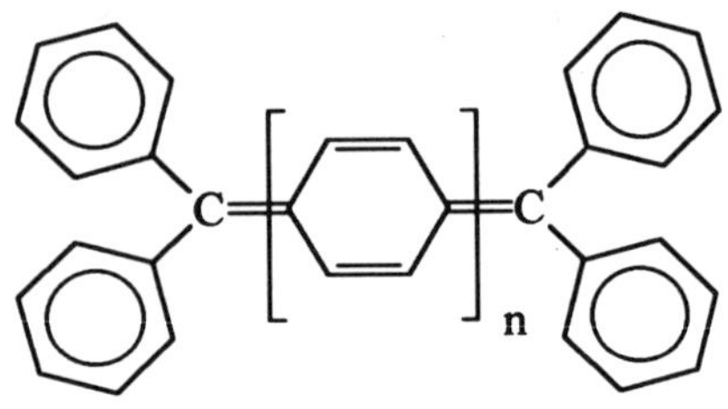

X

3.8 INTERSYSTEM CROSSING

3.8.1 Mixing of Singlet and Triplet States

Intersystem crossing (ISC) [3,24–26], or transitions between singlet and triplet states, cannot occur if the state is a pure singlet or a pure triplet. The wave function of the nominally singlet state has to have some contribution from the triplet wave function, and vice versa. In spectroscopy, ISC makes "forbidden" singlet–triplet absorption bands possible even though they are usually weak. To mix the states there has to be a magnetic field in a direction that exerts a torque able to "flip" an electron spin. There are two mechanisms that fulfill this requirement, spin-orbit coupling and hyperfine coupling with nuclear spins.

Spin-Orbit Coupling. In spin-orbit coupling (SOC), the magnetic field needed to flip the spins and mix the singlet and triplet wave functions is provided by the orbital motion of the electron in the electrical field of the nucleus. Spin-orbit coupling increases approximately with the fourth power of the effective nuclear charge, and is therefore very much more important in diradicals containing oxygen than in hydrocarbon diradicals.

For radicals with localized spins, three additional factors optimize SOC [3]:

1. The singlet wave function should have some zwitterionic character.

2. The two orbitals should be oriented 90° relative to each other and to the axis around which *orbital* angular momentum is being created. For example, in Fig. 3.6 the orbital angular momentum axis happens to be the z axis, the two orbitals are in the xy plane, and they are tilted toward one another to make an angle of about 90°.

3. The two orbitals must be close enough together to overlap to some extent, but not too close, because too much overlap increases the singlet–

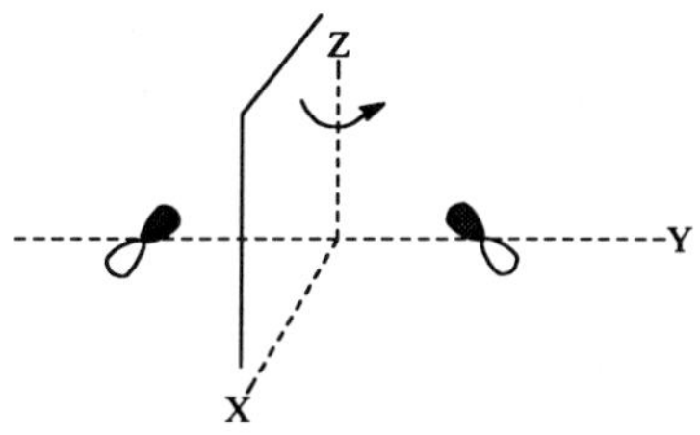

FIG 3.6 Optimum orbital orientation for spin–orbit coupling

triplet energy gap. Diradical **XI**, which is formed in the boat configuration by the sensitized photolysis of 2,3-diazabicyclo[2.2.2]oct-2-ene, is a good example of near optimum configuration for SOC.

XI

Intersystem crossing in this diradical is so efficient that the lifetime of the triplet, estimated from the extent of trapping by O_2, is only about 0.1 ns [5].

Hyperfine Coupling. Another interaction that can mix the spin states is the hyperfine coupling (HFC) of the electron spin with the spin of magnetic nuclei. The importance of HFC is greater in diradicals lacking heavy nuclei. The rate of intersystem crossing by this mechanism *increases* with increasing separation of the diradical ends, because the triplet–singlet energy gap decreases. When HFC is an important mechanism, the ISC rate is more dependent on magnetic field strength.

3.8.2 Energy Matching

The main requirement for the system to make a change from a singlet to a triplet potential energy surface, besides the mixing of the singlet and triplet wave functions, is that the energies must overlap. For example the benzophenone $^1n\pi^*$ singlet is rapidly converted into the $^3\pi\pi^*$ triplet, which is at the same energy. This undergoes efficient "internal conversion" to the lower-energy $^3n\pi^*$ state.

In many cases the requirement of equal energy for the two states is met by a change to a conformation at which the electronic potential energy surfaces intersect. The matching of vibrational energies is not normally a problem since the molecule or molecule plus surrounding solvent will ordinarily have a rather dense manifold of vibrational energy levels in both the singlet and triplet states. For example, the energy of the excited singlet state of a molecule like tetraphenylethylene will be at a maximum when the double bond has been twisted 90° (Fig. 3.7). The triplet state has a minimum at this angle.

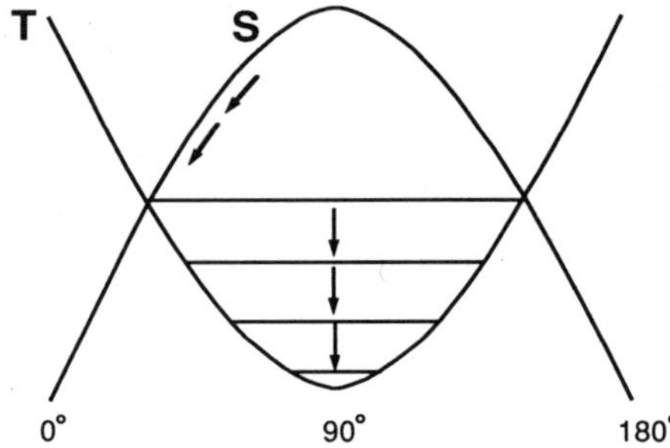

FIG 3.7 Effect of twist angle in tetraphenylethylene.

The two surfaces will intersect at some twist angle between 0 and 90°, and if the molecule remains in this conformation long enough, a transition from the excited singlet to the triplet can occur. Once the transition has been achieved, the molecule will relax vibrationally to the 90°-twisted triplet state.

3.9 FORMATION OF SINGLET PRODUCTS

3.9.1 From Long-Chain Diradicals

The reaction of a long-chain triplet diradical to give a singlet product by cyclization or disproportionation involves three steps (Fig. 3.8), any one of which might be slowest and hence rate determining. First, the diradical must take up a conformation in which the two ends are in a position to react. Second, the nearly cyclic diradical must undergo ISC to become a singlet. The final step is the actual cyclization or disproportionation [27,28].

In the cyclization and disproportionation of 1,6-diphenylhexane-1,6-diyl **XII** the rate of disappearance of the diradical is the rate of ISC to the singlet state, mainly by the HFC mechanism. The cyclization rate of the acyl diradical

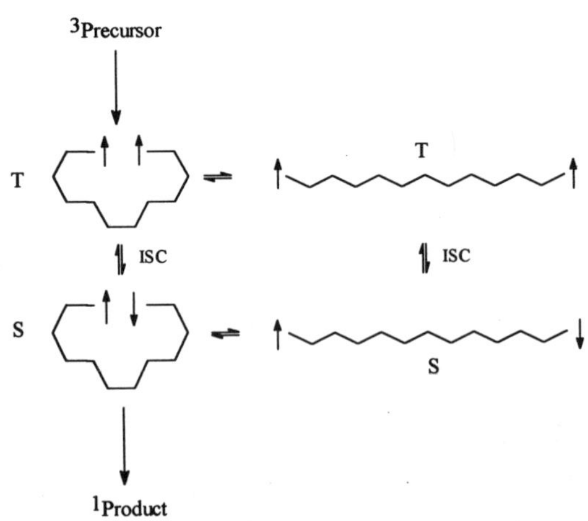

FIG. 3.8 Closure of a long-chain diradical.

XIII is again determined by the ISC rate, but the reaction is faster, and the ISC mechanism is mainly SOC.

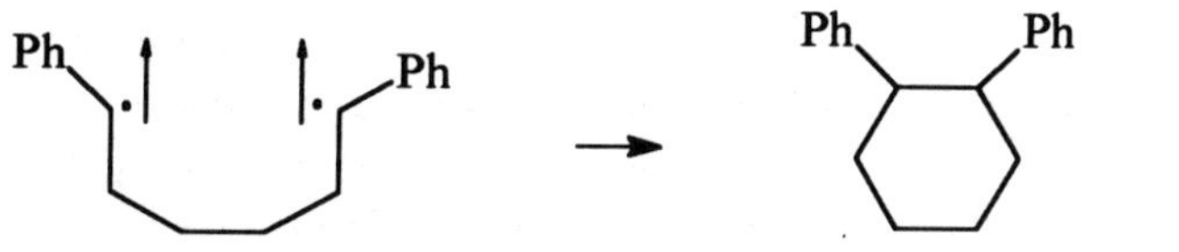

and other products

XII

and other products

XIII

At low temperatures or in viscous solvents the rate-determining step is the change to a nearly cyclic conformation rather than intersystem crossing.

3.9.2 Stereochemical Outcomes

Since a planar or rapidly inverting pyramidal radical can react at either of its two faces, diradicals from chiral precursors can give products in which the initial configuration is retained, partially retained, or lost, depending on the relative rates of ring closure and rotation about the bonds connecting the two radical sites. The decomposition products from meso (**XIV**) and *dl* (**XV**)

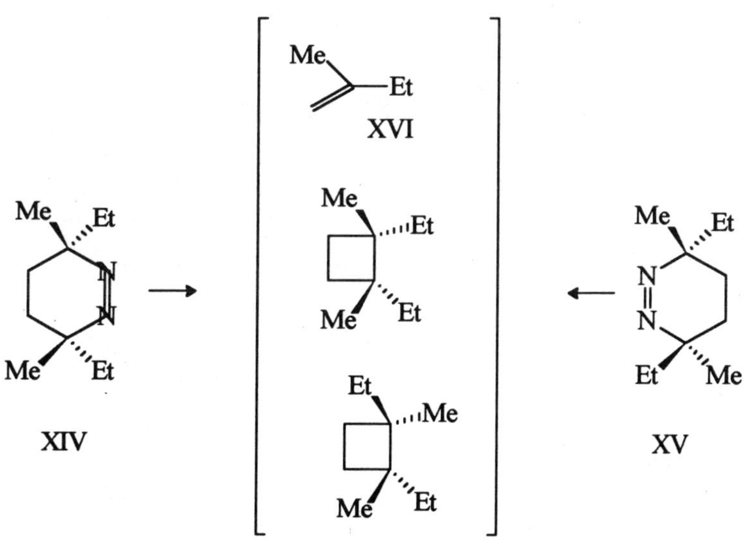

Scheme I

3,6-dimethyl-3,6-diethyltetrahydropyridazines under different conditions (Table 3.1) provide a good example [29–31].

The importance of the multiplicity of the diradical is clear. Thermolysis and direct photolysis, which generate the diradical as a singlet, retain the original stereochemistry in the cyclobutane product to the extent of 95–97%. Under these conditions bond formation is faster than rotation about bonds in the diradical. Sensitized photolysis (with thioxanthone) generates the diradical as a triplet and the need for intersystem crossing makes bond formation slower. As a result, there is much more rotation of the radical face before bonding, and retention of the original stereochemistry decreases to 61–65%. There is also an increase in the yield of the alkene, since that reaction does not have severe conformational requirements either.

3.9.3 Stilbene Isomerization

One of the steps in the photochemical trans–cis isomerization of stilbene (Scheme II) is an example of a reaction in which a diradical has to undergo a change in conformation, although this time no ISC is required. The trans ground state stilbene is excited to a singlet $^1t^*$, which still has the trans configuration. This is then converted to the "perpendicular" singlet $^1p^*$, in which the central bond has been twisted 90°. The perpendicular excited singlet can relax either to cis or trans ground state stilbene [32].

The excited stilbene singlets, in either geometry, can be represented as diradicals by virtue of the weakened central bond. They can also be represented as zwitterions, but this does not imply a polar molecule, since both zwitterionic structures are weighted equally. In the gas phase there is an intrinsic barrier of a few kilocalories per mole to the conversion of $^1t^*$ into $^1p^*$. In solution

TABLE 3.1 Products from a 1,4 Diradical (Scheme I)

	% Yields		
		Cyclobutane	
Conditions and Azo Compound	Alkene **XVI**	meso	*dl*
Thermolysis			
XIV meso	49	43	2.5
XV *dl*	51	3.5	42
Direct photolysis			
XIV meso	61	35	3.5
XV *dl*	60	4	33
Sensitized photolysis			
XIV meso	77	11.5	8
XV *dl*	75	8	12

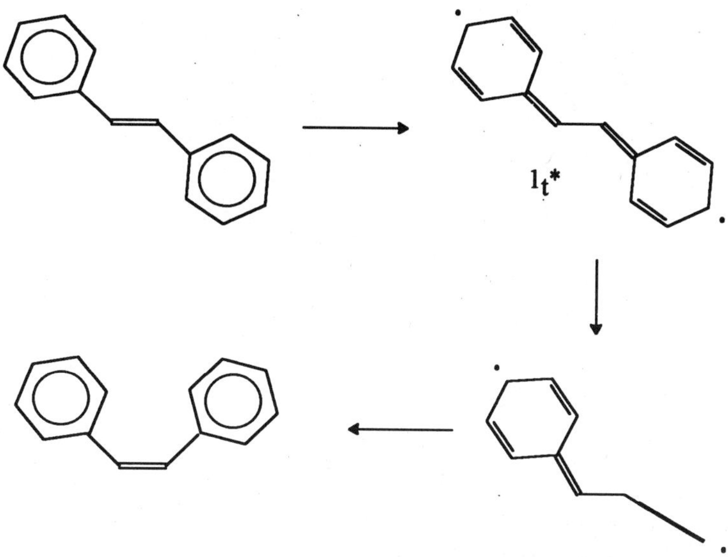

Scheme II

there is an addition to this barrier due to the viscosity of the solvent. The effect of solvent viscosity on the rotation of the C_7H_6 group can be related quantitatively to its effect on the translational diffusion of toluene in the same solvents [33].

3.9.4 Endoperoxide Thermolysis

Endoperoxides of aromatic hydrocarbons decompose into singlet O_2 and the singlet state of the hydrocarbon. The reaction may go by either of two mechanisms, depending on the structure of the endoperoxide. Some endoperoxides, for example, anthracene 1,4-endoperoxides, decompose by a concerted electrocyclic mechanism; others, such as anthracene 9,10 endoperoxides, decompose by a diradical mechanism.

The two mechanisms can be distinguished by their entropies of activation: about zero for the concerted reaction and about $+8$ to $+10$ eu for the diradical mechanism [34]. The diradical mechanism is subject to magnetic field and magnetic isotope effects.

The diradical in (Eq. 3.14) is initially a singlet; if it remains a singlet, the reaction will give singlet hydrocarbon and singlet oxygen. If intersystem crossing to the triplet occurs, the product will be singlet hydrocarbon and triplet oxygen. Because intersystem crossing is catalyzed by the magnetic ^{17}O nucleus, the triplet O_2 is more likely to be formed if the peroxide contained ^{17}O. As a result, the triplet O_2 is selectively enriched in ^{17}O.

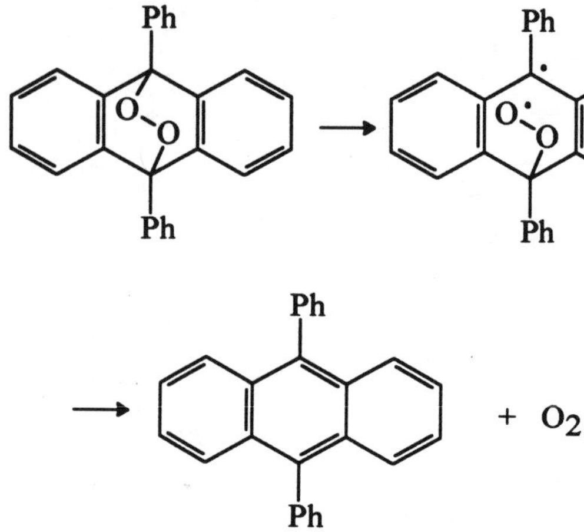

(3.14)

The hydrocarbon helianthrene (HEL) and its enedoperoxide (HELPO) are a special case [35]. HELPO (Eq. 3.15) can be formed not only by the reaction of helianthrene with singlet oxygen, but also by reaction with triplet oxygen.

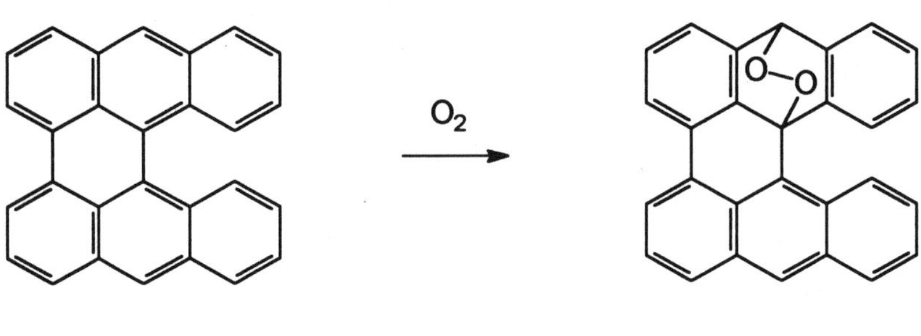

(3.15)

HEL HELPO

The reaction between singlet HEL and triplet oxygen to give singlet HELPO is formally forbidden. The rate is increased both by the use of a more polar solvent and by the use of iodobenzene as solvent. These effects support a mechanism such as that in Scheme III, in which a rate-determining step involves both the creation of a polar intermediate ($^{+}$HEL–O–O^{-}) and intersystem crossing.

$$\mathrm{HEL} + {}^3\mathrm{O}_2 \rightleftharpoons \mathrm{HEL}\cdots{}^3\mathrm{O}_2 \rightleftharpoons {}^{+}\mathrm{HEL{-}O{-}O}^{-} \rightleftharpoons \mathrm{HELPO}$$

Scheme III

The unique reactivity of helianthrene may be a consequence of its nonplanar helical structure [36]. The twisted structure is believed to strain the C=C bonds at the site of the future endoperoxide, a strain that is relieved in the peroxide and probably in the zwitterion.

3.10 TRANSFER OF TRIPLET EXCITATION ENERGY

Because triplets have longer radiative lifetimes than singlets they have a better opportunity to transfer the excitation energy to an acceptor molecule. The process is efficient whenever the acceptor triplet is lower in energy than the donor triplet. The use of benzophenone as a triplet sensitizer in photochemistry is a practical example.

The mechanism of triplet–triplet energy transfer is a simultaneous transfer of two electrons in opposite directions [37,38] and it requires an approach of the donor and acceptor to within about 10–15 Å, the rate decreasing rapidly with distance. In a series of molecules in which a triplet-excited 4-benzophenoneyl donor was separated from a 2-naphthyl acceptor by equatorial cyclohexane or decalin spacers, the triplet–triplet transfer rate decreased as an exponential function of the number of intervening bonds. For example, the rate constant with a 1,3-diequatorial cyclohexane spacer (4 bonds) is $7.7 \times 10^9\ s^{-1}$, but with a 2,6-diequatorial decalin spacer (7 bonds) it is only 3.1×10^6.

In a crystalline solid solution in which the host molecules of the crystal have a sufficiently low-lying triplet state, excitation can be transferred between solute molecules by way of triplet states of the host. The triplet states of the host will be highly delocalized *excitons* assignable to broad energy bands of the crystal lattice rather than to individual molecules. Exciton bands are analogous to the conduction electron bands of semiconductors in that an activation energy must be supplied to promote the spin from the solute triplet to the exciton band. For example, when triplet excitation is transferred from a phenanthrene triplet to a naphthalene molecule in a crystalline biphenyl matrix, experiments at several temperatures showed that the process has an activation energy equal to the difference in energy of the phenanthrene and biphenyl triplet states [39]. A similar process is responsible for the triplet–triplet annihilation of β-methylnaphthalene triplets in a naphthalene host: a triplet exciton enters the naphthalene exciton band and migrates to the site of another β-methylnaphthalene triplet, at which point both triplet excitons are destroyed, forming a β-methylnaphthalene excited singlet in their place [40]. The radiative deactivation of the singlet produces a delayed fluorescence long after the original singlets would have decayed.

REFERENCES

1. Walling, C. *J. Am. Chem. Soc.* **1965,** *87,* 3361.
2. Wagner, P. J. *Acc. Chem. Res.* **1971,** *4,* 168.

3. Turro, N. J. *Modern Molecular Photochemistry,* Benjamin/Cummings: Menlo Park, CA, 1978.
4. Adam, W.: Hannemann, K.; Wilson, R. M. *J. Am. Chem. Soc.* **1986,** *108,* 929.
5. Adam, W.; Grabowski, S.; Wilson, R. M. *Acc. Chem. Res.* **1990,** *23,* 165.
6. Wilson, T. *J. Am. Chem. Soc.* **1966,** *88,* 2898.
7. Adam, W.; Platsch, H.; Wirz, J. *J. Am. Chem. Soc.* **1989,** *111,* 6896.
8. Gordy, W. *Theory and Applications of ESR,* Wiley: New York, 1980.
9. Wasserman, E.; Smolinsky, G.; Yager, W. A. *J. Am. Chem. Soc.* **1964,** *86,* 3166.
10. Hutchison, C. A. J.; Mangum, B. W. *J. Chem. Phys.* **1961,** *34,* 908.
11. Hutchison, C. A. J.; Mangum, B. W. *J. Chem. Phys.* **1958,** *29,* 952.
12. Bersohn, M.; Baird, J. C. *An Introduction to ESR,* Benjamin: New York, 1966.
13. Wasserman, E.; Trozzolo, W. A.; Yager, W. A.; Murray, R. W. *J. Chem. Phys.* **1964,** *40,* 2408.
14. Brandon, R. W.; Closs, G. L.; Davoust, C. E.; Hutchison, C. A. Jr.; Kohler, B. E.; Silbey, R. *J. Chem. Phys.* **1965,** *43,* 2006.
15. Trozzolo, A. M.; Murray, R. W.; Wasserman, E. *J. Am. Chem. Soc.* **1962,** *84,* 4990.
16. Trozzolo, A. M. *Acc. Chem. Res.* **1968,** *1,* 329.
17. Wasserman, E.; Murray, R. W. *J. Am. Chem. Soc.* **1964,** *86,* 4203.
18. Hutchison, C. A. J.; Pearson, G. A. *J. Chem. Phys.* **1967,** *47,* 520.
19. Hutchison, C. A. J. *J. Phys. Chem.* **1967,** *71,* 203.
20. Briere, R.; Dupeyre, R.-M.; Lemaire, H.; Morat, C.; Ramsat, A.; Rey, P. *Bull. Soc. Chim. Fr.* **1965,** 3290.
21. Yang, N. C.; Castro, A. J. *J. Am. Chem. Soc.* **1960,** *82,* 6208.
22. Glarum, S. H.; Marshall, J. H. *J. Chem. Phys.* **1967,** *47,* 1374.
23. Waring, R. K. J ; Sloan, G. J. *J. Chem. Phys.* **1964,** *40,* 772.
24. Salem, L.; Rowland, C. *Angew. Chem. Int. Ed.* **1972,** *11,* 92.
25. Zimmt, M. B.; Doubleday, C. J.; Gould, I. R.; Turro, N. J. *J. Am. Chem. Soc.* **1985,** *107,* 6724.
26. Carlacci, L.; Doubleday, C. J.; Furlani, T. R.; King, H. F.; McIver, J. W. J. *J. Am. Chem. Soc.* **1987,** *109,* 5323.
27. Zimmt, M. B.; Doubleday, C. J.; Turro, N. J. *J. Am. Chem. Soc.* **1986,** *108,* 3618.
28. Wang, J.; Doubleday, C. J.; Turro, N. *J. Am. Chem. Soc.* **1989,** *111,* 3962.
29. Bartlett, P. D.; Porter, N. A. *J. Am. Chem. Soc.* **1968,** *90,* 5317.
30. Engel, P. S. *Chem. Rev.* **1980,** *80,* 99.
31. Casey, C. P.; Boggs, R. A. *J. Am. Chem. Soc.* **1971,** *94,* 6457.
32. Saltiel, J.; Sun, Y.-P. *J. Phys. Chem.* **1989,** *93,* 6246.
33. Sun, Y.-P.; Saltiel, J. *J. Phys. Chem.* **1989,** *93,* 8310.
34. Turro, N. J.; Chow, M.-F.; Rigaudy, J. *J. Am. Chem. Soc.* **1981,** *103,* 7218.
35. Seip, M.; Brauer, H.-D. *J. Am. Chem. Soc.* **1992,** *114,* 4486.
36. Theilacker, W.; Thomas, W. *Liebigs Ann. Chem.* **1960,** *632,* 115.

37. Closs, G. L.; Piotrowiak, P.; MacInnis, J. M.; Fleming, G. R. *J. Am. Chem. Soc.* **1988,** *110,* 2652.
38. Closs, G. L.; Johnson, M. D.; Miller, J. R.; Piotrowiak, P. *J. Am. Chem. Soc.* **1989,** *111,* 3751.
39. Hirota, N.; Hutchison, C. A. J. *J. Chem. Phys.* **1965,** *42,* 2869.
40. Sternlicht, H.; Nieman, G. C.; Robinson, G. W. *J. Chem. Phys.* **1963,** *38,* 1326.

4

RADICAL PAIRS

When a pair of radicals is created in solution, either by the homolysis of a precursor molecule or after coming together by diffusion, a surrounding "cage" of solvent molecules prevents their immediate separation. Radical pairs produced from the same precursor molecule are called *geminate pairs* [1] (Latin, *gemini*, twins) and those formed by diffusion are called *encounter pairs*.

The formation of *geminate products* from the geminate pair requires some constraint on diffusion to prevent the original radicals from losing each other permanently and irreversibly. This constraint can be supplied by a solvent, by a surround zeolite cage, [2] or by adsorption on a silica surface [3–5]. As we shall see, products from the escaped radicals differ in important ways from the geminate products.

4.1 ENCOUNTER PAIRS AND DIFFUSION-CONTROLLED REACTIONS

Any bimolecular reaction involves two steps, the diffusion of the reagents together to form an *encounter pair* or encounter complex and a suitable collision to bring about the reaction. Early students of collision theory were bothered by the fact that the rates of all but a few bimolecular reactions seemed to be completely insensitive to the viscosity of the solvent despite the obvious need for diffusion. This was explained by Rabinowitch and Wood [6,7] after they had done some experiments with a device resembling a shaking billiard table packed with solvent balls and two reagent balls. It turned out that collisions occur in sets. Once the reagents have encountered each other, the surrounding solvent tends to keep them together so that they collide, in a solvent of ordinary viscosity, some 10–100 times. If collisions occur in encounter sets, the rate at which collisions occur is the encounter rate times the number of collisions per encounter (Eq. 4.1).

$$\begin{aligned} k_2 &\propto \text{collisions/sec} \\ &\propto (\text{encounters/sec}) \times (\text{collisions/encounter}) \end{aligned} \tag{4.1}$$

The reason for the insensitivity of the rates of most reactions to the viscosity of the medium is simply that an increase in viscosity decreases the frequency of encounters, but also *increases* the number of collisions per encounter. These effects always cancel unless the reaction is inherently so probable, or the encounter so prolonged, that reaction during the encounter is inevitable. The rates of such *diffusion-controlled* reactions are proportional to the encounter frequency rather than the collision frequency. Diffusion control of the rate occurs in solvents of ordinary viscosity when the rate constant for the reaction of an encounter complex exceeds about 10^{10} sec^{-1}.

The change from encounter-controlled to collision-controlled reactions should occur for activation energies of about 2 kcal mol^{-1}, assuming a normal preexponential factor. Radical recombination or disproportionation reactions tend to be diffusion controlled or on the borderline for diffusion control [8], with some of the radicals reacting during the first encounter and others separating without reaction. The termination reaction (R· + R·) of reactive carbon-centered radicals is close to diffusion controlled even in solvents like isooctane at room temperature. For example, the rate constant for the bimolecular self termination of *n*-pentyl [9] is 2.4×10^{10} M^{-1} s^{-1}; for *t*-butyl [8] it is 7×10^9 M^{-1} s^{-1}; and for benzyl [10] the rate is 5×10^9 M^{-1} s^{-1}.

In some cases termination become diffusion controlled as the viscosity is increased. An example is the almost explosive surge in the rate of radical chain polymerization of methyl methacrylate as the concentration of polymer builds up [11]. This is caused by a slowing down of the radical plus radical chain breaking step in the increasingly more viscous medium. Long-chain growing polymer radicals tend to become entangled with the polymer, but can still grow because the small monomer radicals can diffuse to them.

Ultimately, when the monomer is used up, some long-chain radicals remain stuck in the solid polymer. The stuck radicals can be detected and even titrated but quickly disappear when the polymer is dissolved [12,13].

A highly cross-linked and insoluble material, called popcorn polymer because of its appearance, sometimes forms in fractionating columns used for the distillation of styrene. Popcorn polymer contains stuck radicals and can initiate the polymerization of styrene. Almost any solvent molecule diffusing into a solid polymer will cause the solid to swell, but when styrene monomer diffuses into the popcorn it adds to the stuck radicals and the swelling is irreversible. The swelling also fragments the popcorn, breaking bonds and generating new radicals. Eventually the fractionating column becomes completely clogged by the growth of this kind of polymer.

Other examples of diffusion-controlled reactions are the quenching of fluorescence [14] and very fast proton transfer reactions.

4.1.1 Spin-Statistical Factors

The radicals that diffuse together to form an encounter pair are doublets, each of which might have the spin quantum number m_s equal to $+\frac{1}{2}$ or $-\frac{1}{2}$. The various possible spin states of the resulting encounter complex are, with equal probability, the singlet state ($S = 0$) and the $S_z = 1, 0$, or -1 sublevels of the triplet [14]. As a result, about one quarter of the encounter pairs have a fairly high probability of reacting to form a product molecule (singlet state), and the other three quarters have a fairly high probability of separating without reacting. The ratio is not exactly 1:3 because of intersystem crossings.

4.2 GEMINATE PAIRS

The probability that radicals which have escaped from a geminate cage and diffused apart by more than a few molecular diameters will ever reencounter each other is vanishingly small. Since the same product can often be formed both from a geminate pair and from an encounter pair, the yield of a given product may have a contribution from both processes. The contribution of nongeminate reaction to the yield can be ascertained by running the reaction in the presence of an efficient *scavenger* for the diffusing radicals. The part of the product that cannot be intercepted by the scavenger is geminate product.

Because reaction within the cage may be faster than rotations or conformational changes, the geminate radicals are more likely to retain some of their original structure and stereochemistry in the reaction product.

4.2.1 Scavenging

Some scavengers are stable radicals such as galvinoxyl (**I**) [15], 2,3′,5′,6-tetra-*t*-butylindophenoxyl [16] (BIP, **II**), bisdiphenylenephenylallyl [17] (BDPH, **III**), or the nitroxyl radicals **IV** [18], **V** (TEMPO), or **VI** (TMIO) [19]. These stable radicals trap short-lived free radicals by combining with them. Other scavengers are nitroso spin traps such as 2-methyl-2-nitrosopropane, which gives adducts with recognizable ESR patterns, or atom transfer reagents such as thiophenol.

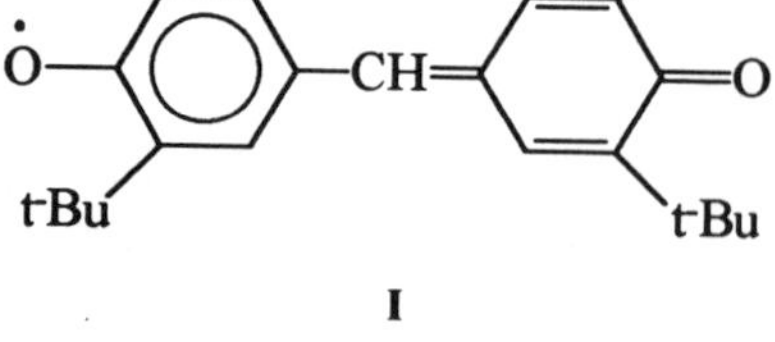

I

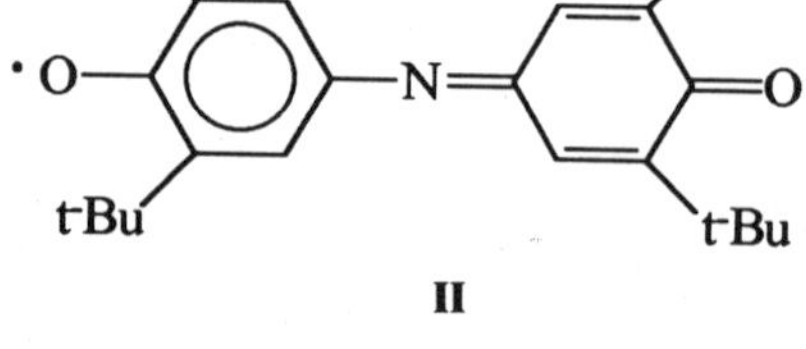

II

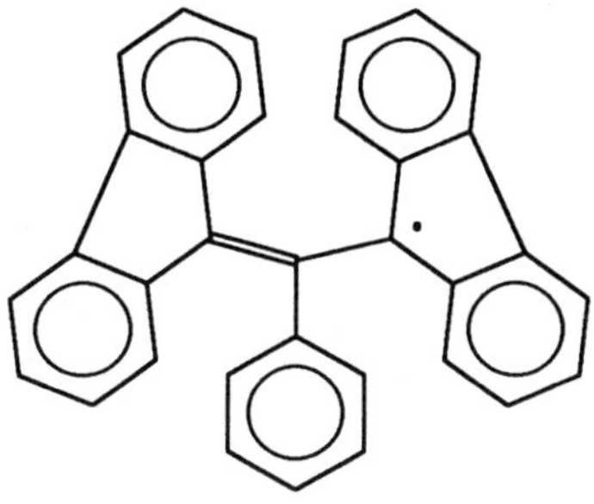

III

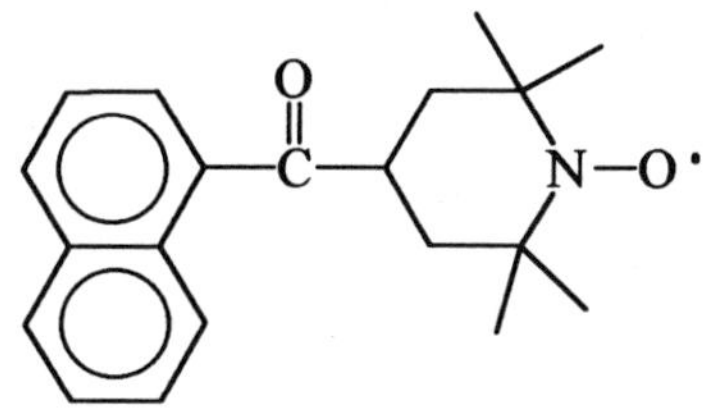

IV

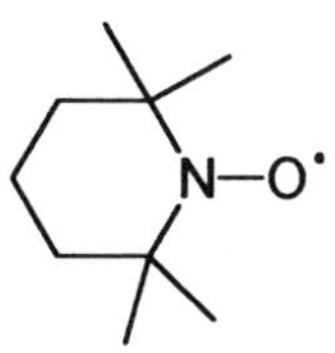

V

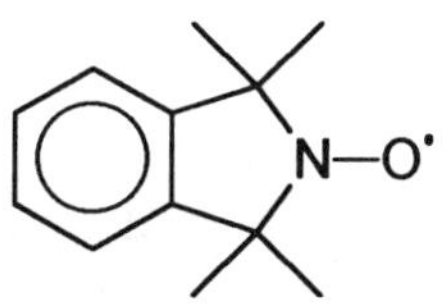

VI

At very low scavenger concentrations trapping is inefficient, but as the concentration of scavenger is increased, the number of trapped radicals rises, eventually levelling off. This is the concentration at which all of the freely diffusing radicals (but not geminate pairs) are being trapped. Because the reaction with the scavenger may be diffusion controlled at high viscosities, a higher scavenger concentration may be needed to trap all of the radicals in viscous solvents [20].

4.2.2 Rates of Production of Diffusing Radicals

The rate constant k_{1s} for the production of pairs of scavengable radicals from the decomposition of an initiator (radical precursor) can be measured by following the disappearance of a low concentration of the scavenger in the presence of a large excess of the initiator. In Eq. 4.2 λ is the slope of the linear plot of optical density against time, P_{av} is the average concentration of the initiator or radical precursor, and ϵ is the extinction coefficient of the scavenger [15].

$$k_{1s} = -\lambda/2P_{av}\epsilon \tag{4.2}$$

Radical Efficiencies and Cage Yields. The fraction of radicals that escape from the cage (the radical efficiency of the initiator) is the first-order rate constant k_{1s} for the production of scavengable pairs of radicals divided by the first-order rate constant for the decomposition of the initiator. The amount of cage reaction, or cage yield, is 1 minus the radical efficiency. A few examples are given in Table 4.1. Some initiators give caged pairs in which

TABLE 4.1 Typical Cage Yields

Initiator	Scavenger	% Cage	Reference
Azobis-isobutyronitrile[a]	Galvinoxyl	34	15
Di-*t*-butylperoxyoxalate[b]	Galvinoxyl	<1	15
Di-*t*-butylmonoperoxyoxalate[c]	Galvinoxyl	51	21
Azobis-α-phenylethane[d]	Nitroso[e]	32	22
t-Butyl triphenylperacetate[f]	None	<11	23

[a]In benzene at 61.65°C.
[b]In benzene at 25.48°C.
[c]In benzene at 25.5°C.
[d]In benzene at 105°C.
[e]2-Methyl-2-nitrosopropane.
[f]In toluene at 25°C, estimated from viscosity effects.

reaction of the pair is hindered by intervening molecular fragments such as N_2 or CO_2, which allow more of the radicals to escape. The less than 1% of cage reaction and greater than 99% radical efficiency for di-*t*-butylperoxyoxalate (DBPO, **VII**), in which the caged radicals are separated by two CO_2 molecules, is an example.

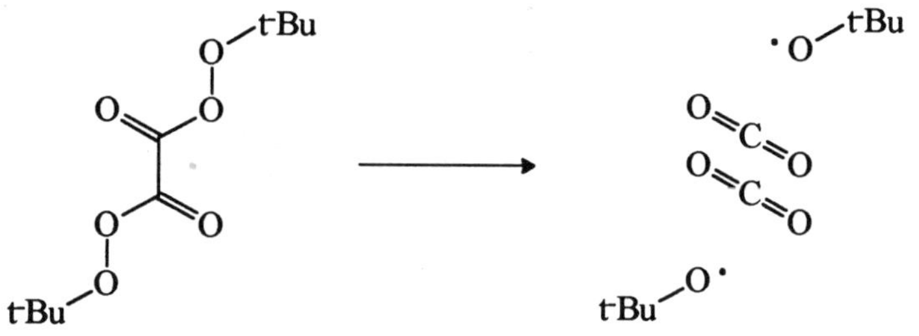

VII

The radical efficiency of di-*t*-butylmonoperoxyoxalate (DBMPO, **VIII**), which forms only one intervening molecule of CO_2, is much less than that of DBPO [21]. Its principal geminate product is di-*t*-butyl carbonate, formed in 35% yield. In the absence of scavenger the carbonate is formed in slightly higher yield because there is also some nongeminate combination of the *t*-butoxy and *t*-butoxycarbonyl radicals.

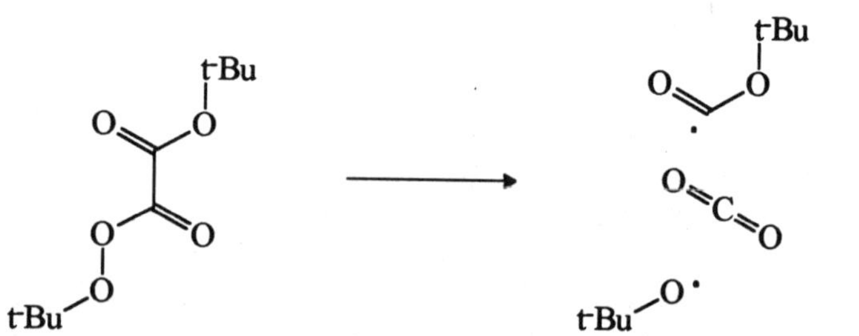

VIII

Effects of Viscosity on Cage Yields. As the solvent is made more viscous, the yield of geminate recombination products increases and the yield of freely diffusing radicals decreases. The formation of di-*t*-butyl peroxide cage product from the decomposition of di-*t*-butyl peroxyoxalate **VII** is a good example. In *n*-pentane at 45°C the yield of di-*t*-butyl peroxide is about 4%, in Nujol, a high molecular weight paraffin oil about 500 times as viscous as pentane, the yield is about 77% [24].

It has been found experimentally for several reactions that the fraction of the radical pairs that escape from the cage is a linear function of $(1/\eta)^n$, with values of n between 0.5 [25] and 1.0 [26]. A theory based on Brownian motion predicts a value of 0.5 for n [27,28], while the Noyes model [29] predicts a value of 1. However, the fit of these relationships for a given value of n is often poor for too great a range in the viscosities and the best value may depend on the type of solvent. Apparently, specific solvation effects must be taken into account [30].

Effects of Multiplicity on Cage Yields. Since the probability of formation of a singlet product depends on the electronic multiplicity of the radical pair, triplet-sensitized photolysis gives less geminate recombination and greater radical efficiency than does thermolysis or singlet-sensitized photolysis.

The homolysis of benzoyl peroxide is an example. Although thermolysis gives very little of the phenyl benzoate cage product, probably because the benzoyloxy radical does not decarboxylate very much during the short lifetime of the cage at thermolysis temperatures, direct photolysis and photolysis sensitized by the singlet excited states of aromatic hydrocarbons gave phenyl benzoate in 8–15% yield. Photolysis with *triplet* sensitizers such as benzophenone gave only a 3% yield of this cage product [31].

Another example is the decomposition of azobis-(3-methyl-1-butene), in which direct photolysis in benzene gave 60% cage product, whereas triplet-sensitized photolysis gave only 7–13%, depending on the sensitizer [32]. However, the difference in cage yield may not be due entirely to the spin correlation effect since the triplet-sensitized photolysis of this azo compound breaks only one bond in the initial step.

The photolysis of azobis-(isobutyronitrile) adsorbed on silica in a benzene slurry gave about 80% geminate recombination in the absence of sensitizer, but about 13% with benzophenone added as a triplet sensitizer [3].

Effects of Magnetic Isotopes. Because intersystem crossing is catalyzed by magnetic nuclei, products formed from escaped radicals have a different isotopic composition than geminate products [33, 34]. When dibenzyl ketone is partially photolyzed in micelles or adsorbed on silica (to enhance the cage effect), the recovered dibenzyl ketone is enriched in ^{13}C relative to the initial ketone [33,35]. Most of the enrichment is at the carbonyl carbon, where the electron–nuclear interaction is greatest. Photolysis of the ketone generates a triplet radical pair. Intersystem crossing in the pairs with ^{13}C converts them into singlets, which have greater probability of recombining to reconstitute the singlet ketone.

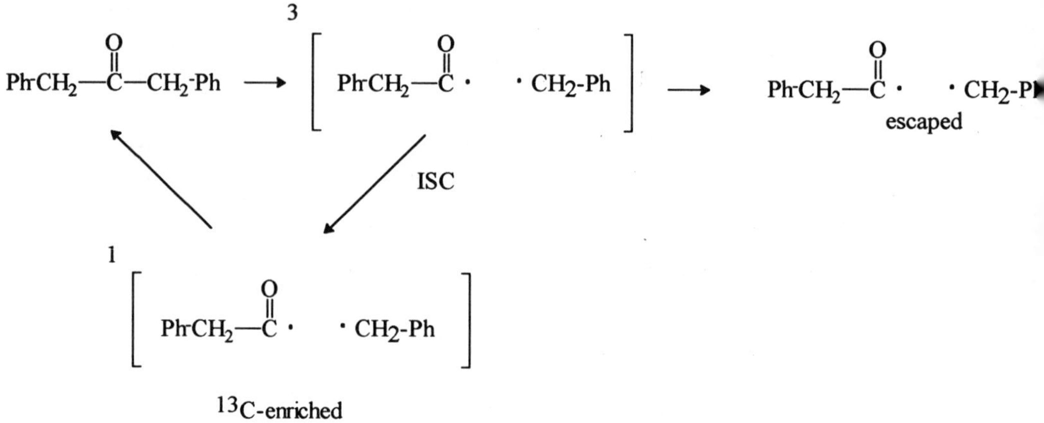

Scheme I

4.2.3 Recombination in the Cage

While the fragment radicals from the homolysis of the initiator or radical precursor are still in the cage, one of the reactions that may occur is recombination to regenerate the original precursor, as in Eq. 4.3.

$$\underset{\text{NAT}}{\mathrm{Ph_3C{-}N{=}N{-}Ar}} \underset{k_{-1}}{\overset{k_1}{\rightleftarrows}} \mathrm{Ph_3C\cdot \ \cdot N{=}N{-}Ar} \xrightarrow{k_D} \text{escape}$$

$$k_{obs} = \frac{k_1 k_D}{k_{-1} + k_D} \tag{4.3}$$

If this is happening, an increase in the lifetime of the cage will decrease the rate constant for the disappearance of the precursor [36,37]. Viscosity effects on the first-order rate constant for the decomposition of *p*-nitrophenylazotriphenylmethane (NAT) are an example. The rate of disappearance of the NAT is given by k_{obs}, which will be less than k_1 unless k_D is $\gg k_{-1}$. Assuming that only k_D is sensitive to changes in the viscosity (η), k_{obs} should obey Eq. 4.4, where A_v is the Arrhenius preexponential factor for viscous flow and A_D is the preexponential factor for k_D. The best fit to Eq. 4.3 is obtained when the empirical parameter α is set at 0.5 [36].

$$\frac{1}{k_{obs}} = \frac{1}{k_1} + \left(\frac{k_{-1}}{k_1 A_D}\right)\left(\frac{\eta}{A_v}\right)^{\alpha} \tag{4.4}$$

The value of k_1 can be obtained from the intercept of the equation.

Multibond Versus One-Bond Cleavage. If the homolysis of an initiator gives not just a pair of radicals but also a fragment molecule during the lifetime of the cage, the probability of recombination to reconstitute the initiator is small and k_{obs} will be independent of the viscosity. For example, the thermolysis of azocumene (**IX**), which gives resonance stabilized α-cumyl radicals, probably breaks both bonds to nitrogen at the same time.

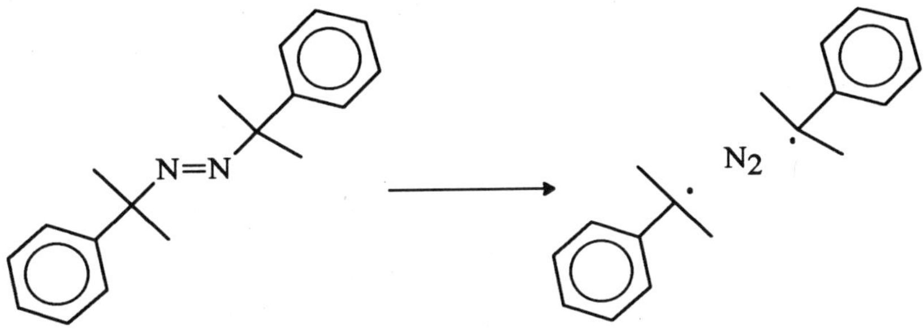

IX

The plot of $1/k_{obs}$ (Eq. 4.4) has zero slope, as would be expected if there is no reformation of the azo compound.

Another example is the thermolysis of *t*-butyl *p*-methylphenylperacetate. Again, the plot of Eq. 4.4 is horizontal, confirming the two-bond cleavage into a *p*-methylbenzyl radical, a molecule of CO_2, and a *t*-butoxy radical, a mode of reaction that is also suggested by the activation parameters [37–39]. In contrast, the thermolysis of *t*-butyl *p*-nitrophenylperacetate is much slower and cleaves only one bond. Although both peresters eventually produce radicals rather than trappable ions, polarized resonance structures can still contribute. With a *p*-methoxy or *p*-methyl substituent the polar structure is important enough to cause two-bond cleavage; with a *p*-nitro substituent it is not.

Acetyl peroxide, despite the fairly easy decarboxylation of the acetoxyl radicals, has been shown to be a one-bond initiator both by ^{18}O scrambling in the partly decomposed peroxide [40] and by kinetic isotope effects [41]. Again, the plot of Eq. 4.4 is horizontal.

Rebonding of Molecular Fragments. Although the probability of reforming the initiator by the reaction of the radical pair with a fragment molecule is very low, addition of a cage radical to a molecule in the cage can occur. A particularly interesting example is a radical switch reaction (Eq. 4.5) in which a molecule of SO_2 is detached from one radical and rebonded to another within the cage [42].

$$\text{Ph}\cdot\ \cdot\text{SO}_2\text{Bu–}t \rightleftharpoons \text{Ph}\cdot\ \text{SO}_2\ \cdot\text{Bu–}t \rightleftharpoons \text{PhSO}_2\cdot\ \cdot\text{Bu–}t \qquad (4.5)$$

A similar rebonding is unlikely for CO_2 and a radical, although CO_2 can react with carbenium ions. Some ambiguity is introduced by the fact that ion-pair structures are part of the valence bond representation of a singlet radical pair. In the polar decomposition of some diacyl peroxides, for example, a polar transition state is believed to give an intermediate of hybrid radical-pair ion-pair character that can generate both ions and radicals. The carboxy-inversion product or carbonic acid ester anhydride formed from one such peroxide has been shown to incorporate a rebonded CO_2 molecule [43,44]. However, the lack of CIDNP signals from such products means that the rebonding must take place after the hybrid pair has lost any radical character. The photolysis of diacyl peroxides, which should give a radical pair accompanied by CO_2, does not give any of the carbonic ester anhydride [45].

4.2.4 Motions In the Cage

Changes in the orientation of the geminate radicals during the lifetime of the cage can affect stereochemistry, ambident bonding, label scrambling, and the ratio of combination to disproportionation.

Stereochemical Effects. When a chiral group is released from a molecule in the form of a freely diffusing radical, it becomes effectively nonchiral, either because it is planar or because of rapid inversion. However, chirality need not be lost in a geminate pair if combination or disproportionation is faster than rotation of the prochiral radical within the cage.

The thermal decompositon of *SS*-(−)-azobis-α-phenylethane (Table 4.1) gives singlet pairs of α-phenylethyl radicals. The recombination of these radicals without rotation in the cage would give non-meso diphenylbutanes exclusively, while randomly oriented radicals are known to give non-meso and meso diphenylbutanes in equal amounts. The cage product, in benzene at 105°C in the presence of a scavenger, consisted of non-meso and meso in a ratio of 1.05. Rotation of the radical is faster than recombination, but not quite fast enough to randomize the configuration entirely.

The thermal rearrangement of the chiral ketimine **X** [46] into the nitrile is a reaction that might, in principle, go either by a concerted [1,3]-sigmatropic rearrangement or by dissociation into a radical pair followed by recombination. The sigmatropic rearrangement with inversion is an allowed process, but in chemistry not everything that is allowed is compulsory [47]. In any case, the observed result in CCl_4 at 60°C in the presence of a scavenger is about 47% net retention, radical efficiency 25%, supporting the radical dissociation mechanism. The steric requirements for the allowed sigmatropic rearrangement may be too severe.

The behavior of 1-methyl-2,2-diphenylcyclopropanecarbonyl peroxide is exceptional among α-branched dialkanoyl peroxides in that its decomposition appears to be entirely a radical reaction. The decomposition of the chiral peroxide in CCl_4 solution at 67°C gives the corresponding 1-methyl-2,2-diphenylpropane, a cage product, with about 67% retention of configuration [48]. Like cyclopropyl itself [49], this radical is probably rapidly inverting

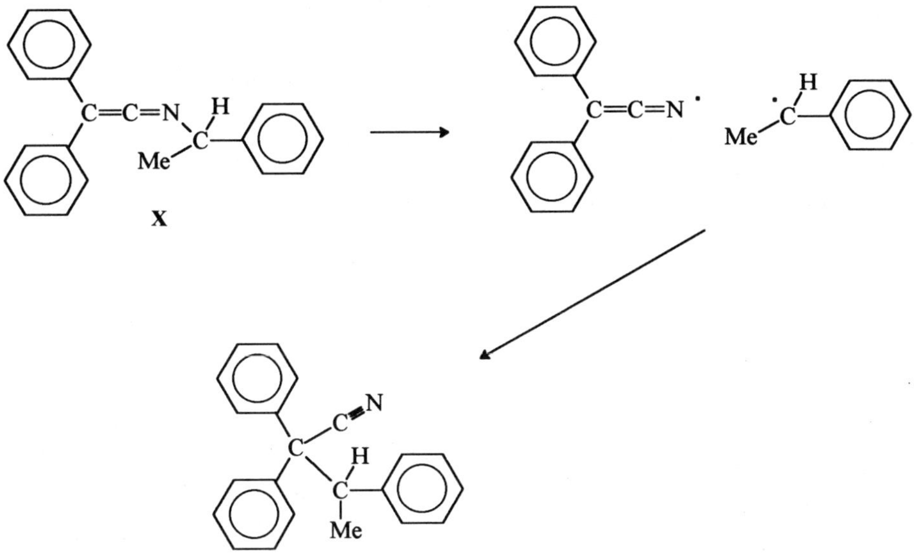

X

rather than planar. Its rotation to present the opposite face evidently occurs at a rate comparable to that of escape from the cage.

Values of the ratios of the rates for rotation of chiral radicals in the cage (k_R) to combination (k_2) in the cage are given in Table 4.2. They depend on the size of the chiral radical, the presence of intervening fragments in the cage, the temperature, and the viscosity. Both rotational diffusion and translational diffusion are dependent on the viscosity of the medium, being ap-

TABLE 4.2 Rotation versus Combination

Radical Pair	Solvent (°C)	k_R/k_2	Ref.
$PhCH_2\cdot$ N_2 $\cdot CH(CH_3)Ph$	Benzene(110)	16	54
$Ph(CH_3)CH\cdot$ N_2 $\cdot CH(CH_3)Ph$	Benzene(105)	15	
$Ph(CH_3)CH\cdot$ $\cdot N{=}NCH_3$	Hexadecane(162)	30	55
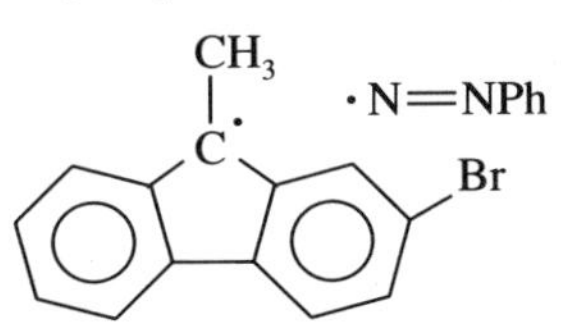	Heptane (80)	0.94	52
	Octadecane	0.30	
$CH_3CH_2(CH_3)CH\cdot$ CO_2 $\cdot OC(CH_3)_3$	Paraffins(101.8)	23[a]–232[b]	51
$Ph(CH_3)CH\cdot$ CO_2 $\cdot OC(CH_3)_3$	Cumene(60)	19	56
$(Ph)_2C{=}C{=}N\cdot$ $\cdot CH(CH_3)Ph$	CCl_4(60)	0.4	46
	CH_3CN(60)	0.8	

[a]Paraffin oil.
[b]Octane.

proximately proportional to $\eta^{-\alpha}$, where α is usually in the range 0.5–0.75 [37,50–52]. In a frozen matrix, which represents the ultimate limit of viscosity, nothing can escape from the cage, and there is also a constraint on rotation. For example the photolysis of meso azobis(2-phenyl-3-methylbutane) **XI** in methylcyclohexane glass gives exclusively the meso isomer of 3,4-diphenyl-2,3,4,5-tetramethylhexane along with disproportionation products [53].

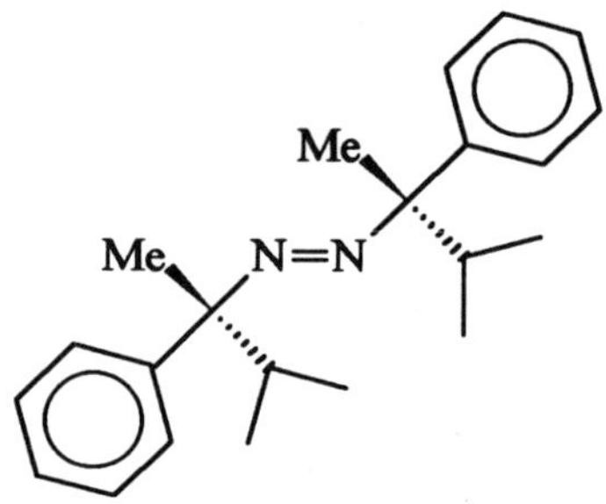

XI

The dissociation of a molecule like (2*S*,3*S*) methyl *threo*-2,3-diphenylbutane-2-carboxylate, followed by recombination in the cage (Scheme II), can give both enantiomers of the diastereomeric erythro product. This product will be racemic only if the two chiral radicals have the same probability of rotating within the cage. The experimental result, an enantiomeric excess of the (+) erythro enantiomer, indicated a large rotational preference for the α-phenylethyl radical over the α-carbomethoxy-α-phenylethyl radical, a factor of 1.8 or greater [57].

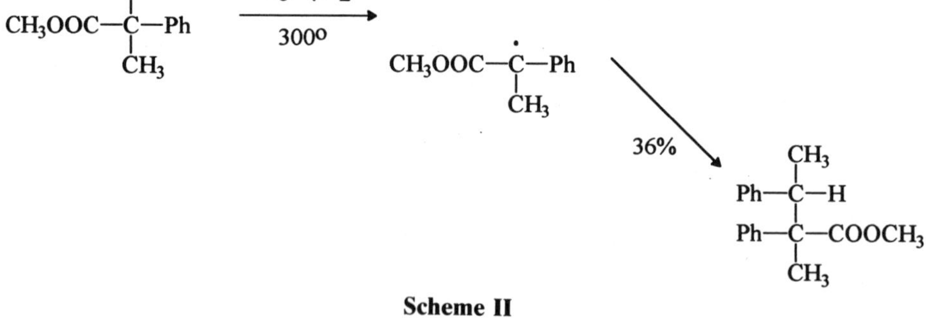

Scheme II

Disproportionation Versus Combination. Radical disproportionation and combination reactions in principle have two steps: rotation of the radicals into an appropriate orientation, and the actual bond formation. In solvents viscous enough to slow down rotations within the cage, rotation can be the rate-determining step, in which case disproportionation/combination ratios (k_d/k_c) become dependent on the viscosity. Geminate and encounter pairs might give *different* ratios, since the initial orientation of the geminate pair is nonrandom, established by the radical precursor.

The behavior of *encounter* pairs of *t*-butyl radicals was studied by photolyzing di-*t*-butyl ketone. Decarbonylation of *t*-butylcarbonyl radicals is slow enough so that there are no geminate pairs of *t*-butyl radicals. The behavior of *geminate* pairs of *t*-butyl radicals was studied by photolyzing 2,2′-azoisobutane in the presence of thiophenol as scavenger. In both cases k_d/k_c for pairs of *t*-butyl radicals increased with viscosity, starting at about 4 or 5 and increasing to about 6 for solvents like *n*-tetradecane. Consideration of models for the reactions suggests a reason for this: the reorientation leading to combination has to move more solvent molecules [58]. The ratio is about the same for viscosities up to about 3 cP, but beyond that, the geminate k_d/k_c is much less sensitive to increase in viscosity, almost leveling off [59]. The geminate pairs, formed from the cis azo compound, are born with a bias toward combination, and at high viscosities rotations in the cage are too slow to remove this bias.

Similar effects might be observable for other azo compounds, depending on whether they are thermolyzed or photolyzed. Most acyclic azo compounds are trans, and thermolysis, at least of symmetrical azo compounds, probably breaks both bonds to nitrogen at the same time [60]. On the other hand, photolysis of most azo compounds probably isomerizes them to the cis isomer, which then gives a radical pair with a different orientation.

The cumyl radicals from the decomposition of azocumene in benzene solution give a k_d/k_c of about 0.05 and the ratio does not appear to depend on whether the reaction was initiated thermally or photochemically, but photolysis in frozen toluene gave k_d/k_c ratios ranging from 1.5 to 3 [61]. When the azocumene was adsorbed on silica, where surface forces act like a viscous solvent in interfering with motion of the radicals, the k_d/k_c ratio was about 0.28 and not significantly dependent on whether the azo compound was thermolyzed or photolyzed [4].

In the decomposition of azobis-(isobutyronitrile) in solution, k_d/k_c for the isobutyronitrile radicals is only about 0.05. In marked contrast to this, photolysis in the solid state gives a k_d/k_c of about 4 [62]. Two kinds of experiment were carried out with deuterated azobis-(isobutyronitrile) in the solid state, labeling different hydrogens within a methyl group, and labelling entire methyl groups [63]. There is a normal isotope effect for H/D abstraction within a given methyl group, because methyl group rotation is too fast to be rate-determining even in the solid. On the other hand, when entire methyl groups were labeled, the isotope effect disappeared: this means that rotation of the entire isobutyronitrile radical is slow enough so that it, rather than atom transfer, is the rate-limiting step.

Ambident Radicals. The products from a pair of ambident radicals will depend not only on the intrinsic reactivity at the several sites, but also on the orientation of the radicals in the pair.

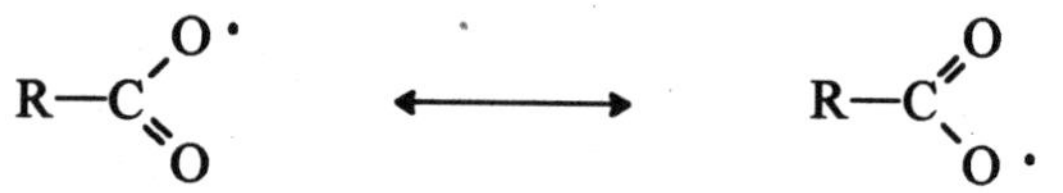

XII

When an acyloxy radical labeled in the carbonyl oxygen reorients itself in the cage before forming a new bond, the label becomes scrambled. For example, in the decomposition of acetyl peroxide in isooctane at 80°C, peroxide recovered after one half-life retains only about 75% of its original carbonyl label, indicating geminate recombination of acetoxy radicals [64]. A lesser amount of scrambling is found in recovered benzoyl peroxide, possibly because bridging by the phenyl group keeps the oxygens distinguishable [65].

Because of steric hindrance, and despite some loss of resonance stabilization, encounter pairs of triphenylmethyl radicals couple at α-para positions rather than just at the α position. The decomposition of *t*-butyl triphenylperacetate forms CO_2 and trityl and *t*-butoxy radicals in about 98% yield. Both the α and para ethers are formed [66]. Since the α ether resists formation by the usual methods, it is probably a cage product. The orientation of radicals generated in the geminate cage is particularly favorable for α bonding, whereas para bonding requires a major rotation.

The cumyl radicals from azodicumyl (Eq. 4.6) also afford alternative bonding sites. In solution about a 2% yield of the α,para dimer can be seen spectroscopically as an intermediate, but it dissociates and eventually gives only the more stable α,α dicumyl and the disproportionation products [60].

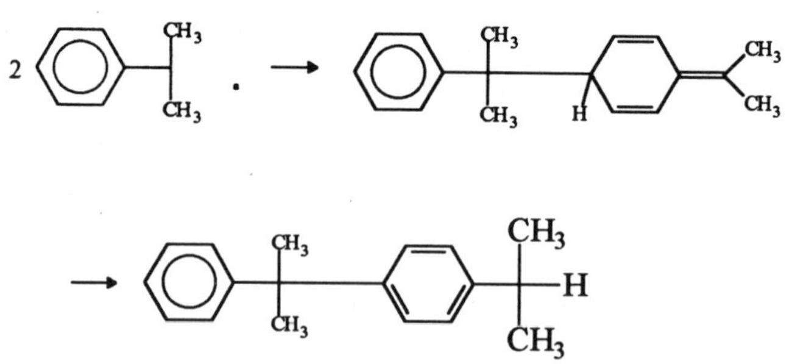

(4.6)

Photolysis of azocumene adsorbed on silica also gives some of the α,para and α,ortho dimers, but on silica these are immediately aromatized [4]. Deuterium labeling showed that the aromatized dimer was entirely a geminate product. The most probable explanation is that the strong adsorption sites catalyze the aromatization of the geminate pairs. Radicals which escape from the strong adsorption site probably also give head-to-tail dimers, but these merely dissociate again.

The thermolysis of azobisisobutyronitrile (AIBN) in CCl_4 at 62.5°C (Scheme III) gives two radical combination products, tetramethylsuccinonitrile (TMSN) and the ketimine (KI) as well as disproportionation products. The combination products KI and TMSN are formed both geminately, within the cage, and nongeminately. The ratio KI/(KI + TMSN) for the geminate part of the reaction is 0.65, but only 0.46 for the noncage reaction [67].

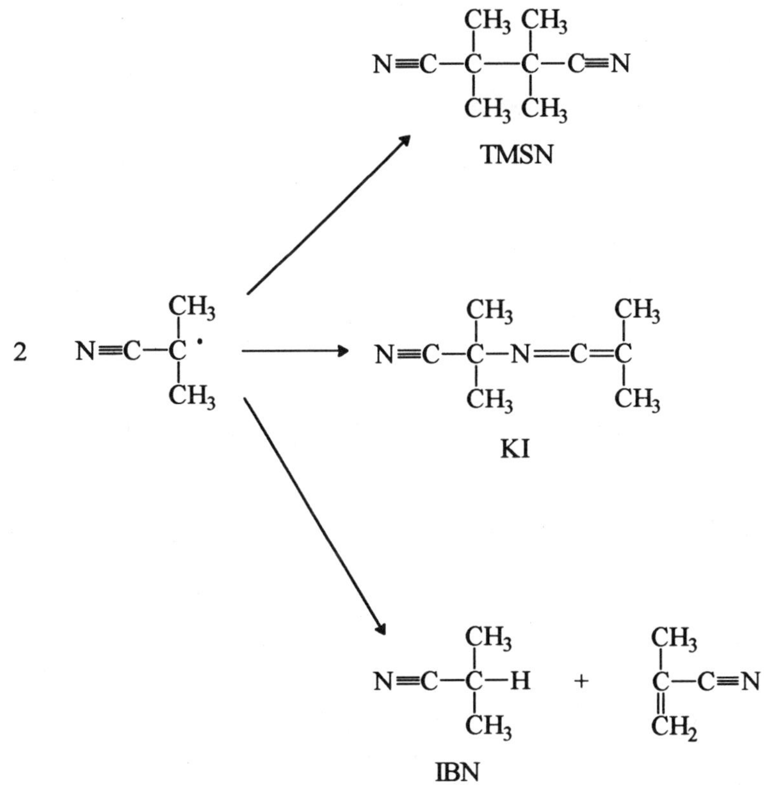

Scheme III

4.3 RATES OF REACTIONS BETWEEN RADICALS

Bimolecular reactions between radicals, involving the formation of an encounter complex and the subsequent reaction of the radical pair, are usually very fast, but there are several methods for measuring the rates.

4.3.1 The Radical Clock Method

The radical clock method gives a rate constant relative to that of a standard "radical clock" reaction, using product ratios.

An example is the reaction of the nitroxyl radical TMIO (**VI**) with 5-hexenyl radical (Scheme IV) with rate constant k_T [68]. The radical clock reaction in this case is the cyclization of the 5-hexenyl radical to cyclopentylmethyl, with rate constant, k_c [69,70]. Products are formed from both of these species, and the product ratio $[H\text{-}T]/[C\text{-}T]$ is related to the rate constant ratio k_T/k_c. When the initial concentration of TMIO is high compared to that of the source of the 5-hexenyl radicals, the ratio of the two rate constants is given by Eq. 4.7.

$$\frac{k_T}{k_c} = \frac{[H\text{-}T]/[C\text{-}T]}{[\text{TMIO}]_i} \tag{4.7}$$

Since the rate constant k_c is known [70,71], k_T can also be evaluated.

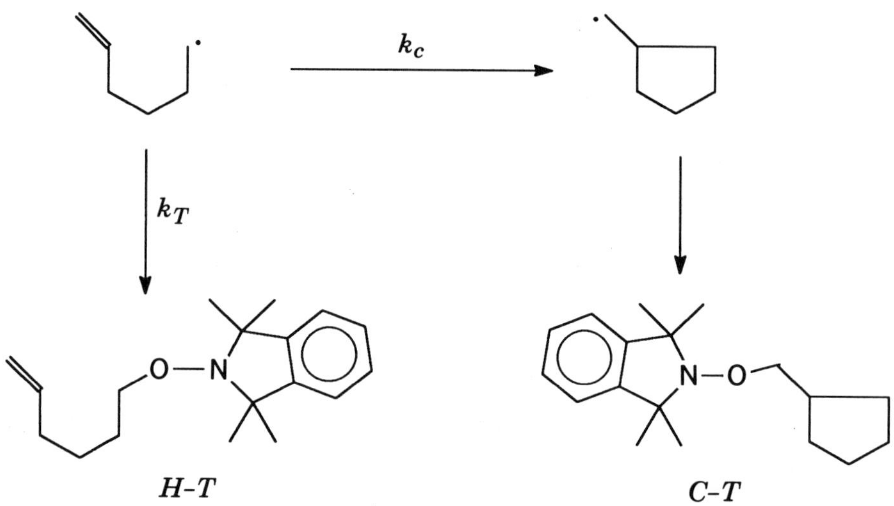

Scheme IV

Some other reactions that have been used as radical clocks are given in Scheme V [68].

4.3.2 Absolute Rates

***Kinetic ESR Spectroscopy* [71,72].** The rate of a radical plus radical reaction (combination plus disproportionation) can be measured by photolysis of the precursor in the cavity of the ESR spectrometer, averaging the rate of a large number of identical radical signal decays. The rate constant for the disap-

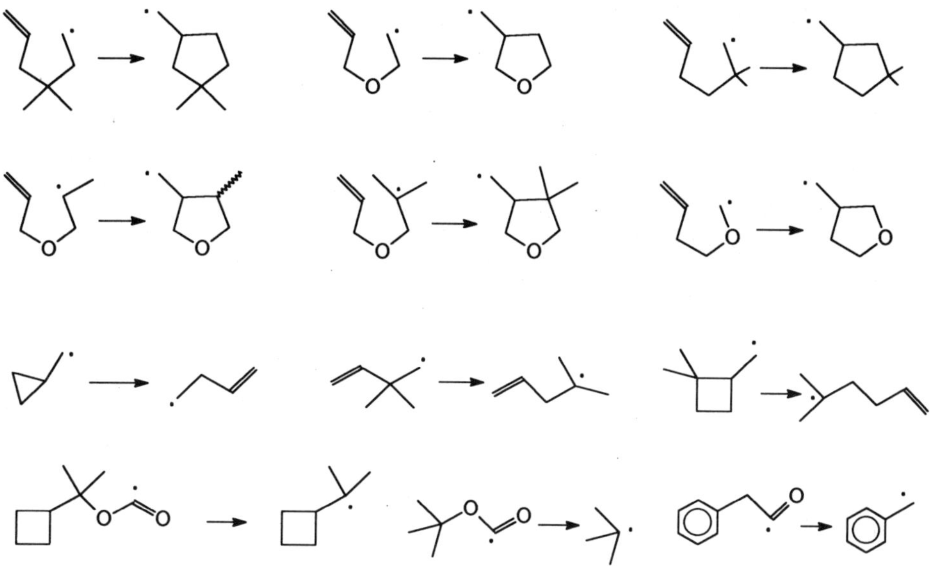

Scheme V

pearance of t-butyl radicals ($2\ k_t$) was determined in this way to be 7.9×10^{-9} dm^3 mol^{-1} s^{-1} in heptane at 294 K [8]. This is within experimental error of the diffusion-controlled rate.

$$t\text{-Bu-CO-}t\text{Bu} \xrightarrow{h\nu} 2\ t\text{-Bu}\cdot \xrightarrow{2\ k_t} t\text{-Bu-}t\text{-Bu} \quad \text{and} \quad t\text{-Bu-H} + CH_2{=}C(CH_3)_2 \tag{4.8}$$

Whenever a unimolecular reaction, such as one of the clock reactions, transforms one radical into another radical, and both are observable by ESR, the rate of the transformation reaction can be obtained by ESR spectroscopy [73]. For example, the phosphoranyl radical (P·) can be generated at a constant concentration by the photolysis of a hydrocarbon solution of di-t-butyl peroxide and triethyl phosphite as in Eq. 4.9. Reaction 4.10 transforms it into the t-butyl radical. Other reactions occurring are given by Eqs. 4.11–4.13.

$$Me_3CO\cdot + P(OEt)_3 \longrightarrow Me_3COP\cdot(OEt)_3 \qquad (P\cdot) \tag{4.9}$$

$$P\cdot \longrightarrow t\text{-Bu}\cdot + O{=}P(OEt)_3 \qquad (k_{10}) \tag{4.10}$$

$$P\cdot + P\cdot \longrightarrow \text{nonradical products} \tag{4.11}$$

$$P\cdot + t\text{-Bu}\cdot \longrightarrow \text{nonradical products} \qquad (k_{12}) \tag{4.12}$$

$$2\ t\text{-Bu}\cdot \longrightarrow \text{nonradical products} \qquad (k_{13}) \tag{4.13}$$

This leads to a rate relationship (Eq. 4.14):

$$\frac{1}{[\mathrm{Me_3C\cdot}]} = \frac{2k_{13}[\mathrm{Me_3C\cdot}]}{k_{10}[\mathrm{P\cdot}]} + \frac{2k_{12}}{k_{10}} \tag{4.14}$$

When the radical concentrations are changed by varying the light intensity, a plot of Eq. 4.14 gives k_{13}/k_{10} from the slope. Since k_{13} is known, this procedure allows k_{10} to be obtained as well.

***Nanosecond Laser Flash Photolysis* [19,74].** Rates of reaction of a radical R· with a stable nitroxide scavenger $(R')_2NO\cdot$ (Eq. 4.15)

$$\mathrm{R\cdot} + \mathrm{(R')_2NO\cdot} \longrightarrow \mathrm{(R')_2NO{-}R} \tag{4.15}$$

can be measured in the following way [74]. Flash pulses generate R· by the photolysis of a precursor such as R_2CO. The decay of the absorption of R· follows pseudo-first-order kinetics (Eq. 4.16), where k_0 is the sum of all the first or pseudo-first-order rate constants for the decay of R· except for its reaction with the nitroxide.

$$k_{\mathrm{exp}} = k_0 + k_{15}[\mathrm{(R')_2NO\cdot}] \tag{4.16}$$

For primary alkyl radicals, which do not have an absorption in the 290–800 nm range, a different technique must be used [75]. The primary alkyl radical adds to 1,1-diphenylethylene (Eq. 4.17), giving a radical which has a strong absorption at about 330 nm. In the presence of both the diphenylethylene and the nitroxide the decay of the 330-nm signal is given by Eq. 4.18.

$$\mathrm{R\cdot} + [\mathrm{CH_2{=}C(C_6H_5)_2}] \longrightarrow \mathrm{R{-}CH_2C\cdot(C_6H_5)_2} \tag{4.17}$$

$$k_{\mathrm{exp}} = k_0 + k_{17}[\mathrm{CH_2{=}C(C_6H_5)_2}] + k_{15}[\mathrm{(R')_2NO\cdot}] \tag{4.18}$$

The value of k_{17} is obtained from a plot of Eq. 4.18 at constant nitroxide concentration. Then a plot at constant diphenylethylene concentration is used to obtain k_{15}.

Some values of k_{15} in isooctane at 20°C are given in Table 4.3.

4.3.3 Slow Combination

The rates at which *n*-nonyl, *t*-butyl, and benzyl radicals are trapped by the nitroxyl scavenger TEMPO (Table 4.3), unlike those of the corresponding alkyl radical plus alkyl radical reactions, are about an order of magnitude lower than the diffusion-controlled rate [19]. The temperature dependence is

TABLE 4.3 Radical plus Nitroxide Combination Rates

Precursor of R·	Method[a]	R·	Nitroxide	$10^{-7}k_{15}$ ($M^{-1}s^{-1}$)
$(RCO_2)_2$	P	$CH_3(CH_2)_7CH_2\cdot$	TEMPO[b]	123 ± 26
$(RCO_2)_2$	P	$(CH_3)_3CCH_2\cdot$	TEMPO	96 ± 22
R_2CO	D	$(CH_3)_3C\cdot$	TEMPO	76 ± 16
R_2CO	D	$(CH_3)_3C\cdot$	TMIO[c]	88 ± 12
R_2CO	D	$C_6H_5CH_2\cdot$	TEMPO	49 ± 8
R_2CO	D	$C_6H_5CH_2\cdot$	TMIO	55 ± 5
$Me_3CO\cdot$ + RH	D	$C_6H_5C\cdot HCH_3$	TEMPO	16 ± 4
$Me_3CO\cdot$ + RH	D	$C_6H_5C\cdot(CH_3)_2$	TEMPO	11.8 ± 0.1
R_2CO	D	$(C_6H_5)_2CH\cdot$	TEMPO	4.63 ± 0.02
$Me_3CO\cdot$ + RH	D	$(C_6H_5)_2C\cdot CH_3$	TEMPO	4.58 ± 0.47
$Me_3CO\cdot$ + RH	D	$(C_6H_5)_3C\cdot$	TEMPO	<0.1
$Me_3CO\cdot$ + RH	D	$1\text{-}C_{10}H_7CH_2\cdot$	TEMPO	8.2 ± 0.2
$Me_3CO\cdot$ + RH	D	$2\text{-}C_{10}H_7CH_2\cdot$	TEMPO	5.7 ± 1.8

[a]P, with diphenylethylene probe; D, direct absorption.
[b]2,2,6,6-Tetramethylpiperidine-*N*-oxyl.
[c]Tetramethylisoindoline-*N*-oxyl.

also different from that expected for a diffusion-controlled reaction. The log *A* values are 9.5–10.5 and *E* is 0.3–1.7 kcal mol^{-1}; the parameters for diffusion-controlled reactions are typically about 11.5 and 3.0. Thus the conversion of the encounter complex to the alkylhydroxylamine appears to be slowed by an entropy barrier. It has been suggested that the entropy barrier is due to the need for localization of the unpaired electron on oxygen (corresponding to structure A below) [19].

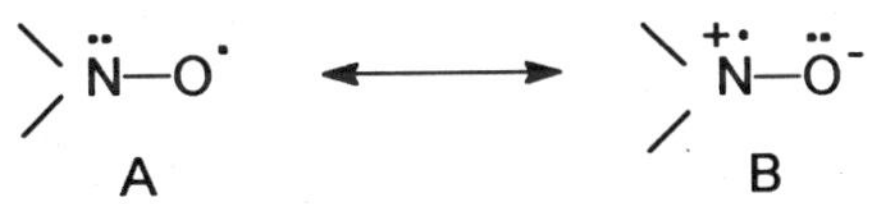

4.4 INTERSYSTEM CROSSING AND CIDNP

Intersystem crossing during the lifetime of a radical pair, catalyzed by hyperfine interactions with nuclear spins, can modify the ratio of pairs reacting to pairs escaping. Since intersystem crossing by the HFC mechanism is more likely in some nuclear spin states than in others, the populations of the nuclear spin states in the products formed by the pair are different from those of the escaping radicals and their products. As a result, NMR spectra taken before the spins have had time to relax are abnormal, some peaks showing unusually strong absorption and others showing emission rather than absorption.

REFERENCES

1. Noyes, R. M. *J. Am. Chem. Soc.* **1955,** *77,* 2042.
2. Turro, N. J.; Lei, X.; Cheng, C.-C.; Corbin, D. R.; Abrams, L. *J. Am. Chem. Soc.* **1985,** *107,* 5824.
3. Johnston, L. J.; de Mayo, P.; Wong, S. K. *J. Org. Chem.* **1984,** *49,* 20.
4. Leffler, J. E.; Zupancic, J. J. *J. Am. Chem. Soc.* **1980,** *102,* 259.
5. Baretz, B.; Turro, N. J. *J. Am. Chem. Soc.* **1983,** *105,* 1309.
6. Rabinowitch, E.; Wood, W. C. *Trans. Faraday Soc.* **1936,** *32,* 1381.
7. Rabinowitch, E.; Wood, W. C. *Trans. Faraday Soc.* **1937,** *33,* 1225.
8. Schuh, H.-H.; Fischer, H. *Helv. Chim. Acta* **1978,** *61,* 2130.
9. Ingold, K. U.; Maillard, B.; Walton, J. C. *J. Chem. Soc. Perkin II* **1981,** 970.
10. Claridge, R. F. C.; Fischer, H. *J. Phys. Chem.* **1983,** *87,* 1960.
11. Schulz, G. V.; Harborth, *G. Makromol. Chem.* **1947,** *1,* 106.
12. Medvedev, S. S. *Acta Physicochim. USSR* **1944,** *19,* 457.
13. Fraenkel, G. K.; Hirshon, J. M.; Walling, C. *J. Am. Chem. Soc.* **1954,** *76,* 3606.
14. Saltiel, J.; Atwater, B. W. In *Advances in Photochemistry,* Interscience: New York, 1988, Chapter 1.
15. Bartlett, P. D.; Funahashi, T. *J. Am. Chem. Soc.* **1962,** *84,* 2596.
16. Bartlett, P. D.; Purrington, S. T. *J. Am. Chem. Soc.* **1966,** *88,* 3303.
17. Koelsch, C. F. *J. Am. Chem. Soc.* **1957,** *79,* 4439.
18. Jones, M. J.; Moad, G.; Rizzardo, E.; Solomon, D. H. *J. Org. Chem.* **1989,** *54,* 1607.
19. Beckwith, A. L. J.; Bowry, V. W.; Ingold, K. U. *J. Am. Chem. Soc.* **1992,** *114,* 4983.
20. Kryger, R. G.; Lorand, J. P.; Stevens, N. R.; Herron, N. R. *J. Am. Chem. Soc.* **1977,** *99,* 7589.
21. Bartlett, P. D.; Gontarev, B. A.; Sakurai, H. *J. Am. Chem. Soc.* **1962,** *84,* 3101.
22. Greene, F. D.; Berwick, M. A.; Stowell, J. C. *J. Am. Chem. Soc.* **1970,** *92,* 867.
23. Lorand, J. P.; Wallace, R. W. *J. Am. Chem. Soc.* **1974,** *96,* 1402.
24. Hiatt, R.; Traylor, T. G. *J. Am. Chem. Soc.* **1965,** *87,* 3766.
25. Koenig, T.; Deinzer, M. *J. Am. Chem. Soc.* **1968,** *90,* 7014.
26. Niki, E.; Kamiya, Y. *J. Am. Chem. Soc.* **1974,** *96,* 2129.
27. Koenig, T. *J. Am. Chem. Soc.* **1969,** *91,* 2558.
28. Frisch, H. L.; Kuivila, H. G. *J. Am. Chem. Soc.* **1977,** *99,* 7200.
29. Noyes, R. M. *Progr. React. Kinet.* **1961,** *1,* 129.
30. Kiefer, H.; Traylor, T. G. *J. Am. Chem. Soc.* **1967,** *89,* 6667.
31. Tokumaru, K.; Ohshima, A.; Nakata, T.; Sakuragi, H.; Mishima, T. *Chem. Lett. (Jpn.)* **1974,** 571.
32. Engel, P. S.; Bishop, D. J.; Page, M. A. *J. Am. Chem. Soc.* **1978,** *100,* 7009.
33. Turro, N. J.; Chow, M.-F.; Chung, C.-J.; Kraeutler, B. *J. Am. Chem. Soc.* **1981,** *103,* 3886.

34. Scaiano, J. C.; Abuin, E. B.; Stewart, L. *J. Am. Chem. Soc.* **1982,** *104,* 5673.
35. Turro, N. J.; Paczkowski, M. A.; Wan, P. *J. Org. Chem.* **1985,** *50,* 1399.
36. Pryor, W. A.; Smith, K. *J. Am. Chem. Soc.* **1970,** *92,* 5403.
37. Neuman, R. C.; Lockyer, G. D. Jr.; Amrich, M. J. *Tetrahedron Lett.* **1972,** 1221.
38. Bartlett, P. D.; Rüchardt, C. *J. Am. Chem. Soc.* **1960,** *82,* 1756.
39. Koenig, T.; Huntington, J.; Cruthoff, R. *J. Am. Chem. Soc.* **1970,** *92,* 5413.
40. Taylor, J. W.; Martin, J. C. *J. Am. Chem. Soc.* **1966,** *88,* 3650.
41. Koenig, T. W.; Brewer, W. D. *Tetrahedron Lett.* **1965,** 2773.
42. Kobayashi, M.; Gotoh, M.; Yoshida, M. *Bull. Chem. Soc. Jpn.* **1987,** *60,* 295.
43. Taylor, K. G.; Govindan, C. K.; Kaelin, M. S. *J. Am. Chem. Soc.* **1979,** *101,* 2091.
44. Sigman, M. E.; Barbas, J. T.; Leffler, J. E. *J. Org. Chem.* **1987,** *52,* 1754.
45. Sheldon, R. A.; Kochi, J. K. *J. Am. Chem. Soc.* **1970,** *92,* 4395.
46. Lee, K.-W.; Horowitz, N.; Ware, J.; Singer, L. A. *J. Am. Chem. Soc.* **1977,** *99,* 2622.
47. Bernardi, F.; Olivucci, M.; Robb, M. A. *Acc. Chem. Res.* **1990,** *23,* 405.
48. Walborsky, H. M.; Chen, C.-J. *J. Am. Chem. Soc.* **1967,** *89,* 5499.
49. Fessenden, R. W.; Schuler, R. H. *J. Chem. Phys.* **1963,** *39,* 2147.
50. Olea, A. F.; Thomas, J. K. *J. Am. Chem. Soc.* **1988,** *110,* 4494.
51. Koenig, T.; Owens, J. M. *J. Am. Chem. Soc.* **1973,** *95,* 8484.
52. Johnson, R. A.; Seltzer, S. *J. Am. Chem. Soc.* **1973,** *95,* 938.
53. Bartlett, P. D.; McBride, J. M. *Pure & Appl. Chem.* **1967,** *15,* 89.
54. Kopecky, K. R.; Gillan, T. *Can. J. Chem.* **1969,** *47,* 2371.
55. Tsolis, A.; Mylonakis, S. G.; Nieh, M. T.; Seltzer, S. *J. Am. Chem. Soc.* **1972,** *94,* 829.
56. Engstrom, J. P.; Greene, F. D. *J. Org. Chem.* **1972,** *37,* 968.
57. Doering, W. von E.; Birladeanu, L. *J. Am. Chem. Soc.* **1986,** *108,* 7442.
58. Schuh, H.-H.; Fischer, H. *Helv. Chim. Acta.* **1978,** *61,* 2463.
59. Tanner, D. D.; Rahimi, P. M. *J. Am. Chem. Soc.* **1982,** *104,* 225.
60. Engel, P. S. *Chem. Rev.* **1980,** *80,* 99.
61. Nelsen, S. F.; Bartlett, P. D. *J. Am. Chem. Soc.* **1966,** *88,* 137.
62. Jaffe, A. B.; Skinner, K. J.; McBride, J. M. *J. Am. Chem. Soc.* **1972,** *94,* 8510.
63. McBride, J. M. *J. Am. Chem. Soc.* **1971,** *93,* 6302.
64. Taylor, J. W.; Martin, J. C. *J. Am. Chem. Soc.* **1967,** *89,* 6904.
65. Martin, J. C.; Hargis, J. H. *J. Am. Chem. Soc.* **1969,** *91,* 5399.
66. Lorand, J. P.; Bartlett, P. D. *J. Am. Chem. Soc.* **1966,** *88,* 3294.
67. Hammond, G. S.; Wu, C.-H. S.; Trapp, O. D.; Warkentin, J.; Keys, R. T. *J. Am. Chem. Soc.* **1960,** *82,* 5394.
68. Beckwith, A. L. J.; Bowry, V. W.; Moad, G. *J. Org. Chem.* **1988,** *53,* 1632.
69. Kochi, J. K.; Krucic, P. J. *J. Am. Chem. Soc.* **1969,** *91,* 3940.
70. Chatgilialoglu, C.; Ingold, K. U.; Scaiano, J. C. *J. Am. Chem. Soc.* **1981,** *103,* 7739.

71. Griller, D.; Ingold, K. U. *Acc. Chem. Res.* **1980,** *13,* 317.
72. Griller, D.; Ingold, K. U. *Acc. Chem. Res.* **1980,** *13,* 193.
73. Watts, G. B.; Griller, D.; Ingold, K. U. *J. Am. Chem. Soc.* **1972,** *94,* 8784.
74. Chateauneuf, J.; Lusztyk, J.; Ingold, K. U. *J. Org. Chem.* **1988,** *53,* 1629.
75. Paul, H.; Small, R. D. J.; Scaiano, J. C. *J. Am. Chem. Soc.* **1978,** *100,* 4520.

5

CIDNP AND CIDEP

5.1 SPIN POPULATIONS

Chemically induced dynamic nuclear polarization (CIDNP) is a phenomenon that causes the NMR lines of the products of radical reactions to have abnormal intensities [1]. *Polarization* refers to a nonequilibrium distribution of the molecules among the nuclear spin energy levels. At equilibrium the molecules of the total population N are distributed among the levels of energy ϵ_i according to the Boltzmann equation:

$$N_i = Ae^{-\epsilon i/kT}, \tag{5.1}$$

where $A = N/\Sigma\epsilon_i/kT$.

The interaction of electromagnetic radiation with any collection of molecules can equally well produce net absorption or net emission. If the population of the lower energy level exceeds that of the upper energy level, as is normally the case, the result is absorption and promotion of some of the molecules from the lower to the higher energy level. If the sample is polarized in such a way that the population of the lower level is higher than it would be at equilibrium, the absorption will be abnormally strong. Similarly, if the sample contains an excess of molecules in the upper energy level, interaction with the radiation will produce a net emission in proportion to the excess population of the upper level.

The polarization of the spin populations need not be very great in order to produce a strongly enhanced absorption or emission. In NMR the equilibrium population of the upper level is less than that of the lower level by a Boltzmann factor of $\exp(-g_N\beta_N H/kT)$ that is only very slightly less than unity. For a field strength of 10,000 gauss the excess population of the lower spin state is only about 0.00068%. Increasing this to 0.68% would involve less than 1% of the nuclei, but would increase the strength of the NMR signal by a factor of 1000. Since the fraction of spins that needs to be changed to cause a drastic change in the spectrum is so small, the mechanism that produces the spin polarization need not be very efficient and may involve very subtle effects not otherwise noticeable.

Chemically induced dynamic electron polarization (CIDEP) is a similar phenomenon in electron spin resonance spectroscopy in which electron spin levels have abnormal populations [2].

The phenomenon of CIDNP was discovered independently in 1967 by Joachim Bargon [3] and by Ward and Lawler [4,5]. Bargon observed an abnormal NMR signal from the protons of benzene formed during the decomposition of benzoyl peroxide in cyclohexanone at 110°C. The benzene ^{1}H singlet began to appear in *emission*, increasing in strength to a maximum at about 4 minutes into the run, decreasing and passing through zero at about 8 minutes, and reaching its normal constant absorption intensity after about 12 minutes. Ward and Lawler observed an abnormal NMR spectrum for the 1-butene formed by reactions 5.2 and 5.3.

$$n\text{-}C_4H_9Br + n\text{-}C_4H_9Li \longrightarrow LiBr + 2\ n\text{-}C_4H_9\cdot \tag{5.2}$$

$$2\ n\text{-}C_4H_9\cdot \longrightarrow CH_2{=}CHCH_2CH_3 + n\text{-}C_4H_{10} \tag{5.3}$$

The multiplets corresponding to the protons at C1 and C2 showed both emission and enhanced absorption lines.

Since it is possible to maladjust an NMR spectrometer to show absorption lines upside down (i.e, as emissions), or as the first derivative of the absorption, CIDNP may well have been observed earlier and dismissed as an artifact.

5.2 CLASSIFICATION OF CIDNP SPECTRA

The CIDNP spectra may be classified into two types, those showing a *net effect* and those showing a *multiplet effect*.

Net effect spectra are obtained from molecules whose normal NMR spectrum contains a line that results from a group of equivalent nuclei not coupled to other nuclei. This line may appear as enhanced absorption (A) or, as in the case of the benzene spectrum observed by Bargon, as emission (E).

Multiplets appear in CIDNP spectra with some of the component lines enhanced (A) and an equal number as emission (E). If the first line on the left, that is, the one at lower field or greater δ, appears as A, the signal is designated as an A/E multiplet. If the first line appears as E, the signal is called an E/A multiplet.

The polarization responsible for the net spectra is an *energy polarization* because there has to be a net change in the total Zeeman energy of the spin system. For example, in the case of emission, the enhanced population of the higher energy level clearly requires a greater total Zeeman energy. Correspondingly, enhanced absorption requires a greater population of the lower level and hence a lower total Zeeman energy.

In the multiplet effect nuclear spins are merely sorted rather than changed. Since each enhanced absorption line is accompanied by an emission line, the total Zeeman energy is about the same as in a sample at equilibrium; excess population at some energy levels is balanced by lower population at other energy levels. Because the total Zeeman energy is merely distributed differently and not changed, this sorting of the spins is called *entropy* polarization as opposed to energy polarization.

Both a net effect and a multiplet effect are sometimes seen in the same spectrum, energy polarization and entropy polarization occurring together. The multiplets show both A and E, but the intensities do not balance exactly, with either the absorption lines or the emission lines being stronger.

5.3 THE CKO RADICAL PAIR THEORY

As discussed in Chapter 4, if the radical pair is not already a singlet, the formation of a geminate product requires intersystem crossing to the singlet state. Similarly, *escape* from geminate reaction requires the radical pair to be a triplet. Polarization of the nuclear spins in the geminate product or in the escaped radicals occurs because the probability of intersystem crossing is different for different nuclear spin states. According to the radical pair model, intersystem crossing occurs as a result of motions of the radicals of the pair [6–9]. Another mechanism, restricted to photochemical reactions, depends on intersystem crossing from a triplet excited state of a molecule. This mechanism will be discussed in the section on electron spin polarization.

5.3.1 The High Field Approximation: T_0-S Transitions

Many reactions leading to CIDNP are carried out in high magnetic fields (i.e., 10^3 gauss or greater), often in the probe of the NMR spectrometer. Under these conditions there is a very great simplification of the theory because only the transition between the singlet and the T_0 level of the triplet needs to be considered [9]. There are no transitions between the singlet and the T_+ or T_- sublevels of the triplet.

One of the requirements for mixing of the electronic spin states is matching of the energies. At high field the singlet and triplet levels are well separated in energy as long as the radicals are close together. The left side of Fig. 5.1 shows a large separation in energy (J_{ab}) when the distance r between the radicals a and b is small. As the radicals begin to diffuse apart to greater distances, J_{ab} decreases and the levels come closer together. In fact, the singlet level actually crosses the T_{-1} level at a critical distance r_c, but so abruptly

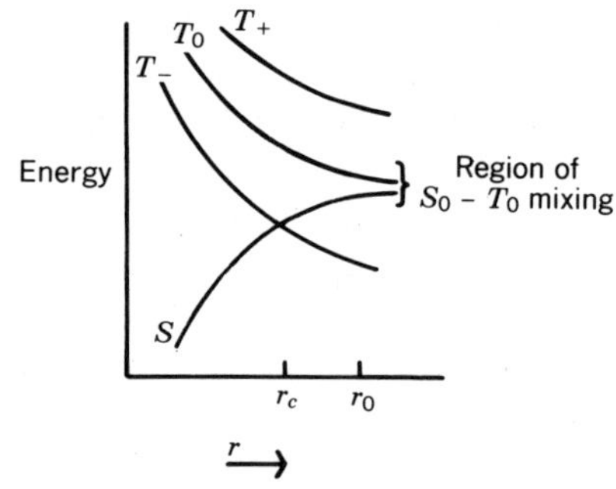

FIG. 5.1 Requirement for $S_0 - T_0$ mixing. Reprinted with permission from R. Kaptein, *J. Am. Chem. Soc.* **1972,** *94,* 6251. Copyright 1972 American Chemical Society.

that the singlet and triplet energies are close together for too short a time for any mixing of the singlet and T_{-1} states to occur. However, at a still greater distance r_0 the singlet level and the triplet T_0 ($S = 1$, $m_s = 0$) come close and stay close together *over a large range of r*. Because the two electronic levels remain close together during a large part of the diffusion process, there is time for the singlet and T_0 levels to become mixed. Under a perturbation caused by the interaction of the electron and nuclear spins, an originally singlet pair of radicals becomes a quasi-singlet with some triplet character, and vice versa.

The value of r_0 is such that the radicals are no longer within the original solvent cage, but are still close enough so that a geminate collision is highly probable. Recombination of such a pair is sometimes called secondary recombination to distinguish it from reaction within the original cage.

Now that we have seen how the singlet and T_0 levels might get close enough in energy for intersystem crossing to occur, it remains to be seen what part the nuclear spins play in this process. Because the electron spins are loosely coupled, they precess about the z axis independently. When first formed, the T_0 vectors look like Fig. 5.2*a*. The z components of the vectors 1 and 2 cancel, giving a zero component in the z direction. When first formed, the singlet looks like Fig. 5.2*b*: the two vectors 1 and 2 are directly opposed, so that all of their components cancel, not just the z component.

The electron spin angular momentum vectors precess about the z axis. If

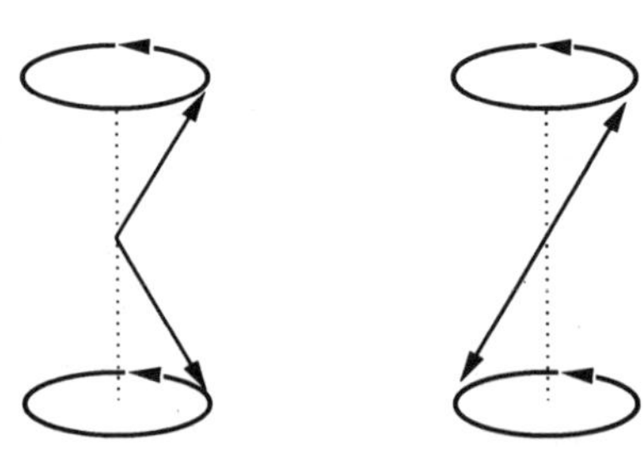

FIG. 5.2 Singlet and triplet spin vectors.

they precess at the same rate, they will stay in phase and the diagram will continue to look like Fig. 5.2*a* for the triplet and 5.2*b* for the singlet. However, if they precess at different rates, the relative orientations of the two vectors will change, at times corresponding to the singlet state (Fig. 5.2*b*) and at times corresponding to the triplet (Fig. 5.2*a*); the state oscillates between singlet and triplet. More precisely, the quantum mechanical probability of the state being a singlet or a triplet oscillates.

In the absence of magnetic fields other than the external magnet field H_0, the electron spin angular momentum vectors would precess about the z axis with a Larmor angular frequency given by Eq. 5.4.

$$\omega = g\beta_e\hbar^{-1}H_0 \tag{5.4}$$

The actual z component of the magnetic field experienced by the electrons is a sum of the external field H_0, the field caused by electron orbital motion, and the fields due to the spins of nearby magnetic nuclei. Because of this, the precession frequencies can be different for radicals in different nuclear spin states. The difference between the two precession frequencies ω_1 and ω_2 is given by Eq. 5.5 [11]. The frequencies are in radians per second.

$$\omega_1 - \omega_2 = \beta_e\hbar^{-1}\Delta g H_0 + \Sigma a_{1i}m_{1i} - \Sigma a_{2j}m_{2j} \tag{5.5}$$

The Δg $(= g_1 - g_2)$ in the first term comes from spin-orbit coupling and will be different from zero when the g values of the two radicals are different. The second and third terms come from the ESR hyperfine splitting constants a_{1i} and a_{2j} for magnetic nuclei in the two radicals and the nuclear magnetic quantum numbers m_{1i} and m_{2j}.

For a typical pair of radicals the frequency difference is $10^{-7} - 10^{-8}$ sec^{-1}, which is fast enough to allow mixing of the singlet and triplet states within the lifetime of the radical pair.

The polarization can be illustrated most simply with a system in which one radical ($HR_1\cdot$) has only one proton coupled to the electron and the other radical ($R_2\cdot$) has none. The radical pair with the proton in the nuclear spin state $|\alpha\rangle$, spin $+\frac{1}{2}$, will obey Eq. 5.6, while the pair with the proton in the $|\beta\rangle$ state will obey Eq. 5.7.

$$\omega_1 - \omega_2 = \beta_e\hbar^{-1}\Delta g H_0 + (\tfrac{1}{2})a \tag{5.6}$$

$$\omega_1 - \omega_2 = \beta_e\hbar^{-1}\Delta g H_0 - (\tfrac{1}{2})a \tag{5.7}$$

To make a specific prediction we must know the signs of Δg and the coupling constant a. Assuming that both of these are positive, a radical pair with proton spin $|\beta\rangle$ and initially in the singlet state will take longer to reach the triplet state. During this time it will have a chance to give a singlet reaction product: this product will have a higher than normal spin $|\beta\rangle$ population, and will give emission. Correspondingly, the pairs with the proton in the $|\alpha\rangle$ state will have a greater chance of changing to the triplet state and escaping as free radicals. The products from these radicals will have an enhanced population with spin $|\alpha>$ and will show enhanced absorption. For a radical pair initially in the triplet state, the predictions are just the reverse.

5.3.2 Kaptein's Rules

Handy rules for making qualitative predictions of the polarization have been given by Kaptein [10] in terms of the signs of two quantities Γ_{ne} and Γ_{me}. These rules are reliable for *most* reactions run at high field, for example, in the field of the NMR magnet. Some exceptions will be discussed later. When Δg is large compared to the hyperfine constants of the coupled nuclei, the net effect predominates and Eq. 5.8 applies.

$$\Gamma_{ne} = \mu\epsilon\Delta g a_i \tag{5.8}$$

A positive Γ_{ne} predicts enhanced absorption (A) and a negative value predicts E.

When the hyperfine coupling constants a_i and a_j are large compared to Δg, the multiplet effect predominates and Eq. 5.9 is applicable.

$$\Gamma_{me} = \mu\epsilon a_i a_j J_{ij}\sigma_{ij} \tag{5.9}$$

A positive Γ_{me} predicts E/A and a negative value predicts A/E.

Only the signs of the parameters are needed. The parameter $\Delta g = g_1 - g_2$, where g_1 is the g value for the radical containing the magnetic nucleus in question, and g_2 is the g value for the other radical. The ESR splitting parameters a_i and a_j have their usual meaning; J_{ij} is the NMR coupling constant of the nuclei i and j. The other parameters are assigned values of $+1$ or -1 according to the rules in Table 5.1.

5.3.3 The Rules Exemplified

The thermolysis of acetyl trichloroacetyl peroxide in CCl_4 in the presence of iodine gives methyl iodide and trichloroethane [11]. The NMR spectra of the products before and after the spins have relaxed are shown in Fig. 5.3.

TABLE 5.1 The Parameters in Kaptein's Rules

Symbol	Plus	Minus
μ	Triplet pair or encounter pair	Singlet pair
ϵ	Geminate product	Scavenged product
σ_{ij}	The nuclei i,j belong to the same radical	The nuclei i,j belong to different radicals

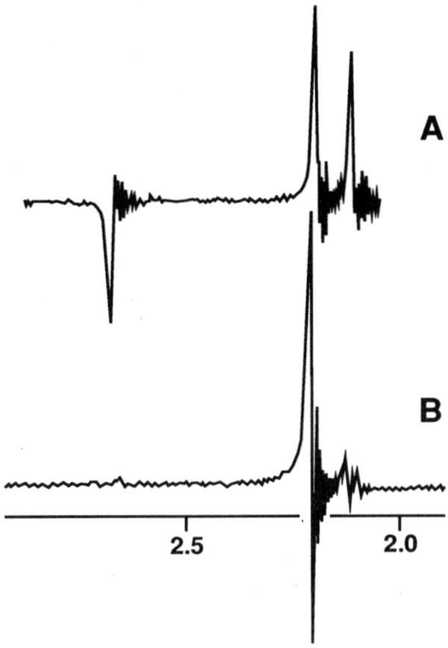

FIG. 5.3 (*A*) CIDNP spectrum of trichloroethane, acetyltrichloroacetyl peroxide, and methyl iodide. (*B*) After relaxation of the spins. Reprinted with permission from H. R. Ward, *Acc. Chem. Res.*, **1972,** *5,* 19. Copyright 1972 American Chemical Society.

The emission at δ2.7, fading to a weak absorption in the relaxed spectrum, is due to the protons in trichloroethane. The enhanced absorption at δ2.1 is due to methyl iodide. The line at about δ2.2 is due to the remaining peroxide.

The methyl iodide enhanced absorption at δ2.1 is a net effect. Since the methyl radical–trichloromethyl radical pairs were generated thermally, $\mu = (-)$. Since the methyl iodide is formed from radicals that escaped geminate reaction and were later trapped by the iodine scavenger, $\epsilon = (-)$. Since the methyl radical has a lower g value than the trichloromethyl radical, Δg is negative. The hyperfine splitting constant of the protons in the methyl radical is negative. Putting all of these into the expression for the net effect gives $\Gamma_{ne} = (-)(-)(-)(-) = (+)$, hence the signal is an enhanced absorption.

The emission from the trichloroethane protons at δ2.7 is also a net effect. All of the factors except for ϵ are the same as before, but ϵ is $(+)$ because the trichloroethane is the product of geminate combination. The net sign is therefore $(-)$ and the result is *E*.

The thermolysis of dipropionyl peroxide in hexachloroacetone gave the ethyl chloride NMR spectrum shown in Fig. 5.4 [12]. In the decomposition of this peroxide the propionyloxy radicals either decarboxylate very rapidly or are bypassed entirely, so that the pair is a pair of ethyl radicals. Since Δg is zero, there is no net effect and the polarization is a pure multiplet effect. The ethyl chloride is formed by escaped ethyl radicals which abstract chlorine from the hexachloroacetone.

The quartet at δ3.52 due to the methylene protons and the triplet at δ1.47 due to the methyl protons both show an *A*/*E* multiplet effect. The parameters

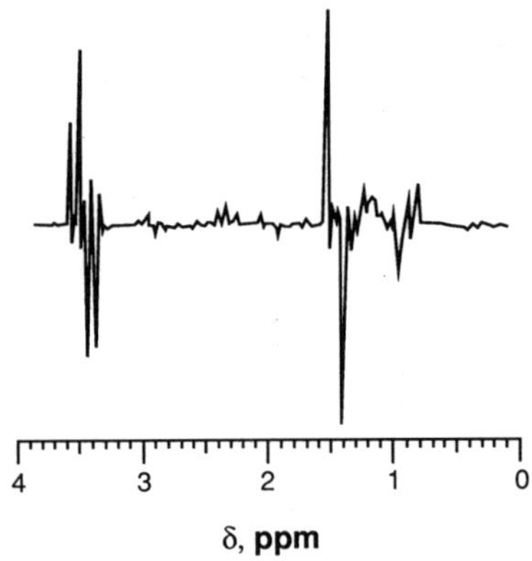

FIG. 5.4 The decomposition of dipropionyl peroxide in hexachloroacetone [10].

for Γ_{me} are μ $-$, ϵ $-$, a_1 $-$, a_2 $+$, J_{12} $+$, and σ $+$, giving $(-)(-)(-)(+)(+)(+) = (-)$, predicting A/E.

The benzhydryl benzyl radical pair **I** in the scheme shown below can be made in three ways [13]:

1. Diphenylmethylene **II**, which has a triplet ground state, abstracts hydrogen from toluene to give triplet **I**.
2. Thermolysis of the azo compound **III** gives nitrogen and singlet **I**.
3. Decomposition of benzoyl peroxide in the presence of a mixture of diphenylmethane and toluene abstracts hydrogen to give a mixture of freely diffusing benzyl and benzhydryl radicals.

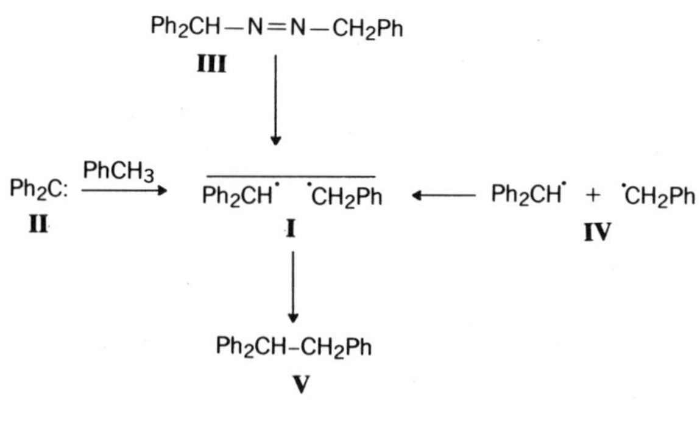

Geminate combination gives triphenylethane **V**.

When the precursor radical pair is the triplet (from **II**), $\Gamma_{me} = (+)(+)(+)(+)(+)(-) = (-)$, hence the CIDNP spectrum of the aliphatic protons of **V** is A/E (Fig. 5.5*a*). When the precursor radical pair is the singlet, from the thermolysis of the azo compound, the factor μ becomes negative and the product is $(+)$ and an E/A spectrum (Fig. 5.5*b*) results. When the precursor consists of freely diffusing radicals (Fig. 5.5*c*) the result is E/A

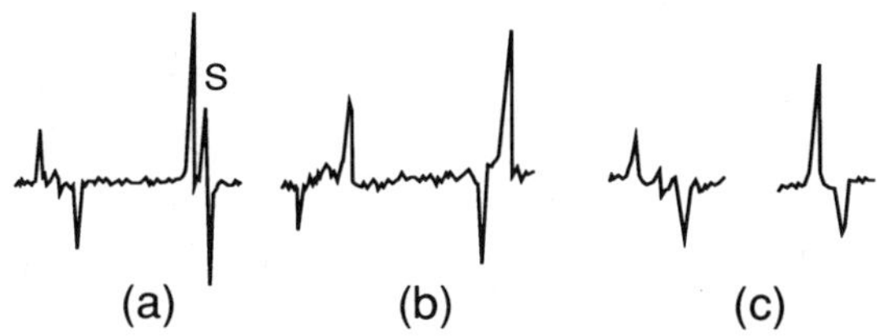

FIG. 5.5 CIDNP spectra from 1,1,2-triphenylethane **V**. From (*a*) triplet pairs, (*b*) singlet pairs and (*c*) freely diffusing radicals. From G. L. Closs, *Chemically Induced Dynamic Nuclear Polarization,* in *Advances in Magnetic Resonance,* J. S. Waugh, Ed., Academic, New York, 1974, p. 181.

again. Encounters between the freely diffusing radicals give triplet and singlet pairs in a 3:1 ratio, so that the effect is like that from an initial triplet, but weaker.

Since the two radicals have almost the same *g* factors, Δg and Γ_{ne} are close to zero and there is almost no net effect.

5.3.4 Some Limitations of the Kaptein Rules

Even when the radical pair theory is applicable, exceptions to the Kaptein rules may occur when the magnitudes of Δg and the hyperfine coupling constants are not sufficiently different [14]. The exact quantitative theory can still be used, however. For greater ease in generalizing the results it is convenient to calculate the relative line intensities (plus for *A*, minus for *E*) as a function of a dimensionless parameter $G = \Delta g H_0/a_A$, in which a_A is the hyperfine coupling constant of the nucleus in question.

Consider for example a two-proton system. In general, the net effect rule will be valid for any value of *G* if the two hyperfine constants are the same. If the two hyperfine constants a_x and a_H are not the same, it will also be valid whenever $|G| \geq |a_x/a_A|$, but will fail when $|G| < |a_x/a_A|$. An example is the CIDNP from the photolysis of benzoin (Eq. 5.10). One of the products is the regenerated starting material, a geminate product.

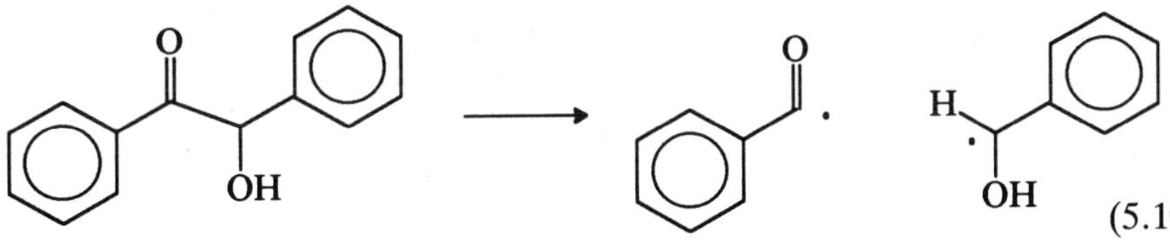

(5.10)

The signal from the benzylic proton of the geminate benzoin is correctly predicted from $\mu,\epsilon,\Delta g > 0 < a_H$ to be a net emission. However, when the carbonyl is labeled with ^{13}C the signal from this proton is split into a doublet [15]. The doublet, showing *A*/*E*, has a slightly stronger *A* than *E*; there is a small but definite net A effect in violation of the rule. In this example the hyperfine constants differ substantially, since a_c is 130 gauss and a_H only 14.9 gauss. The inequality $|G| < |a_x/a_A|$ will hold even up to relatively high values of G.

The multiplet rule will remain valid when $|G| < |a_x/a_A|$ and fail when $|G| > |a_x/a_A|$. An example of its failure is the ^{1}H CIDNP from the photolysis of benzoin labeled, this time, with ^{13}C in the hydroxybenzyl position. An E/A multiplet, $J_{CH} = +145$ Hz, is found for the benzylic proton. The result predicted by the rule is A/E on the basis of ϵ,μ,a_c,σ, $J > 0 > a_H$. Another product is benzaldehyde, formed when an escaped hydroxybenzyl radical loses its hydroxyl proton to an acceptor. The aldehyde proton signal, split by the ^{13}C, shows an A/E doublet, while the rule would predict E/A on the basis of μ,a_c,σ, $J > 0 > a_H,\epsilon$. The ^{13}C hyperfine constant for the hydroxybenzyl radical, 23 G, is less than twice as large as the ^{1}H hyperfine constant, -14.9 G. For a field strength H_0 of 21 kG, G for this radical pair $= -3.5$, hence $|G| > |a_X/a_A|$, and the rule is violated.

Still other exceptions to the Kaptein rules are found when the radical pair theory is not really applicable. Examples are the formation of a product molecule in its triplet state [16], and reactions involving electron transfer [17].

5.3.5 Low Fields: T_+ and T_- to S Transitions

At high magnetic field strengths, after the radicals have separated by a few angstroms, the energies of the S and T_0 electronic levels become degenerate, while the T_+ and T_- levels remain distinct and can be neglected. At low fields, however, the energies are close together even at low interradical distances and mixing of all four states must be considered. This complicates the theory to such an extent that predictions of the CIDNP spectra have to be computed numerically [18–21]. Transitions can occur when the changes in electron spin quantum number Δm_s and nuclear spin quantum number Δm_I add up to zero, that is, if $\Delta m_s + \Delta m_I = 0$ [21]. For example, the state $|S,\alpha>$ mixes with the state $|T_+,\beta>$. There are several differences from the high-field polarization rules:

1. The sign of Δg affects the polarization only at high fields.
2. Escape products have opposite polarizations from geminate products at high field, but the same polarization at low fields.
3. For a one-proton radical pair the polarization becomes zero at zero field.

Figure 5.6 shows the observed and computed intensities of the CIDNP signal from the proton in $CHCl_3$ formed from the 2-propyl trichloromethyl radical pair VI [20]. The radical pair was generated by the photolysis of diisopropyl ketone in CCl_4 at various field strengths, then the sample was transferred rapidly to a 60-mHz NMR spectrometer. Assuming a singlet pair,

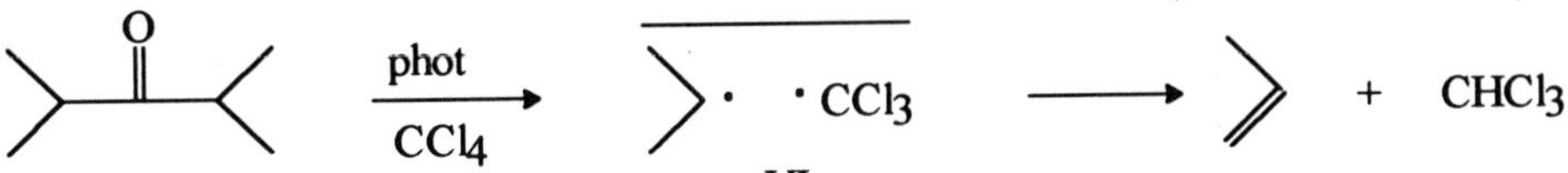

VI

$CHCl_3$

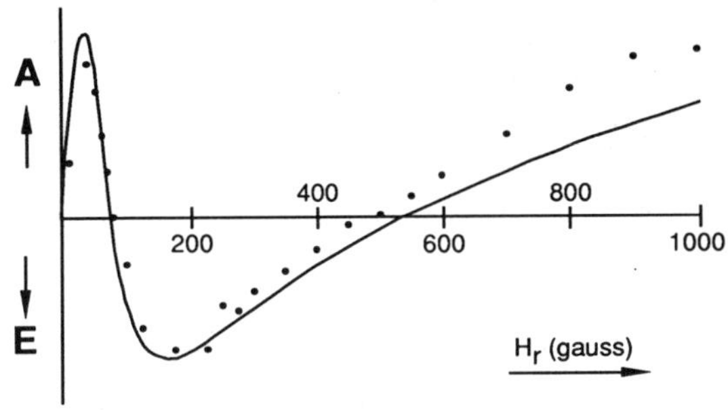

FIG. 5.6 Signals from $CHCl_3$ formed by photolysis of diisopropyl ketone. Reprinted with permission from R. Kaptein and J. A. den Hollander, *J. Am. Chem. Soc.* **1972,** *94,* 6269. Copyright 1972 American Chemical Society.

a negative splitting factor, a positive Δg, and a positive ϵ, the high field signal should be an enhanced absorption, as seems to be the case. At very low fields the signal is again A, but at intermediate fields it is E. The curve is the theoretical prediction scaled to the experimental points.

5.4 CIDNP FROM BIRADICALS

In the photolysis of cycloalkanones with rings larger than six carbons the excited singlet undergoes intersystem crossing to the triplet, which then cleaves one of the bonds at the carbonyl to give a triplet diradical. The photolysis of cycloheptanone is an example [22]. The products are the regenerated cycloheptanone and hept-5-ene-1-al. Because the radical ends never get as far apart as the r_0 of Fig. 5.1, the usual high-field T_0 to S intersystem crossing does not occur, and the relevant theory is like that for radical pairs at low field. Even though the reaction is run in the high field of the spectrometer, all of the signals are in emission, regardless of the signs of the coupling constants. The interpretation is that the ground state of the biradical is a singlet, there is $T_- - S$ mixing, and J is negative. It will be recalled that in radical pairs there was no $T_- - S$ mixing because the system did not remain long enough at the critical distance where the two surfaces crossed. In a suitable diradical the ends are constrained to remain longer at the critical distance.

In the case of cyclohexanone the unsaturated aldehyde product and the regenerated ketone have opposite polarizations, possibly because the aldehyde is formed from the singlet and the ketone is formed from the triplet by way of a spin-orbit coupling.

5.5 PAIR AND CAGE SUBSTITUTION

Pair substitution refers to the transformation of one of a pair of radicals into a different radical. For example, the pair RCOO· ·R might be transformed into the pair R· ·R by decarboxylation, or the pair R· ·R might be transformed into a pair R· ·R′ if one of the radicals rearranges. These transformations are often fast enough so that the nuclei of the new pair retain the polarization that they had in the original pair, a phenomenon called the memory effect.

An example is the formation of ethane from the thermolysis of acetyl peroxide [23,24]. The ethane CIDNP spectrum shows emission. The polarization responsible for this net effect could not have occurred in the $CH_3\cdot\ \cdot CH_3$ pair, because Δg for that pair is zero. It must have occurred in the $CH_3\cdot\ \cdot OOC$-CH_3 pair and persisted in the methyl radical pair.

Cage substitution refers to the formation of a new pair of radicals when one of the radicals of the first formed geminate pair is intercepted by a scavenging molecule. An example is the transformation of an initial isopropyl radical pair, formed by the thermolysis of isobutyryl peroxide, into an isopropyl/trichloromethyl pair by reaction with a molecule of CCl_3Br (Eq. 5.11) [24]

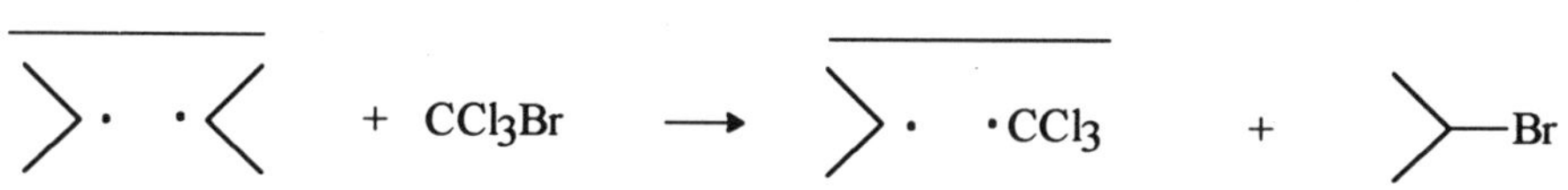

(5.11)

Chloroform, formed by the disproportionation of the isopropyl/trichloromethyl pair, shows enhanced absorption when a high concentration of CCl_3Br is used. At lower concentrations of the scavenger, emission is observed because most of the chloroform is formed from encounters between escaped isopropyl and trichloromethyl radicals.

Another example is the conversion of a benzoyloxy/phenyl radical pair, formed by the thermolysis of benzoyl peroxide, into a benzoyloxy/ethyl pair by reaction with ethyl iodide [25]. The *E*/*A* polarization of the ethyl benzoate is consistent with its formation from the singlet benzoyloxy/ethyl pair.

5.6 OTHER NUCLEI

Chemically induced dynamic nuclear polarization is possible with any magnetic nucleus although the examples discussed so far have all involved ^{1}H or ^{13}C.

5.6.1 CIDNP from ^{13}C

Because the polarization is much larger than for ^1H, CIDNP can be observed even from samples with only the 1.1% natural abundance of ^{13}C, although it is necessary to accumulate the signal from a large number of scans [26]. An example is the decomposition of benzoyl peroxide in cyclohexanone at 110°C, which gives benzoic acid, CO_2, biphenyl, and phenyl benzoate (Eq. 5.12).

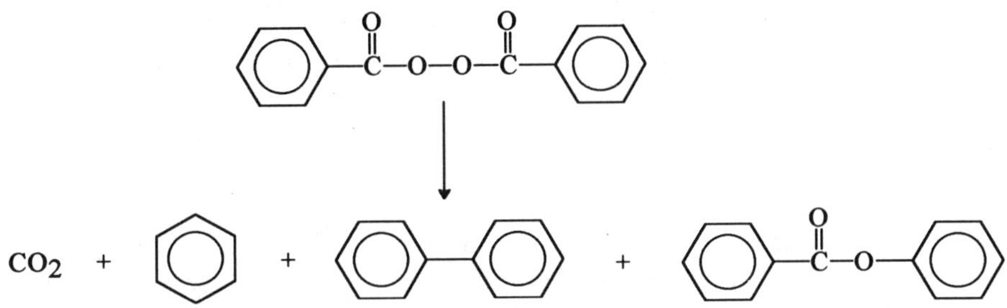

(5.12)

The phenyl benzoate, a rather minor product formed by geminate recombination, gave strong emission peaks from the carbonyl and unprotonated carbons. An A/E doublet from the benzene carbons collapses to a single line when the hydrogen is decoupled by double irradiation. In general, only the carbons at or next to the reaction center are polarized. The polarization is positive (A) in recombination products and negative in cleavage products such as CO_2 or in hydrogen abstraction products such as benzene.

Another example is the thermolysis of acetyl benzoyl peroxide in tetrachloroethylene at 110°C, which also gives a strong emission from the CO_2 and strong absorptions from the methyl and carbonyl carbons of methyl benzoate. Without proton decoupling the methyl carbon of the ester appears as an enhanced quadruplet. The ^{13}C signal from methane is seen as a quintet in emission.

The ^1H CIDNP spectrum of methyl chloride from the decomposition in hexachloroacetone of acetyl peroxide 55% labeled with ^{13}C in the methyl groups has an $A/E + A$ doublet from the $^{13}CH_3Cl$ flanking the signal from the unlabeled CH_3Cl [23].

5.6.2 Deuterium

Since the ratio of the nuclear g factors g_D/g_H is 0.154, the coupling constants a_D and J_{HD} are smaller by a factor of 6.5 than the corresponding a_H and J_{HH}. The decomposition of dideuterioacetyl peroxide in hexachloroacetone gives CH_2DCl and DCH_2CH_2D. The ^2D spectrum (15.4 MHz, H_0 23.5 kG) shows an $A + A/E$ triplet for the chloride and an $E + E/A$ triplet for the ethane [23].

5.6.3 Nitrogen

The decomposition of azo compounds R—N=N—R′ can in principle lead to the radicals R· and R·′ either directly or by way of an intervening diazenyl radical.

$$\text{R—N=N—R}' \longrightarrow \text{R}\cdot + \text{R}\cdot' + \text{N}_2 \tag{5.13a}$$

$$\text{R—N=N—R}' \longrightarrow \text{R—N=N}\cdot + \text{R}\cdot' \tag{5.13b}$$

The intervention of the diazenyl radical can be detected by the use of ^{15}N CIDNP. Because of the negative gyromagnetic ratio of ^{15}N, an additional minus sign must be introduced into the Kaptein rule, making it $\Gamma_{ne} = -\mu\epsilon\Delta g a_i$ [27].

The decomposition of *cis*-phenylazocumene gives enhanced absorption from the regenerated azo compound and emission due to $^{29}N_2$, confirming that the phenyldiazenyl radical is an intermediate, as expected on other grounds [28]. An alkyl diazenyl radical is also an intermediate in the decomposition of the cis norbornyl azo compound, so the "one-bond" cleavage mechanism is not limited to arylazo compounds (Scheme I).

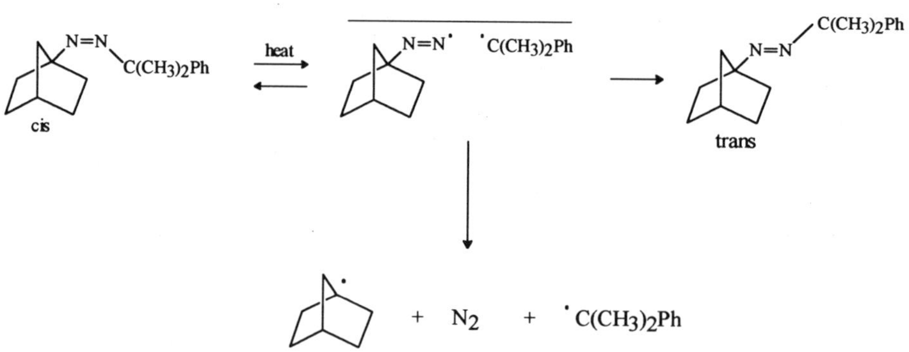

Scheme I

The ^{15}N signal from the escaped $^{29}N_2$, a product from escaped radicals, is seen as *E*, and those from the geminate reconstituted trans and cis azo compounds are *A*.

5.6.4 ^{19}F

The photolysis of bis(pentafluorobenzoyl) peroxide in CCl_4 gives pentafluorophenyl pentafluorobenzoate and chloropentafluorobenzene [29]. The fluorophenoxy moiety of the ester, a geminate product, showed strong enhanced absorption from the ortho and meta fluorines and emission from the para fluorine. These signals are consistent with positive a_F for the ortho and meta positions and negative a_F for the para position. The signals from the chloropentafluorobenzene, from escaped radicals, had the opposite polarization.

5.7 SELECTED APPLICATIONS OF CIDNP

5.7.1 Detection of Intermediates

Finding a CIDNP signal for a reaction product or for a particular intermediate means that some of the reaction took a corresponding radical path. But because CIDNP is so sensitive, it may be detecting only a minor mechanism, not the main one. On the other hand, if the mechanism detected can explain all of the facts, it should be taken seriously.

Interconversion of Ion and Radical Pairs by Electron Transfer. The thermolysis of diacyl peroxides appears to involve both radical pairs and ion pairs, or possibly a hybrid or rapidly equilibrating ion pair/radical pair [30–32]. The thermolysis of *t*-butylacetyl *m*-chlorobenzoyl peroxide (Scheme II) in *o*-dichlorobenzene at 120°C gave the carboxyinversion compound (CIN), the corresponding ester (E), and the hydrocarbons neopentane, 2-methyl-1-butene, 2-methyl-2-butene, and 1,1-dimethylcyclopropane [33].

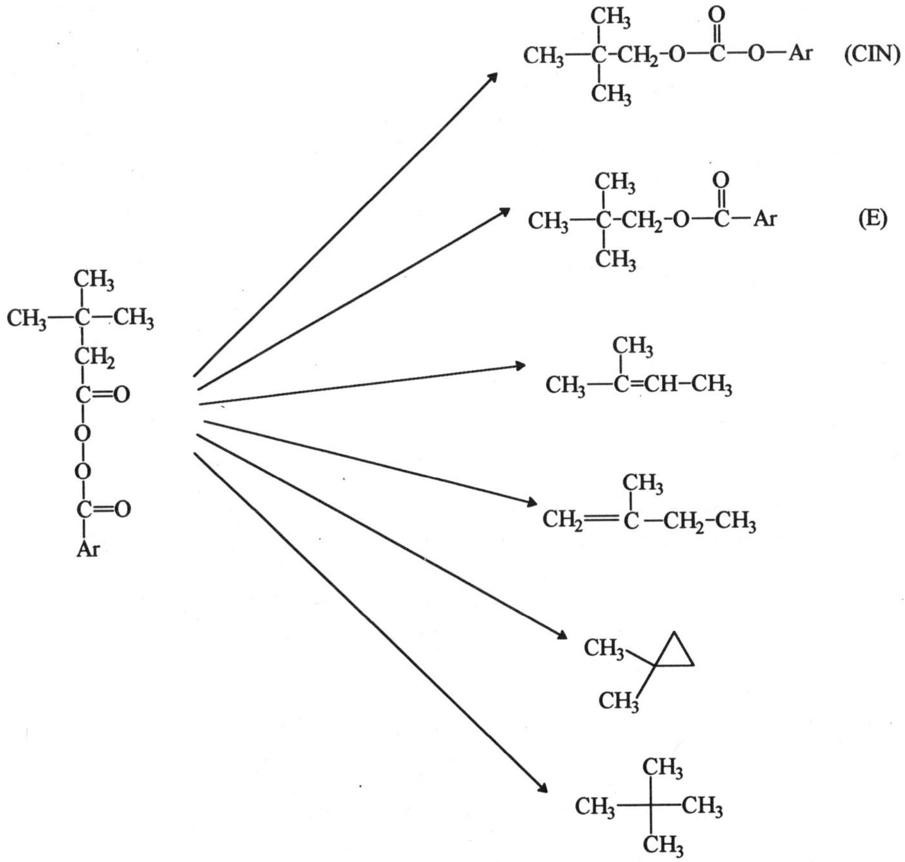

Scheme II

Formation of the alkenes and the cyclopropane requires a rearrangement of the carbon skeleton, which must have occurred in a neopentyl cation, since neopentyl radicals rearrange much too slowly, if at all. The CIDNP emission signals from these compounds are consistent with polarization of the terminal CH_2 of the neopentyl radical (Scheme III). Electron transfer converted this into the neopentyl cation, which rearranged. The polarized hydrogens in the products, indicated by underlines, correspond to the polarized terminal methylene of the neopentyl radical. Carboxyinversion compounds (CIN) are formed from ion pairs initially and are never polarized [34–36], and photolysis of diacyl peroxides gives radicals exclusively and no carboxyinversion products.

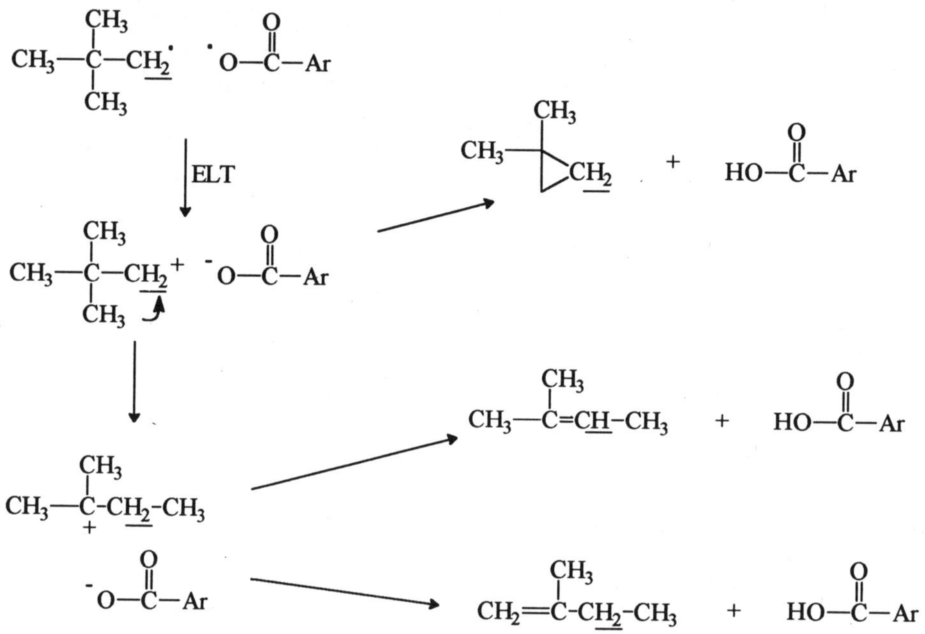

Scheme III

Acetophenone Enol. Photolysis of acetophenone in the presence of phenol (Scheme IV) gives 1,2-dibenzoylethane and acetophenone pinacol. The CIDNP emissions are observed from the methyl group of regenerated acetophenone and the hydroxyl and methylene groups of the enol [37]. These signals are consistent with polarization arising in a triplet pair formed by the reaction of triplet acetophenone with the phenol.

An α-Lactone Diradical. The Huček perester, discussed in more detail in Chapter 7, undergoes both radical and base-induced decomposition reactions in which an attacking radical removes a hydrogen atom from the phenolic OH or an attacking base removes a proton [38]. The base-initiated reaction is believed to give an open-chain triplet form of the α-lactone (Scheme V).

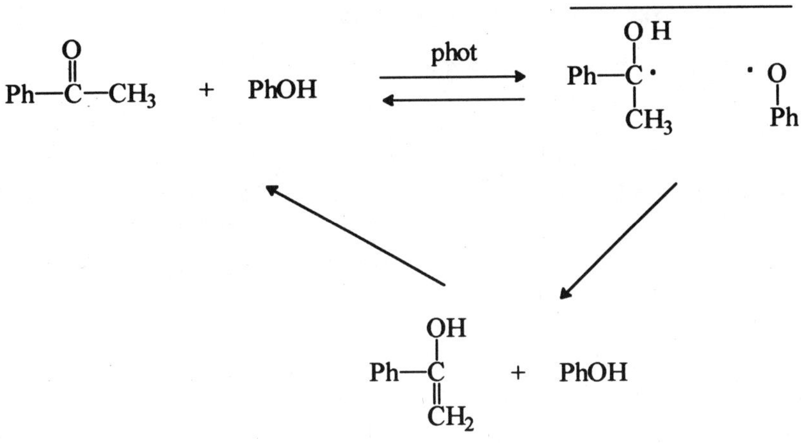

Scheme IV

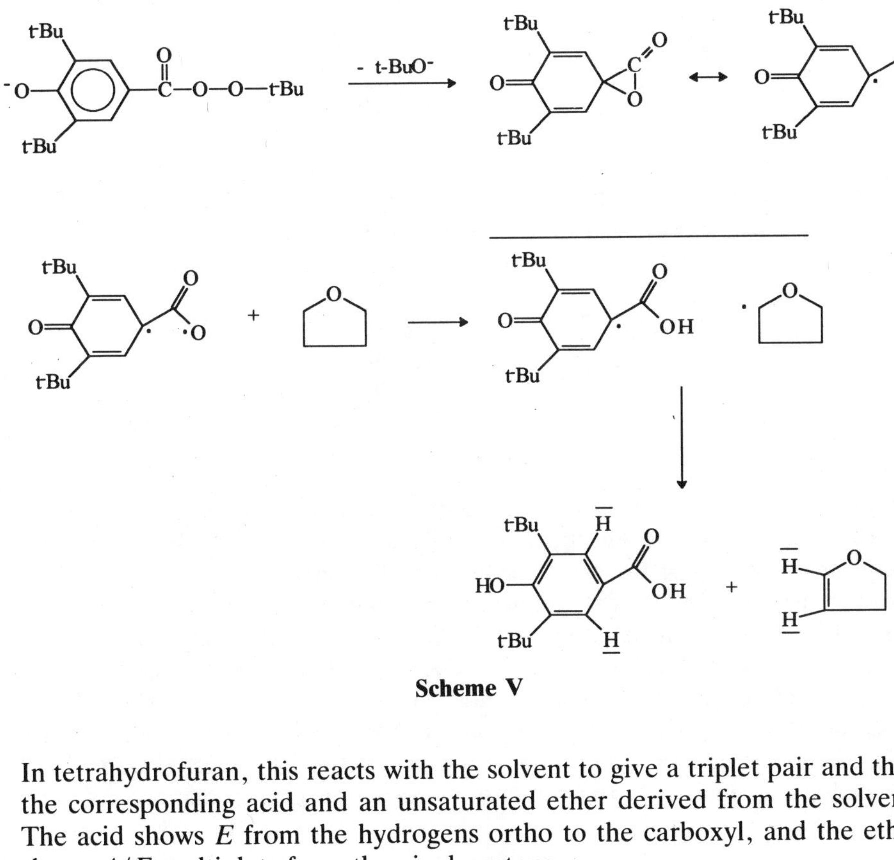

Scheme V

In tetrahydrofuran, this reacts with the solvent to give a triplet pair and then the corresponding acid and an unsaturated ether derived from the solvent. The acid shows *E* from the hydrogens ortho to the carboxyl, and the ether shows *A*/*E* multiplets from the vinyl protons.

5.7.2 Alternatives to Sigmatropic Rearrangement

In chemistry, as in society, not everything that is allowed is compulsory. An example is the 1,3 migration shown here of a dimethylaminomethyl group in a triene [39]. Although it is allowed as a concerted sigmatropic rearrangement, either as a suprafacial migration with inversion or an antarafacial migration with retention, it appears to go at least in part by a stepwise radical pair process (Eq. 5.14). The benzylic protons give an *E*/*A* multiplet.

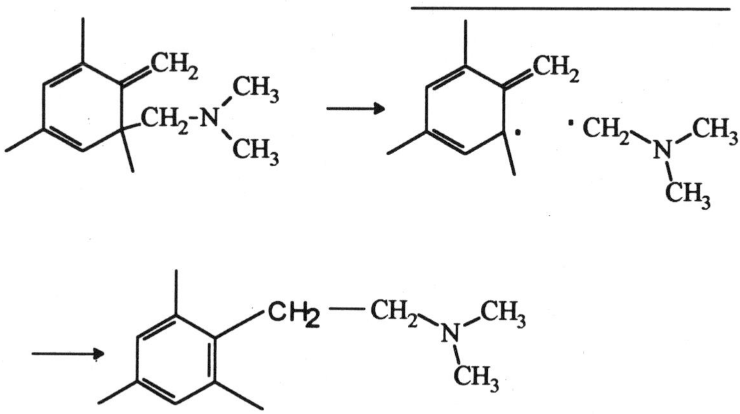

(5.14)

Other examples are found in some Stevens rearrangements of ylides (Eq. 5.15). The mechanism probably varies from case to case.

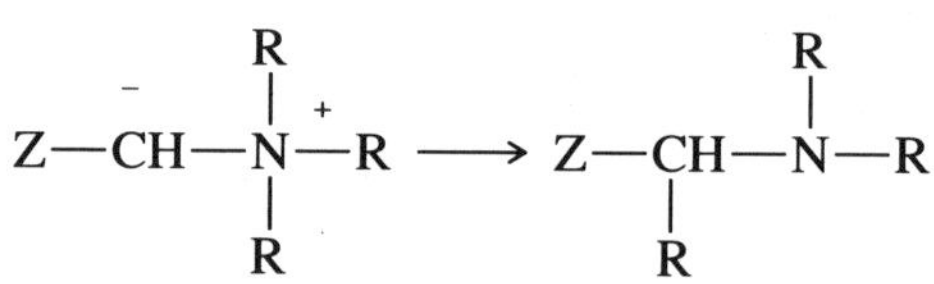

(5.15)

The plausible mechanisms involve a sigmatropic concerted reaction with inversion of the migrating group, or heterolysis of the bond to the migrating group, or homolysis of the bond to the migrating group.

In the rearrangement of the benzyl compound shown here (Eq. 5.16) the radical pair mechanism is supported by the isolation of a small amount of dibenzyl as a byproduct as well as an *E*/*A* signal from the benzylic protons [40]. In a related reaction in which a chiral α-phenylethyl group migrated, the configuration was 89% retained rather than inverted. Although racemization is expected for pairs that separate to the optimum distance for $S - T_0$ mixing, most of the recombination probably occurs from more intimate pairs.

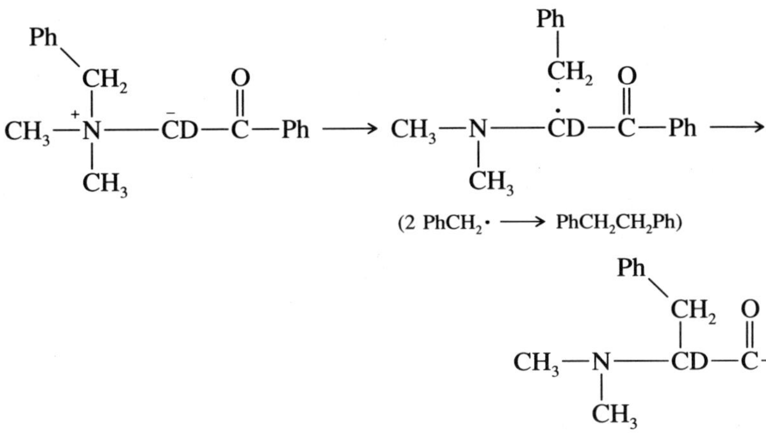

(5.16)

5.7.3 Methylene Multiplicity

A methylene CH_2: can be generated either as a singlet by photolyzing diazirine directly or as a triplet by photolyzing it with a triplet sensitizer (Scheme VI) [41]. In $CDCl_3$ solution, besides various free radical products, the direct (singlet) reaction gives $Cl_2DC–CH_2Cl$ by abstraction of Cl and recombination; the triplet sensitized reaction gives $Cl_3C–CDH_2$ by abstraction of D and recombination. The ^1H signal from the singlet reaction product is a triplet A split by the adjacent deuterium. The ^1H signal from the triplet reaction product is an unresolved *E*.

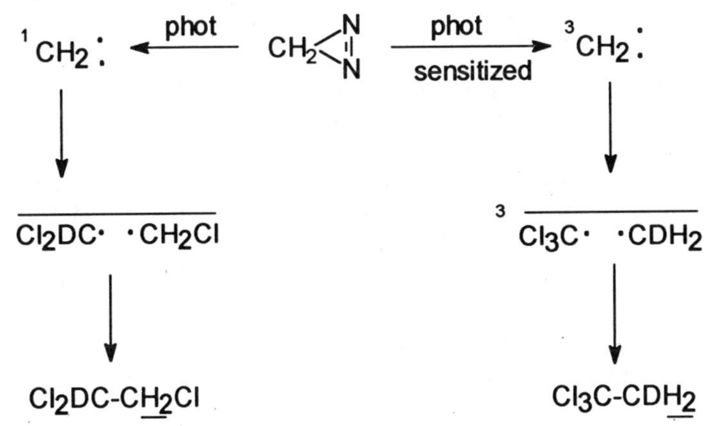

Scheme VI

The benzophenone-sensitized photolysis of diazomethane in toluene gives triplet methylene (Eq. 5.17). Reaction with the solvent then gives ethylbenzene showing an *A*/*E* signal from the methyl hydrogens [42]. In contrast, ethylbenzene formed in a similar experiment without any triplet sensitizer shows no CIDNP, presumably because the reaction of the carbene with the toluene involves direct insertion rather than an abstraction–recombination sequence [43].

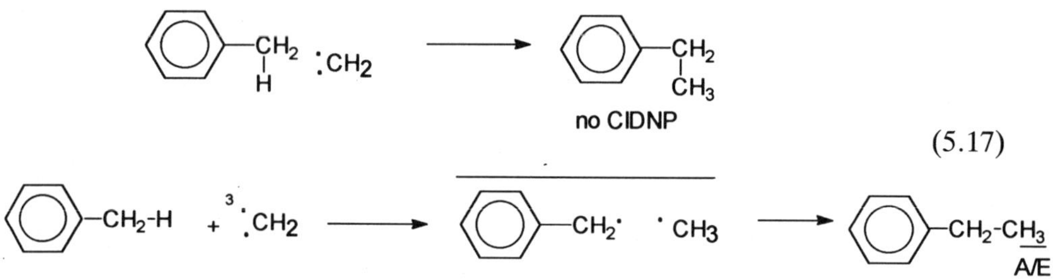

(5.17)

5.8 CIDEP

Chemically induced dynamic electron polarization (CIDEP) refers to enhanced absorption or emission in electron spin resonance spectra. Unlike NMR signals, ESR signals come only from the radicals that escape geminate recombination, but the high field $T - S_0$ radical pair sorting mechanism still applies. Another polarization mechanism, called initial polarization, or the triplet mechanism, occurs only in photochemical reactions. "Triplet" in this context refers to the triplet state of a molecule rather than a triplet radical pair [44]. The triplet mechanism affects both the CIDNP and CIDEP spectra.

5.8.1 The Triplet or Initial Polarization Effect

Polarization from the triplet effect [44] depends on the fact that the rates of intersystem crossing from the excited singlet to each of the three triplet sublevels depends on the nuclear spin state. If the spin-polarized triplet reacts, for example by H atom transfer or electron transfer, with spin conservation and within the longitudinal relaxation time ($^3T_1 \sim 10^{-8} - 10^{-9}$ s), the differences in populations of the T_{+1} and T_{-1} states will persist in the radicals being formed. Suppose, for example, that a molecule in the T_+ state with electron spins $|\alpha\alpha>$ gives one of its spin $|\alpha>$ electrons to a diamagnetic molecule with electron spins $|\alpha\beta>$. The triplet will become a spin $|\alpha>$ radical and the diamagnetic molecule will become a radical with spin $|\alpha\alpha\beta>$, which abbreviates to $|\alpha>$. Similarly, a molecule in the T_{-1} state will give two radicals with spin $|\beta>$.

5.8.2 Photoreduction of Pyruvic Acid

A reaction that shows CIDNP as well as CIDEP, and which will serve to exemplify both of the electron spin polarization mechanisms, is the photoreduction of pyruvic acid by hydrogen donors (Scheme VII) [45].

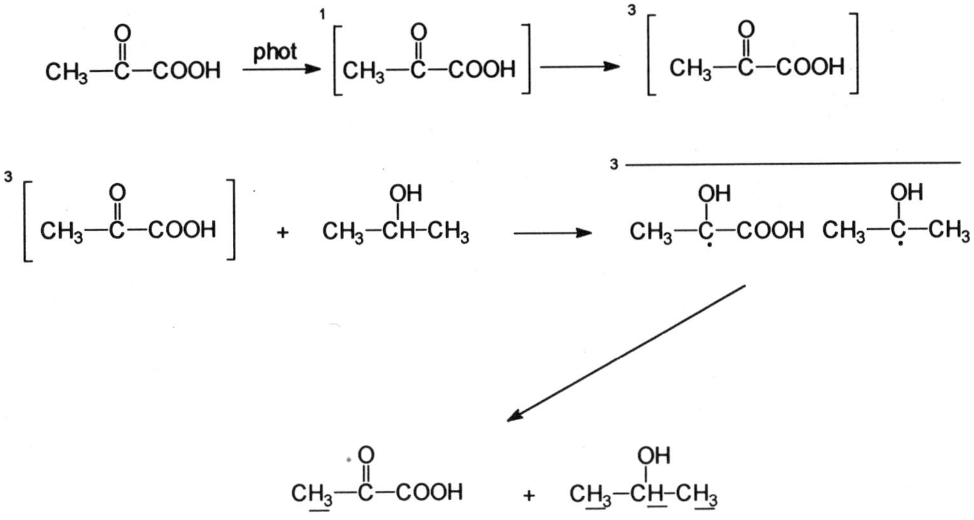

Scheme VII

Irradiation of a mixture of pyruvic acid and 2-propanol in CCl_4 gives triplet pyruvic acid by intersystem crossing. This abstracts hydrogen from the 2-propanol to give a triplet radical pair. Back transfer of the hydrogen atom then regenerates the pyruvic acid and 2-propanol. The methyl protons of the 2-propanol give an *A*/*E* CIDNP signal as predicted by the radical pair model. The methyl protons of the pyruvic acid are expected on the basis of the radical pair model to give an A CIDNP signal, but instead the signal from these protons is a net emission. This result is the effect of the triplet polarization mechanism.

Figure 5.7 is the ESR spectrum from the $CH_3C{\cdot}(OH)COOH$ radical formed during the photoreduction of pyruvic acid by lactic acid. It shows a quartet

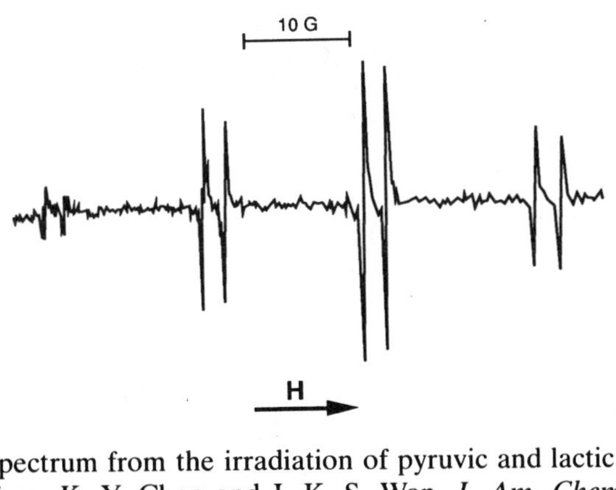

FIG. 5.7 ESR spectrum from the irradiation of pyruvic and lactic acids. Reprinted with permission from K. Y. Choo and J. K. S. Wan, *J. Am. Chem. Soc.*, **1975,** *97,* 7127. Copyright 1975 American Chemical Society.

of doublets, A_{CH_3} = 17.0 G, A_{OH} 2.2 G. The normal spectrum would have a 1:3:3:1 intensity distribution in the quartet. In the observed spectrum the low field lines are weaker than the high field lines: the radicals with these proton spins are fewer than expected, a multiplet effect. A careful analysis of the spectrum shows a small enhanced absorption superimposed on the multiplet effect. This is due to a contribution from the triplet effect added to the dominant radical pair effect.

REFERENCES

1. Lepley, R.; Closs, G. L., Eds. *Chemically Induced Magnetic Polarization,* Wiley: New York, 1973.
2. Fessenden, R. W.; Schuler, R. H. *J. Chem. Phys.* **1963,** *39,* 2147.
3. Bargon, J.; Fischer, H.; Johnsen, U. *Z. Naturforsch.* **1967,** *22,* 1551.
4. Ward, H. R. *J. Am. Chem. Soc.* **1967,** *89,* 5517.
5. Ward, H. R.; Lawler, R. G. *J. Am. Chem. Soc.* **1967,** *89,* 5518.
6. Closs, G. L. *J. Am. Chem. Soc.* **1969,** *91,* 4552.
7. Kaptein, R.; Oosterhoff, J. L. *Chem. Phys. Lett.* **1969,** *4,* 195.
8. Kaptein, R.; Oosterhoff, J. L. *Chem. Phys. Lett.* **1969,** *4,* 214.
9. Closs, G. L., *Adv. Magn. Res.* **1974,** *7,* 157.
10. Kaptein, R.; Verheus, F. W.; Oosterhoff, L. J. *Chem. Commun.* **1971,** 877.
11. Ward, H. R. *Acc. Chem. Res.* **1972,** *5,* 18.
12. Kaptein, R. *Chem. Phys. Lett.* **1968,** *2,* 261.
13. Closs, G. L.; Trifunac, A. D. *J. Am. Chem. Soc.* **1970,** *92,* 2186.
14. Hutton, R. S.; Roth, H. D.; Bertz, S. H. *J. Am. Chem. Soc.* **1983,** *105,* 6371.
15. Bertz, S. H. *Chem. Commun.* **1980,** 831.
16. Closs, G. L.; Czeropski, M. S. *J. Am. Chem. Soc.* **1977,** *99,* 6127.
17. Closs, G. L.; Czeropski, M. S. *Chem. Phys. Lett.* **1978,** *53,* 321.
18. Adrian, F. J. *Chem. Phys. Lett.* **1971,** *10,* 70.
19. Morris, J. I.; Morrison, R. C.; Smith, D. W.; Garst, J. F. *J. Am. Chem. Soc.* **1972,** *94,* 2406.
20. Kaptein, R.; den Hollander, J. A. *J. Am. Chem. Soc.* **1972,** *94,* 6269.
21. Closs, G. L., Ed. In *Chemically Induced Magnetic Polarization,* D. Reidel: Ubrino, 1977.
22. Closs, G. L.; Doubleday, C. E. *J. Am. Chem. Soc.* **1972,** *94,* 9248.
23. Kaptein, R.; Brokken-Zijp, J.; de Kanter, F. J. J. *J. Am. Chem. Soc.* **1972,** *94,* 6280.
24. Kaptein, R. *J. Am. Chem. Soc.* **1972,** *94,* 6262.
25. Cooper, R. A.; Lawler, R. G.; Ward, H. R. *J. Am. Chem. Soc.* **1972,** *94,* 552.
26. Lippmaa, E.; Pehk, T.; Buchachenko, A. L.; Rykov, S. V. *Chem. Phys. Lett.* **1970,** *5,* 521.
27. Green, J. G.; Dubay, G. R.; Porter, N. A. *J. Am. Chem. Soc.* **1977,** *99,* 1264.

28. Porter, N. A.; Green, J. G.; Dubay, G. R. *Tetrahedron Lett.* **1975,** 3363.
29. Roth, H. D.; Kaplan, M. L. *J. Am. Chem. Soc.* **1973,** *95,* 262.
30. Leffler, J. E. *J. Am. Chem. Soc.* **1950,** *72,* 67.
31. Greene, F. D.; Stein, H. P.; Chu, C.-C.; Vane, F. M. *J. Am. Chem. Soc.* **1964,** *86,* 2080.
32. Walling, C.; Waits, H. P.; Milovanovic, J.; Pappiaonnou, C. G. *J. Am. Chem. Soc.* **1970,** *92,* 4927.
33. Lawler, R. G.; Barbara, P. F.; Jacobs, D. *J. Am. Chem. Soc.* **1978,** *100,* 4912.
34. Taylor, K. G.; Govindan, C. K.; Kaelin, M. S. *J. Am. Chem. Soc.* **1979,** *101,* 2091.
35. Cooper, R. A.; Lawler, R. G.; Ward, H. R. *J. Am. Chem. Soc.* **1972,** *94,* 545.
36. Walling, C.; Lepley, A. R. *J. Am. Chem. Soc.* **1972,** *94,* 2007.
37. Rosenfeld, S. M.; Lawler, R. G.; Ward, H. R. *J. Am. Chem. Soc.* **1973,** *95,* 946.
38. Huček, A. M.; Barbas, J. T.; Leffler, J. E. *J. Am. Chem. Soc.* **1973,** *95,* 4698.
39. Baldwin, J. E.; Brown, J. E. *J. Am. Chem. Soc.* **1969,** *91,* 3647.
40. Schöllkopf, U.; Ludwig, U.; Ostermann, G.; Patsch, M. *Tetrahedron Lett.* **1969,** 3415.
41. Roth, H. D. *J. Am. Chem. Soc.* **1971,** *93,* 4935.
42. Roth, H. D. *J. Am. Chem. Soc.* **1972,** *94,* 1400.
43. Roth, H. D. *J. Am. Chem. Soc.* **1972,** *94,* 1761.
44. Wan, J. K. S.; Elliot, A. J. *Acc. Chem. Res.* **1977,** *10,* 161.
45. Choo, K. Y.; Wan, J. K. S. *J. Am. Chem. Soc.* **1975,** *97,* 7127.

6

THE SMALL RADICALS

6.1 THE SOLVATED ELECTRON

In 1864 Weyl discovered that metallic sodium dissolves readily in liquid ammonia to give a blue solution [1]. Kraus, in 1908, electrolyzed this solution and found that the species responsible for the blue color carried a negative charge. Since it did not give gas or any other product when it discharged at the anode, he concluded that it was an electron. The electron in liquid ammonia solution is quite stable. Although $\cdot e^{-}_{NH_3}$ eventually reacts with the ammonia to give sodamide and hydrogen, this process takes about 15 h at the boiling point of ammonia [2]. Unlike metallic sodium, in which the electrons are paired in conduction bands, the solution in liquid ammonia is paramagnetic.

Solutions of solvated electrons, apparently in equilibrium with diamagnetic solvated electron-pairs and ion pairs, can also be prepared by dissolving an alkali metal in other solvents that can solvate the alkali metal cation [3]. These include amines, 1,2-dimethoxyethane (DME), crown ethers, and even alcohols and water [4,5]. The reaction of potassium with very pure water or alcohols produces only a transient, localized blue color.

A better way of producing and studying solvated electrons is to use ionizing radiation in which an electron is detached either from a solvent molecule or from an aromatic hydrocarbon solute [6–10].

6.1.1 The Structure of the Solvated Electron

The structure of the solvated electron has been studied by means of its optical absorption spectrum [8,10] and by ESR [7,11] both in liquids and in gasses.

In 2,5-dimethyltetrahydrofuran solutions of sodium tetraphenylboron, the optical absorption spectrum of the (e^-, Na^+) pair, obtained by nanosecond pulse radiolysis, has two peaks, at 575 and 735 nm, and a shoulder at longer wavelengths [10]. This solvent was chosen because it is sterically hindered from tightly solvating the sodium ion. In 2,2,5,5-tetramethyltetrahydrofuran the peaks are even better resolved. The structure of the (e^-, Na^+) pair is uncertain, but could be spherically symmetrical and hydrogen-like, with the sodium ion in the center and the electron delocalized over the solvation shell.

Theoretical models for the solvated electron put the electron in an orbital centered on a cavity of the solvent [12–14] and held by a force field provided by surrounding dipoles. This may also be compared with the electron in the hydrogen 1*s* ground state, which resides in an orbital whose properties are the result of the central nuclear charge. The cavity orbital resembles an atomic 1*s* orbital mixed with 3*s* Rydberg-like orbitals of N or O atoms of the surrounding solvent [15].

6.1.2 The Hydrated Electron

Hydrated electrons can be formed by the irradiation of aqueous solutions of aromatic hydrocarbons. The excited state of the hydrocarbon either loses its excitation energy as fluorescence or transfers an electron to the solution, giving $\cdot e^-_{aq}$ [16].

Another method is to inject the electron by two-photon UV ionization of a water molecule rather than ionization of an added solute [9]. Using this method it was found that a 100-femtosecond (10^{-15} sec) pulse produced two different kinds of absorbing species. The first species absorbs in the infrared and appears with a characteristic time of 110 fs. This "dry" electron is formed much too fast for the surrounding water molecules to relax to the equilibrium cage orientations characteristic of the hydrated electron, $\cdot e^-_{aq}$. The hydration time is about 240 fs.

6.1.3 Reactions of Aqueous Electrons with H_2O and H^+

Two of the most important reactions of the aqueous electron are with the hydronium ion and with water to give H·, which may be thought of as the conjugate acid of the electron (Eq. 6.1 and 6.2).

$$\cdot e^-_{aq} + H_3O^+ \longrightarrow H\cdot + H_2O \tag{6.1}$$

$$\cdot e^-_{aq} + H_2O \longrightarrow H\cdot + HO^- \tag{6.2}$$

Reaction 6.1 is fast and diffusion-controlled, with a rate constant [17] of $2.1 \times 10^{10}\ M^{-1}\ sec^{-1}$.

The reaction with H_2O (Eq. 6.2) has a rate constant of only 890 sec^{-1}, compared to $1.8 \times 10^7\ M^{-1}\ sec^{-1}$ for the reverse reaction [6,17]. The ratio of these rates corresponds to a pK_a for the hydrogen atom in water of about 9.6.

The activation energy for reaction 6.2 is 4.5 ± 1 kcal/mol [18]. The activation energy of the reverse reaction is 6.3 ± 0.6 kcal/mol [19].

From the values given above it was possible to derive several thermodynamic properties of $\cdot e^-_{aq}$: the heat of formation $\Delta H^\circ_f = -32.6 \pm 1.6$ kcal/mole, the heat of hydration of the gas phase electron $\Delta H_s = -34 \pm 1.6$ kcal/mole, the standard entropy $S^\circ = 16.7 \pm 5.4$ cal/(mol deg), and the entropy of hydration of the gas phase electron $\Delta S_s = 11.7 \pm 5.4$ cal/(mol deg) [19]. The positive entropy of hydration means that the effect of breaking the structure of the water solvent outweighs the effect of orienting the water molecules about the electron.

Both $\cdot e^-_{aq}$ and $\cdot e^-_{NH_3}$ can react bimolecularly to give H_2 and the corresponding lyate ion [6].

$$2\ \cdot e^-_{aq} \longrightarrow H_2 + 2\ HO^-\ (6 \times 10^9\ M^{-1}\ sec^{-1}) \tag{6.3}$$

The reaction of $\cdot e^-_{NH_3}$ is very much slower than that of $\cdot e^-_{aq}$

$$2\ \cdot e^-_{NH_3} \longrightarrow H_2 + 2\ H_2N^- \tag{6.4}$$

6.1.4 Addition to Benzene

Although both $\cdot e^-_{aq}$ and H· (in water) ultimately give the same product, the addition of $\cdot e^-_{aq}$ to benzene in water is two orders of magnitude slower than the addition of H· [6]. To make the reaction with $\cdot e^-_{aq}$ (Eq. 6.5) observable at all against the background of the H· reaction (Eq. 6.6), the equilibrium between $\cdot e^-_{aq}$ and H· (Eq. 6.1) is shifted by adding a large excess of HO^-.

$$C_6H_6 + \cdot e^-_{aq} \longrightarrow C_6H_6^{\cdot -} \qquad (k = 1.4 \times 10^7\ M^{-1}\ sec^{-1}) \tag{6.5}$$

$$\Big\downarrow\ H^+$$

$$C_6H_6 + H\cdot \longrightarrow C_6H_7\cdot \qquad (k = 1.1 \times 10^9\ M^{-1}\ sec^{-1}) \tag{6.6}$$

Subsequent steps convert the benzene anion-radical or the cyclohexadienyl radical to 1,4-cyclohexadiene.

6.1.5 $S_{RN}2$ Substitution

The anion-radical intermediates in the $S_{RN}2$ substitution reaction [20] are formed by electron transfer from various reagents to the substrate. The solvated electron $\cdot e^-_{aq}$ an act as one of these reagents [21,22].

$$\cdot e^-_{aq} + RI \longrightarrow RI^{-}\cdot \longrightarrow R\cdot + I^- \tag{6.7}$$

$$\cdot e^-_{aq} + C(NO_2)_4 \longrightarrow \cdot C(NO_2)^- \longrightarrow \cdot NO_2 + (O_2N)_3C^- \tag{6.8}$$

6.1.6 Addition to Carbonyl Compounds

The addition to carbonyl compounds gives ketyls, some of which are rather stable anion–radicals.

$$\cdot e^-_{aq} + R_2C{=}O \longrightarrow R_2C^{\cdot}{-}O^- \tag{6.9}$$

Dimerization of the conjugate acid of the ketyl gives the corresponding pinacol. Many reactions of synthetic utility involve the reduction of a carbonyl compound in the presence of a metal. In principle the transfer of the electron from the metal to the carbonyl compound could be indirect by way of the solvated electron, but more probably takes place at the surface of the metal. Examples are the preparation of pinacol from acetone and magnesium and the reduction of esters to acyloins by sodium.

6.1.7 Addition to Oxygen

Both $\cdot e^-_{aq}$ and H· react with O_2 at diffusion-controlled rates to give the superoxide anion radical $O_2\cdot^-$ and its conjugate acid HOO· [23].

6.2 THE HYDROGEN ATOM

6.2.1 Formation

Because of the equilibrium with $\cdot e^-_{aq}$ (Eqs. 6.1 and 6.2), the hydrogen atom is found wherever $\cdot e^-_{aq}$ is found, particularly in acidic solutions.

In the gas phase, atomic hydrogen can be made by passing H_2 diluted with argon through a 10-MHz radiofrequency discharge [24–26] or through a microwave field [27], or by passing H_2 over a hot tungsten filament [28–30]. Since the lifetime of atomic hydrogen in the gas phase is quite long, it can be delivered to a reagent in a liquid phase by passing the gas through the liquid, or to a solid reagent by diffusion.

One such example of the transport of atomic hydrogen from the gas phase is the polymerization of styrene, induced by bubbling the carrier gas through the liquid [27]. Another example is the formation of isopropyl radicals when H·, formed from H_2 by contact with a hot tungsten wire in the gas phase, reacts with propylene adsorbed on silica [28]. The isopropyl radicals subsequently dimerize or disproportionate to give propylene and propane; the propane to dimer ratio varies with the fraction of the surface covered by the adsorbed propylene. Reaction of H· with *cis*-2-butene gives a mixture of 1-butene, *trans*-2-butene, and butane [31].

Lasers have been used to generate H· or D· by photolysis of HI, HCl, H_2S, NH_3, PH_3, and acetylene [32]. The translational energy of the atom can be varied by changing the wavelength at which a precursor HX is photolyzed [33]. Another advantage of lasers is that they can be used to measure the yield of H· or D· by the fluorescence of the atom in its $2p$ state [33]. For example, the photolysis of deuteroformaldehyde (CHDO) gives both H· and D·. The H·/D· ratio depends on the photolysis wavelength, dropping sharply after the energy threshold for cleavage of the C–D bond is reached [34].

Hydrogen atoms can be generated directly in condensed phases by photolysis of sulfhydryl compounds [35,36] or *t*-butyl performate [37].

$$C_6H_5\text{—S—H} \xrightarrow{h\nu\ 300\ \text{nm}} C_6H_5\text{—S}\cdot + H\cdot \tag{6.10}$$

$$t\text{-BuO}_3\text{CH} \xrightarrow{h\nu\ 300\ \text{nm}} t\text{-Bu—O}\cdot + CO_2 + H\cdot \tag{6.11}$$

Mercury in the gas phase absorbs at 254 nm to give $Hg(^3P_1)$, an electrophilic excited state that forms exciplexes with donor C–H bonds and causes homolysis into alkyl radicals and hydrogen atoms [38]. A practical application is the synthesis of dehydrodimers from alkanes [39].

6.2.2 Abstraction or Displacement Reactions

Hydrogen atoms will abstract both halogen [40] and hydrogen [30,35].

$$H\cdot + RX \longrightarrow HX + R\cdot \tag{6.12}$$

$$H\cdot + RSH \longrightarrow H_2 + RS\cdot \tag{6.13}$$

$$H\cdot + RH \longrightarrow H_2 + R\cdot \tag{6.14}$$

$$H\cdot + H_2 \longrightarrow H_2 + H\cdot \tag{6.15}$$

$$H\cdot + D_2 \longrightarrow HD + D\cdot \tag{6.16}$$

The energy of the transition state for reaction 6.16 is expected to depend sharply on the angle of approach [33,42–45]. For a collinear transition state the calculated barrier is 8 kcal mole^{-1}, rising steeply to about 50 kcal mole^{-1} for a sideways approach.

Reaction 6.17 is the disproportionation reaction of H· with the ethyl radical [46]. The H·–H–C angle at the saddle point of the potential energy surface is calculated to be 162° rather than 180°. This potential energy saddle point, however, is not the transition state if the transition state is defined as the position of maximum free energy along the reaction coordinate. The transition state is actually somewhat on the product side of the potential energy saddle point. Although the potential energy begins to decrease when the saddle point is passed, the entropy is becoming increasingly more negative. The decrease in potential energy is compensated by the increase in the contribution of the $T\Delta S^‡$ term to $\Delta G^‡$, which equals $\Delta H^‡ - T\Delta S^‡$.

$$H\cdot + CH_3CH_2\cdot \longrightarrow H_2 + CH_2{=}CH_2 \tag{6.17}$$

Photolysis of *t*-butyl performate at high concentrations gives considerable induced decomposition, with evolution of H_2 as indicated by Eq. 6.18 [37].

$$H\cdot + H{-}(CO){-}OO\text{-}t\text{-}Bu \longrightarrow H_2 + CO_2 + \cdot O{-}t\text{-}Bu \tag{6.18}$$

Abstraction of aldehydic hydrogen by H· is two to three orders of magnitude faster than abstraction of hydrogen from a *t*-butyl group [47]. However, it is interesting that H· is able to attack the formyl hydrogen rather than the peroxide function. No production of H_2 occurs in the *thermolysis* of *t*-butyl performate [48]. Apparently formyloxy radicals, unlike acetoxy, do not decarboxylate readily, and therefore do not provide any H·.

Relative rates of abstraction of H by H· in various molecular environments have been established by competition experiments, usually with a standard substrate such as a thiol [35,37]. Perhaps the most reliable method, even though it requires the estimation of corrections due to isotope effects, is the photolysis of a deuterated thiol in the presence of the substrate [35]. Some results are shown in Table 6.1.

The selectivity of H· in abstraction reactions is about the same as those of methyl and phenyl radicals.

6.2.3 Addition Reactions of H·

Hydrogen atoms add to ethylene to give ethyl radicals [30] and to propylene to give isopropyl radicals [28]. Addition to 1-propyne gives the 1-methyl-1-

TABLE 6.1 Relative Reactivities (per H) for Abstraction of H by Atomic Hydrogen[a]

Alkanes		Alcohols		Ethers	
Relative to $R\mathbf{CH}_3$ = [1.]		Relative to $\mathbf{CH}_3OH$ = [1.]		Relative to $\mathbf{CH}_3OH$ = [1.]	
$R_2\mathbf{CH}_2$	5.	$CH_3\mathbf{CH}_2OH$	4.2	Dioxane	2.7
$R_3\mathbf{CH}$	42.	$(CH_3)_2\mathbf{CH}OH$	14.1	THF (α)	16.5
		$(\mathbf{CH}_3)_3OH$	0.024	$(CH_3\mathbf{CH}_2)_2O$	9.0
				$((CH_3)_2\mathbf{CH})_2O$	17.4

[a]The position is shown by the boldface letter.

vinyl radical (Eq. 6.19). Addition to 1,3-butadiene takes place at the end carbon to give methylallyl [30].

$$\mathrm{H\cdot + HC{\equiv}C{-}CH_3 \longrightarrow H_2C{=}\dot{C}{-}CH_3} \tag{6.19}$$

$$\mathrm{H\cdot + CH_2{=}CH{-}CH{=}CH_2 \longrightarrow CH_3{-}CH{=}CH{-}CH_2\cdot} \tag{6.20}$$

The homogeneous pyrolysis of 1,3-cyclohexadiene at around 400° is a chain reaction in which hydrogen atoms abstract H from the cyclohexadiene, and also add to it, to give an allylic radical [49].

Addition of H· to benzene gives the cyclohexadienyl radical [30]. The tritiation of monosubstituted benzenes in solution at 40°C has been studied by photolyzing mixtures of the substrate with thiophenol-*t* [36]. Addition of the T· to the benzene ring (Eq. 6.21), followed by abstraction of H by another atom or radical (Eq. 6.22), gives a mixture of ortho, meta, and para derivatives.

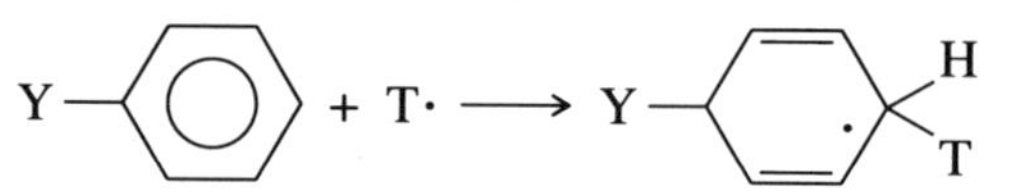

(6.21)

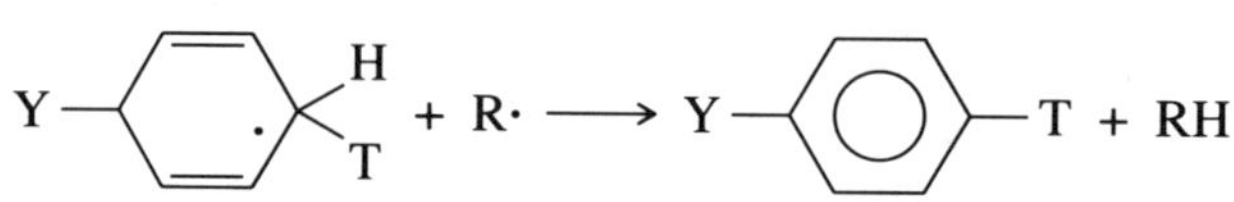

(6.22)

Tritiation at the ortho positions has been found to range from 42 to 67%, which is significantly greater than the 40% expected statistically.

Assuming that reaction 6.22 is fast enough to make Eq. 6.21 essentially irreversible, then the rate of reaction 6.21 for the meta position (for example) is equal to the total tritiation rate times the fraction of meta-tritiated product. Rates calculated in this way, called meta and para partial rate factors, can be plotted against the Hammett sigma constant for the substituent Y. The resulting plots show considerable scatter, perhaps because the different methods used in the determination of the total tritiation rates are subject to different errors, or because the assumption, that Eq. 6.21 is rate-limiting, is not completely valid. The slope (rho) is about -0.25, indicating that T· is very slightly electrophilic [36].

6.3 HALOGEN ATOMS

6.3.1 Halogenation Chain Mechanisms

Free radical halogenation is a chain reaction consisting of initiation, chain-carrying, and chain-terminating steps. Chains can be initiated by any process that generates radicals able to abstract hydrogen from the substrate being halogenated. In photochemically initiated chlorination or bromination, using the halogen itself rather than a "positive halogen" compound, the initiating step is homolysis of the halogen molecule itself.

Initiation:

$$X_2 \xrightarrow{h\nu} 2\,X\cdot \tag{6.23}$$

In the absence of complicating factors, the remaining steps can be as simple as Eqs. 6.24–6.27.

Chain carrying:

$$X\cdot + R\text{—}H \xrightarrow{k_{abs}} HX + R\cdot \tag{6.24}$$

$$R\cdot + X_2 \longrightarrow R\text{—}X + X\cdot \tag{6.25}$$

Termination:

$$2\,X\cdot \longrightarrow X_2 \tag{6.26}$$

$$R\cdot + X\cdot \longrightarrow R\text{—}X \text{ or alkene} + HX \tag{6.27}$$

Mechanisms for halogenation by positive halogen compounds such as *t*-butyl hypochlorite or *N*-bromosuccinimide are similar. Initiation is by photolysis or thermolysis of the positive halogen compound. The chain-carrying steps include reactions like Eq. 6.28 and Eq. 6.29.

$$R\cdot + t\text{-BuOCl} \longrightarrow RCl + t\text{-BuO}\cdot \tag{6.28}$$

$$t\text{-BuO}\cdot + RH \longrightarrow t\text{-BuOH} + R\cdot \tag{6.29}$$

Rates. The rate constant [50] for the chain-carrying step (Eq. 6.24), with Cl· as the abstracting atom and ethane as the substrate, in the gas phase at room temperature, is $6.10 \pm 0.11 \times 10^{-11}$ cm^3 molecule^{-1} sec^{-1}, or 3.67×10^8 M^{-1} sec^{-1}. The rate constant for chain breaking by a reaction in which Cl· abstracts hydrogen from the ethyl radical (Eq. 6.27) is $1.21 \pm 0.04 \times 10^{-11}$ cm^3 molecule^{-1} sec^{-1}.

Selectivity. Given the uncomplicated mechanism of Eqs. 6.23–6.27, the yield of each isomeric substitution product R–X or R′–X is proportional to the rate constant for the corresponding hydrogen abstraction step.

$$k_{abs}/k'_{abs} = [\text{Yld R–X}]/[\text{Yld R}'\text{–X}] \tag{6.30}$$

Selectivity for abstraction of hydrogens in *different* substrate molecules R–H and R′–H is given by a similar equation modified only by the inclusion of the substrate concentrations.

$$(k_{abs}/k'_{abs})([\text{R–H}]/[\text{R}'\text{–H}]) = [\text{Yld R–X}]/[\text{Yld R}'\text{–X}] \qquad (6.31)$$

Several reactions not included in the simple mechanism can obscure the true selectivity of the halogen atom. One such reaction is the abstraction by the alkyl radical of hydrogen from the hydrogen halide, the reverse of the chain-carrying step (Eq. 6.24) [51]. In chlorination, this is not a serious difficulty because the reverse reaction is slightly endothermic, or at worst, nearly thermoneutral. This means that it is too slow to compete with the highly exothermic reaction of R· with Cl_2. For bromination, the reverse of Eq. 6.24 is exothermic, but can still be neglected if the Br_2 concentration is kept high enough.

Another complication is the possibility that R· might rearrange, giving a halogenated product with a structure different from that of the radical formed by the initial hydrogen abstraction. This is not a problem in the case of unsubstituted alkyl radicals, because migration of alkyl or hydrogen is an electronically forbidden process. Migration of aryl, although not forbidden, is usually not fast enough to interfere. However, if the substrate already contains a bromine substituent, migration of this substituent can be competitive [51,52].

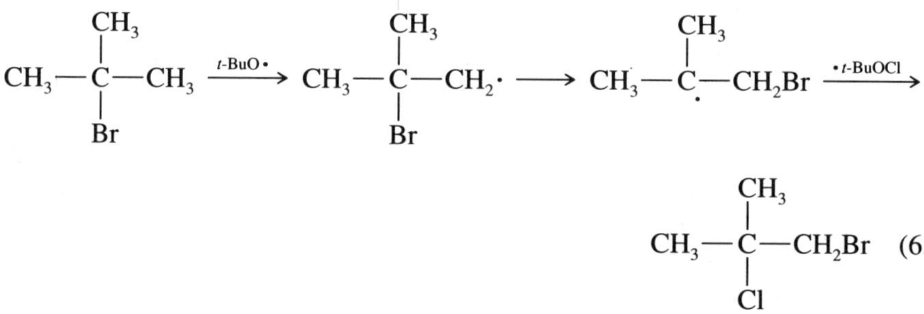
(6.32)

If the concentration of the monohalogenated products is allowed to become too great, the further halogenation of these compounds can obscure the interpretation of the results. This difficulty is minimized by using relatively high concentrations of the substrate and stopping the reaction at low extents of conversion.

In the gas phase, and in relatively inert solvents such as CCl_4, Cl· is rather unselective in its abstraction of hydrogen atoms because the reactions are so fast as to be almost encounter controlled, that is, with rate constants exceeding $10^{10}\ M^{-1}\ s^{-1}$. This is especially true of the reaction of Cl· with particularly reactive compounds such as ethanol, aniline, and hexamethylbenzene [53–55].

Although tertiary hydrogens are preferred over secondary hydrogens, secondary over primary, and benzylic over nonbenzylic, the selectivities in inert solvents are small factors of less than 10. For example in the chlorination of 2,3-dimethylbutane in CCl_4, the selectivity for abstraction of tertiary over primary hydrogens, corrected for the difference in number, is 6.5 [56]. The selectivity for the benzylic alpha hydrogens of ethylbenzene over the beta hydrogens is even less, only about a factor of 2, for reasons that will become clear [57]. For various aliphatic tertiary, secondary, or primary hydrogens in inert solvents such as CCl_4 average values of the relative rates are about 4.5:4:1 [56].

The bromine atom is less reactive than Cl·, as indicated by the dissociation energies, $D(\text{Br–H}) = 87 \text{ kcal mole}^{-1}$, $D(\text{Cl–H}) = 103 \text{ kcal mole}^{-1}$. It is also more selective. For example, the relative rates for tertiary, secondary, and primary aliphatic hydrogens are estimated to be about 2000:50:1 [58]. Lower reactivity is often, though not always, accompanied by a higher selectivity. A rationale for this is that the transition state for the reaction of a less energetic reagent will be more like the product [59,60]. The difference between forming a stable tertiary radical rather than a primary radical will then be more important. It is not unexpected that the bromine atom is very much more selective.

6.3.2 Medium Effects

In aromatic solvents and in carbon disulfide, in which chlorine atoms are strongly solvated, they are less reactive and much more selective. For example, the preference for abstraction of tertiary rather than primary hydrogen from 2,3-dimethylbutane is (on a per hydrogen basis) a factor of 32 in benzene, and a factor of 225 in carbon disulfide, but only a factor of 6.5 in carbon tetrachloride [56,61].

The reason for the greater selectivity is the conversion of the reactive Cl· into a less reactive complex [53,62]. In benzene this species is a charge-transfer pi complex with benzene (**I**). It has been detected as a transient absorbing at 490 nm during pulse radiolysis of benzene in CCl_4 [63]. The uncomplexed Cl· in CCl_4 absorbs at 330 nm [64].

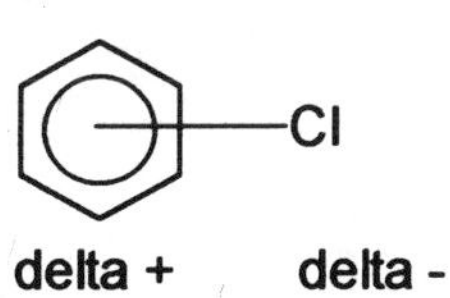

I

The rate constant for the formation of **I** in CCl_4, measured by laser flash photolysis [53], is $6.0 \times 10^9 \text{ M}^{-1} \text{ s}^{-1}$ at room temperature. The rate constant for the reaction of the free chlorine atom with 2,3-dimethylbutane (abstraction of H from both positions) is 2.6×10^9. The rate constant for the abstraction of the tertiary H by the complexed reagent **I** is much lower than this, only 4.6×10^7. The tertiary/primary selectivity of **I** exceeds that of the uncomplexed chlorine atom by more than an order of magnitude.

For chlorination by Cl· (plus complexed Cl·) in a series of substituted benzene solvents, log (tert/prim) selectivity for the reaction with 2,3-dimethylbutane is an approximately linear function of the Hammett meta sigma of the substituent on the substituted benzene, decreasing as the substituent is made more electron-withdrawing. A more meaningful linear free-energy relationship is found between log (tert/prim) and the basicity of the solvent toward HCl [56]. Both relationships support the pi complex structure for Cl· in aromatic solvents.

When the basicity of the benzene is very high, as in the case of benzenes with three to five methyl substituents, the selectivity falls again [65]. For example, under the same standard conditions (22 ± 2°, [arene] = 2.0 M, [DMB] 0.15 M, and CCl_4 as the inert diluent) the per hydrogen tert/prim selectivity is 32 when benzene is the arene, rising to 37 for *m*-xylene, but then dropping to 12 for 1,2,3,5-tetramethylbenzene. The probable reason is that the arene–chlorine complex becomes almost unreactive, leaving the free Cl·, though low in concentration, as the principal reagent.

Diluting neat 2,3-dimethylbutane with substituted benzenes also increases the selectivity (Fig. 6.1) [56,57]. The rise in selectivity with increase in concentration of the aromatic additive is more rapid if the diluent has an electron-releasing substituent.

When an alkyl benzene is chlorinated in an inert solvent, the selectivity between the alpha and beta side-chain positions depends strongly on the concentration of the alkyl benzene. The alkyl benzene substrate has an effect

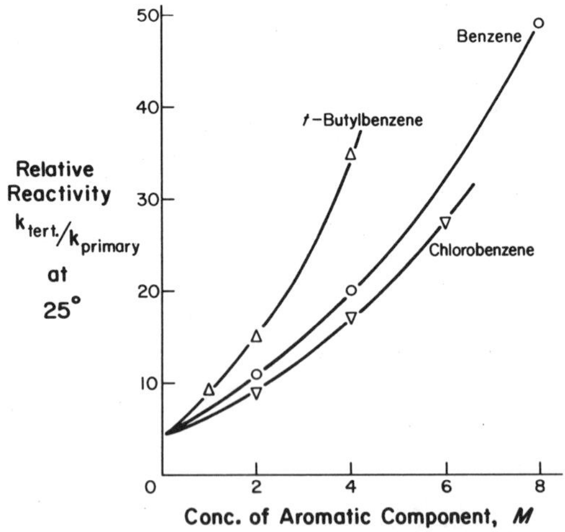

FIG. 6.1 Effects of aromatic diluents on the selectivity of Cl· for tertiary versus primary hydrogens [56].

on its own chlorination resembling the effect of added benzenes on the chlorination of 2,3-dimethylbutane. At low concentrations of the substrate the chlorinating reagent is the free Cl·. At higher concentrations the chlorinating reagent is increasingly the pi complex between Cl· and a molecule of the substrate. The selectivity, alpha/beta, for chlorination of ethylbenzene, adjusted for the difference in number of hydrogens, declined from 14.5 in the neat (8.1 molar) substrate, to 5.65 in a 5-M solution in cyclohexane, to 2.70 in a 1.5-M solution [57].

In a similar series of experiments in which the competing hydrogens were the aliphatic hydrogens of cyclohexane and the alpha hydrogens of ethylbenzene, the results were quite different: in the range of 4.7 to 1.5 M ethylbenzene concentration the alpha ethylbenzene/cyclohexane selectivity changed only from 1.58 to 1.24. Using toluene in place of the ethylbenzene, the results were even more surprising: the selectivity actually *increased* slightly at the lower toluene concentrations. The probable reason for these effects is that most of the encounters between complexed Cl· and the aromatic substrate result in nothing more than transfer of the Cl· from one complexing host to the other. In the case of the *intra*molecular alpha/beta selectivity in the halogenation of ethylbenzene, the effect of this kind of inefficiency is felt equally at both the alpha and the beta positions, and the expected greater stabilization of a benzylic radical can begin to make a difference. The experimental results already described for the chlorination of 2,3-dimethylbutane in complex-forming solvents were from experiments in which the substrate concentration was fairly high, *in the range of 1 to 5 M,* and the tert/prim selectivity seemed to be independent of the substrate concentration and dependent only on the concentration of the complexing solvent component.

In experiments in which the dimethylbutane concentration was varied from 5 to 0.05 M or less, and the concentration of the complex-forming component (benzene or carbon disulfide) was held constant at 4 M, something different happened [66]. The selectivity *increased* as the initial substrate concentration was decreased. The probable explanation of this anomaly is that the equilibrium between free and solvated chlorine atoms is not completely established; some of the newly formed chlorine atoms attack the dimethylbutane before they are complexed by the benzene. At lower dimethylbutane concentrations the chances of the benzene complex being formed first are better and the selectivity rises. This hypothesis was confirmed by laser flash photolysis measurements of the relevant rate constants [67].

6.3.3 Sigma* Complexes of Cl·

Like benzene, pyridine forms a selective complex with Cl·, but the complex is a σ* complex at the nitrogen rather than a pi complex [68,69].

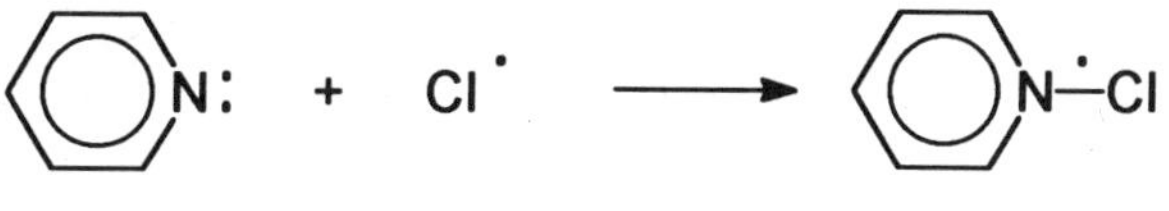

(6.33)

The bond between the N and the Cl is a 3-electron $(\sigma)^2(\sigma^*)^1$ bond involving the occupied sigma orbital on nitrogen and a Cl $3p_\sigma$ singly occupied orbital. According to an ab initio calculation the N–Cl distance is about 2.27 Å, which is 135% of a normal N–Cl bond length and about one third of the van der Waals contact distance [68]. Other 3-electron bonded molecules that have been studied by ESR include both neutral radicals such as $(C_6H_5)_3PCl\cdot$ and charged species such as $Cl_2\cdot^-$ and $R_2SSR_2\cdot^+$ [69,70].

The increased selectivity of Cl· complexed to a pyridine nitrogen has been put to a practical application in the template chlorination of steroids. The Cl· is attached to the pyridine N of a suitable nicotinate ester of the steroid [68].

In the chlorination of 2,6-dimethylquinoline [71] the complexing with the nitrogen reverses, the usual preference for radical abstraction of hydrogen from the methyl at position 6 and gives preferential abstraction from the adjacent methyl group at position 2 by a factor of about 1.7.

6.3.4 Substituent Effects

In the abstraction of hydrogen from the side chain of an *n*-alkylbenzene by an electronegative radical, there are two substituent effects. The rate is enhanced at the alpha position, where the developing radical is stabilized by resonance with the aromatic ring. The other important factor is the electron-withdrawing inductive effect of the phenyl group, greatest in the alpha position and diminishing steadily along the chain. Intramolecular selectivities for the chlorination of *n*-butylbenzene are shown in Table 6.2. Those in the top line, for uncomplexed Cl·, were found by extrapolation to an infinitely dilute solution in CCl_4 [57]. The selectivities for the less reactive, pi-complexed Cl·, from experiments with neat *n*-butylbenzene, are shown in the second line, and those from experiments in benzene on the bottom line. All of the selectivities are greater with the less reactive, pi-complexed chlorine atom.

The selectivities in the alpha position are the result of a resonance effect partly offset by the electron-withdrawing inductive effect of the phenyl group. The inductive effect is important because of the electronegativity of Cl, which favors canonical structures like **II**.

$$R^+ \ H\cdot \ Cl^-$$

II

TABLE 6.2 Chlorination of *n*-Butylbenzene

Position	α	β	γ	δ
Dilute in CCl_4	5.9	2.7	4.0	[1]
Neat	13.9	7.3	9.2	[1]
In benzene	20	10	13	[1]

TABLE 6.3 Chlorination of *n*-Butyl Chloride

Position	α	β	γ	δ
Percent attacked with Cl_2	7.8	23.5	49.9	19.5
Percent attacked with *t*-BuOCl	21.1	19.4	43.9	15.6

Despite the difference in electronegativity of Cl· and pi-complexed Cl·, the alpha/beta selectivities do not differ very much between the two reagents. The resonance effect should be most important, relative to the inductive effect, for the less reactive complexed chlorine atom. On the other hand, the complexed atom may be more sensitive to the steric effect of the phenyl group. All three series of reagents show a dip in selectivity at the beta position.

The known pattern of relative reactivity (selectivity) of Cl· has been used to identify the chain-carrying step in chlorination reactions. For example, in the chlorination of 1-chlorobutane by means of Cl_2 and light or *t*-BuOCl and light [72] (Table 6.3), the percentages of attack at the alpha position, where the inductive effect is greatest and the selectivity should be most sensitive to the electronegativity of the reagent, are quite different for these two reagents. The abstracting species in the *t*-BuOCl reaction is probably *t*-BuO·.

Correlations with Substituent Parameters. Selectivities for the abstraction of alpha hydrogen from substituted toluenes by solvated chlorine atoms have been determined by competitive photochlorination of the substituted toluene and toluene–α–d_1 without any other solvent [73]. The rate is increased by electron-donating substituents, as expected for substrates reacting with an electronegative radical. The logarithm of the selectivity of the substituted toluene relative to toluene itself, at 70°C, plotted against the Hammett sigma value for the substituent, gives a rho of -0.76.

The bromination of toluene and substituted toluenes has been studied both with bromine as the reagent and with *N*-bromosuccinimide. In benzene solution at 80°C the selectivity obeys the Hammett relation using sigma$^+$ [74]. Both bromine and several *N*-bromoamides gave points falling near the same line, suggesting either that the abstracting species is the same (Br·) in each case, or that the contributions from other abstracting species were not sufficient to cause deviation from the bromine line. It should be noted, however, that a later study has shown that, in bromination with *N*-bromosuccinimide, the abstracting radical can be either Br· or the succinimidyl radical, or both, depending on the reaction conditions [75–77].

The rho value for the bromination of the substituted toluenes in benzene, equal to -1.34, is much larger than that found for chlorination. This greater selectivity indicates a less reactive reagent than the solvated chlorine atom.

The selectivity or rho value for the abstraction of an alpha hydrogen should depend not only on the reactivity of the abstracting radical, but also on the reactivity of the substrate [78,79]. For example, Br· should be more selective in abstracting an alpha hydrogen from substituted toluenes X-C_6H_4-CH_3, than from substituted cumenes X-C_6H_4-$CH(CH_3)_2$, in which the hydrogen being abstracted is tertiary rather than primary. It is legitimate and convenient to define the inherent or mean reactivity for each series as the ratio of the rate constant for the reaction of C_6H_5-R and the rate constant for the reaction of C_6H_5-CH_3, adjusting as usual for any differences in the number of alpha hydrogens. These quantities are selectivities, but since the reference compound is the same in each case, they are also measures of the reactivity. Besides the results for toluene, data are also available for the reaction, with *N*-bromosuccinimide, of substituted cumenes [80,81], benzyl methyl ethers [82], α,α-dimethyoxytoluenes [82], ethylbenzenes [81,83,84], diphenylmethanes [78,79], benzyl phenyl ketones [79], allylbenzenes [79,85], and dibenzyl ethers [86]. Although these data are from different laboratories, cover a range of about 10 degrees in temperature, and suffer from the uncertainty, already discussed, about the role of abstracting radicals other than Br·, they do show the expected inverse correlation between the selectivities (ρ) and the inherent reactivities of the different alpha hydrogens. The data are shown in Table 6.4, where it can be seen that the inverse correlation is in fact obeyed. The relationship can also be illustrated by plotting the ρ values against the logarithm of the reactivities. When this is done, all of the points, except that for toluene itself, are found to fall near a single straight line. The fit of the points to this line is not very good, but the experimental uncertainty in the ρ values is also considerable. Sigma$^+$ rather than sigma values were used in obtaining the ρ values in Table 6.4. Although there are theoretical reasons for using sigma$^+$, the reactivity/selectivity correlation based on these is no better than that based on the usual sigmas.

TABLE 6.4 Selectivity and Reactivity in the Bromination of the Alpha Position of Substituted Benzenes

Substrate *m*- or *p*-X	Reactivity per **H**	Rho (for σ+)
X-$PhCH_3$	[1]	−1.38
X-$PhCH_2Ph$	18	−0.62
X-$PhCH_2(CO)Ph$	22	−0.72
X-$PhCH_2CH_3$	24	−0.86
X-$PhCH_2CH{=}CH_2$	25	−0.68
X-$PhCH(CH_3)_2$	54	−0.29
X-$PhCH(OCH_3)_2$	53	−0.38
X-$PhCH_2OCH_3$	54	−0.35
X-$PhCH_2OCH_2Ph$	79	−0.12
(X-Ph)$_2CHOCH_3$	99	0

TABLE 6.5 Isotope Effects in the Halogenation of Toluenes

	k_H/k_D				
Reagent (Substrates)	*p*-Cl-$C_6H_4CH_2D$	$C_6H_5CH_2D$	*p*-CH_3O-$C_6H_4CH_2D$	C_6H_5-$CHDCH_3$	C_6H_5-$CD(CH_3)_2$
NBS	5.08	4.86	3.22	2.67	1.81
Br_2	5.22	4.59			
Cl_2	1.44	1.30			

6.3.5 Isotope Effects

The k_H/k_D isotope effects for the halogenation of toluene and substituted toluenes (Table 6.5) also show a relationship between selectivity and reactivity [87]. As expected, the isotope effect in reactions with the more reactive chlorine atom is lower than that in bromination. The isotope effect also decreases in the bromination of a series of increasingly reactive substrates, *p*-chlorotoluene, toluene, *p*-methoxytoluene, ethylbenzene, and cumene.

6.3.6 Reaction with Ozone

Chlorine atoms are believed to destroy ozone in the Antarctic stratosphere by a reaction that gives ClO· and O_2 [88]. Since O_3 absorbs ultraviolet, depletion of ozone increases the amount of ultraviolet light reaching the surface, and should also increase the incidence of skin cancers.

6.3.7 Addition Reactions

The well-known peroxide effect on the direction of addition of HBr to alkenes gives anti-Markovnikov product because the first step is the addition of a bromine atom rather than a proton. The direction of addition is such as to give the more stable radical, usually the more substituted radical.

The ESR spectra of β-chloro and β-bromoalkyl radicals [89] show strong splitting by the halogen, indicating that the carbon–halogen bond lies in the same plane as the *p*-orbital containing the odd electron. When the radical reacts with a halogen molecule, this conformation favors the trans dihalide as the predominant adduct.

Chlorine atoms in CCl_4 can add to the benzene ring at positions already occupied by certain other substituents, leading to ipso substitution [90]. The reaction with *p*-nitrochlorobenzene, for example, gives *p*-dichlorobenzene. Both chlorine and ^{82}Br atoms can displace bromine from bromobenzene [91].

6.3.8 Iodine Atoms

Iodine atoms can be generated by the photolysis of I_2 [92], by the photolysis of organic iodides [93,94], and by the thermolysis of I_2 [95].

The rate constant [96] for the thermal dissociation of I_2 in hexachlorobutadiene has been estimated to be about 1.4×10^{-12} sec^{-1} at 36°C, and for the reverse reaction about 7×10^{9} M^{-1} sec^{-1}. Although the abstraction of hydrogen by I· is hopelessly endothermic (33 kcal $mole^{-1}$ for methane), I· is able to abstract iodine atoms from benzyl iodide with a rate constant of 7.1×10^{4} M^{-1} sec^{-1} at 36°C. The rate constant for the back reaction is about 3.3×10^{6} M^{-1} sec^{-1}. The subsequent abstraction of iodine from I_2 by the benzyl radical results in an exchange of labeled iodine.

Iodine atoms also add reversibly to carbon–carbon double bonds, catalyzing cis–trans isomerizations [97]. The exchange of radioactive iodine in I_2 with cis and trans 1,2-diiodoethylene and with aryl iodides involves formation of an ipso intermediate by the addition of I·. This reaction can be initiated thermally at 140–190°C in nitrobenzene or hexachlorobutadiene [95]. The formation of an ipso-substitution intermediate by reaction with I· is supported both by the kinetics (one-half order with respect to I_2) and by the observation that the rates are correlated with the Hückel localization energies.

6.4 THE HYDROXYL RADICAL

6.4.1 Generation of HO·

This important radical is formed by the radiolysis of water, by photolysis of hydrogen peroxide [98], and by the reaction of atomic hydrogen with NO_2 [95,99], but is more conveniently generated by Fenton's reagent or one of its variants. One way of generating HO· in nonaqueous solvents is by reaction 6.34 [100].

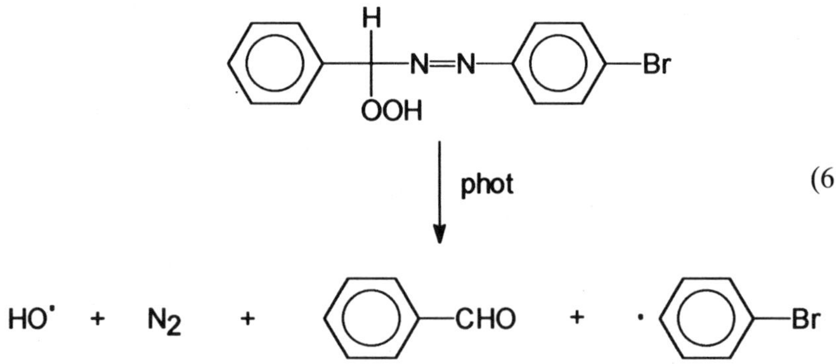

(6.34)

Fenton Reaction. The single-electron reduction of hydrogen peroxide by ferrous iron (Fenton reaction) [101,103], or reduction by any of a number of other metal ions such as Ti^{3+} [104,105] gives hydroxide ion and hydroxyl radical.

$$H_2O_2 + Fe^{2+} \longrightarrow HO^- + HO\cdot + Fe^{3+} \quad (6.35)$$

The hydroxyl radicals subsequently either undergo further reduction by Fe^{2+},

$$HO\cdot + Fe^{2+} \longrightarrow HO^- + Fe^{3+} \qquad (6.36)$$

or react with themselves,

$$2\ HO\cdot \longrightarrow H_2O_2 \qquad (6.37)$$

with hydrogen peroxide,

$$HO\cdot + H_2O_2 \longrightarrow H_2O + HOO\cdot \qquad (6.38)$$

or with an organic compound.

The reactions with organic compounds include both hydrogen abstraction and addition. The resulting organic radicals can dimerize, be reduced by Fe^{2+},

$$R\cdot + Fe^{2+} + H^+ \longrightarrow R\text{-}H + Fe^{3+} \qquad (6.39)$$

or be oxidized by Fe^{3+}. The latter reaction carries on a chain by regenerating Fe^{2+}, which can react with more hydrogen peroxide.

The solvent for reactions with Fenton's reagent is usually water, but an organic solvent can be used if the metal ion is solubilized in the form of an acac chelate [106].

Photolysis. Photolysis of hydrogen peroxide [98] or hydroperoxides requires a short wavelength (254 nm), a disadvantage since the radiation tends to be absorbed by the substrate rather than the peroxide [107]. Phthalimide hydroperoxides can generate HO· at longer wavelengths, but with low efficiency [108].

Potassium Peroxynitrite/Potassium Nitrate. A solid solution of potassium peroxynitrite ($O{=}N{-}OO^-\ {}^+K$) in potassium nitrate can be prepared by irradiating potassium nitrate at 254 nm [109]. This solid, added to an aqueous solution at pH 7.0 generates HO· by the homolysis of peroxynitrous acid (Eq. 6.39a).

$$O{=}N{-}OO^- + H^+ \longrightarrow O{=}N{-}OOH \longrightarrow NO_2\cdot + HO\cdot \qquad (6.39a)$$

Applied to DNA this reagent produces nonspecific strand cleavage. It also oxidizes and degrades proteins, a damage like that from radiation.

6.4.2 Abstraction Reactions

In reactions of HO· with alcohols [104] or with tetrahydrofuran, alpha hydrogens are more readily abstracted than hydrogens in beta or gamma po-

TABLE 6.6 Electrophilic Character of HO·

Substrate	*k* (Relative, per **H**, 30°)	Substrate	*k* (Relative, per **H**, 30°)
Fe^{2+}	[1.][a]	$H(CO)N(\mathbf{CH}_3)_2$	0.97
$HOC\mathbf{H}_2CH_2CH_3$	3.36	$HOC\mathbf{H}_2CCl_3$	0.48
$HOCH_2C\mathbf{H}_2CH_3$	1.34	Acetone	0.054
Tetrahydrofuran		Acetic acid	0.030
alpha	1.45		
beta	0.23		

[a]Rates relative to that of Eq. 6.35, estimated to be about 3×10^8.

sitions. For alpha or beta hydrogens the relative reactivity increases in the usual sequence primary < secondary < tertiary [101], reflecting differences in C–H bond dissociation energies.

The electrophilic character of HO· is apparent from the representative rates in Table 6.6. Abstraction is faster for hydrogens alpha to the oxygen of alcohols or ethers [101] or alpha to amide nitrogen [110]. It is slower for C–H next to electron-withdrawing groups such as CCl_2 or carbonyl.

The ultimate nonradical products of the abstraction reaction depend on the relative rates of oxidation, reduction, or dimerization of the intermediate radical [101]. Isopropyl alcohol gives 93% acetone and only 0.7% of the dimer hexan-2,5-diol. With added Cu^{2+}, however, 11.2% propan-1,2-diol is formed.

The acetonyl radical from the reaction with acetone can be trapped in part as the dimer, 2,5-hexanedione, if the Fe^{2+} is added slowly to the H_2O_2 plus substrate, rather than adding the peroxide to ferrous ion plus substrate.

6.4.3 Addition Reactions

The addition of HO· to multiple bonds, aromatic as well as aliphatic, is very fast, having a rate constant of 10^9 or 10^{10}. With acetylene, the product is hydroxyvinyl radical. This intermediate is reduced by ferrous ion to give acetaldehyde [111] and oxidized by cupric (but not ferric) ion to give hydroxyacetaldehyde.

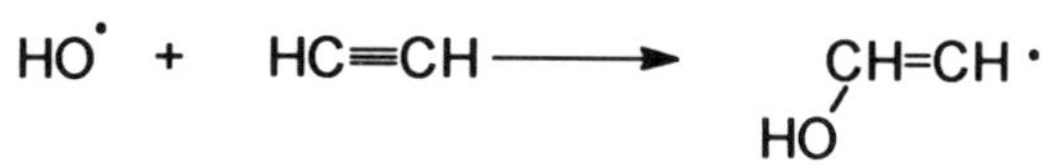

(6.40)

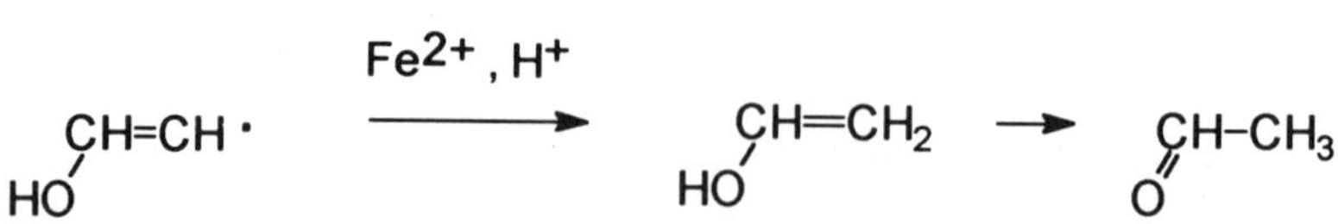

(6.41)

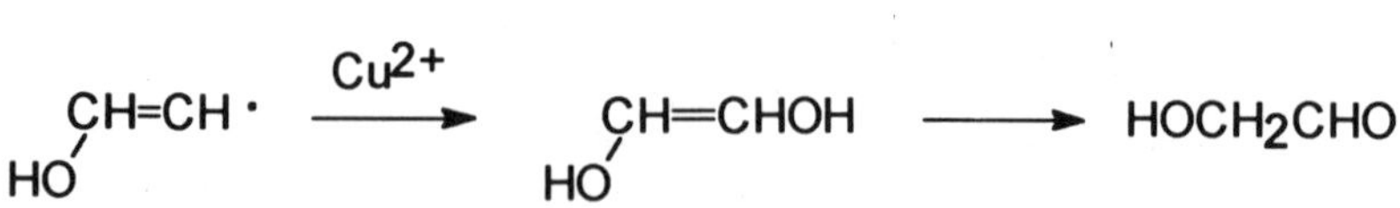

(6.42)

The addition to maleic acid gives a radical which is reduced by the ferrous ion to malic acid [112,113].

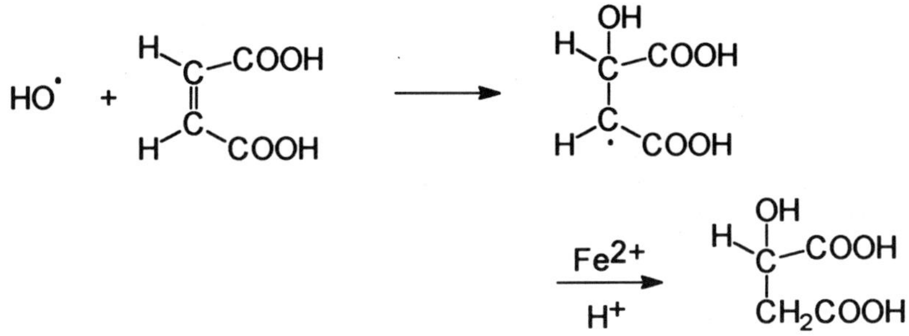

(6.43)

The hydroxyl radical will also add to dimethyl sulfoxide, which then undergoes a beta cleavage reaction to give methanesulfinic acid and a methyl radical [114].

6.4.4 Reactions with Aromatic Compounds

The initial reaction of HO· with aromatic compounds is also addition, even when a side chain with reactive hydrogens is present. The substituted cyclohexadienyl radicals are detectable by their absorption spectra and by ESR [115]. With benzene, the final products are phenol and biphenyl, as shown in Eqs. 6.44–6.46.

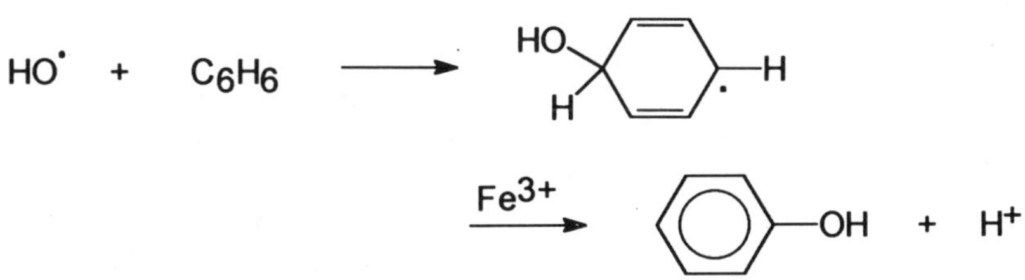

(6.44)

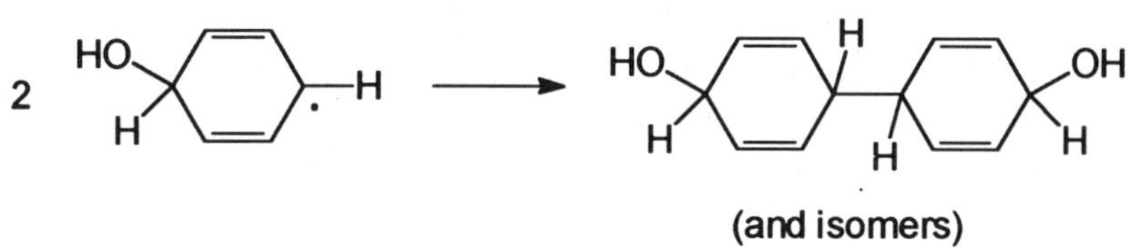

(6.45)

(6.46)

With substituted benzenes, the structure of the intermediate seen by ESR in the reaction with hydrogen peroxide, titanous ion, and various concentrations of sulfuric acid, depends on the pH [105,116,117]. For example, in the oxidation of phenylacetic acid in sulfuric acid concentrations above 0.02 M, the spectrum corresponds to benzyl radical, and at lower concentrations to a mixture of the hydroxycyclohexadienyl radical **III** and its isomers.

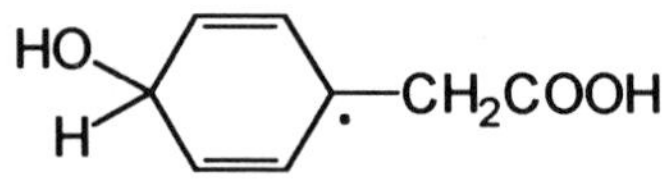

III

The nature of the ultimate nonradical products (from the reaction with hydrogen peroxide, ferrous ion, 0.05 M perchloric acid, and various amounts of ferric and cupric ions) suggests the following mechanism (Eqs. 6.47–6.49) [101,118].

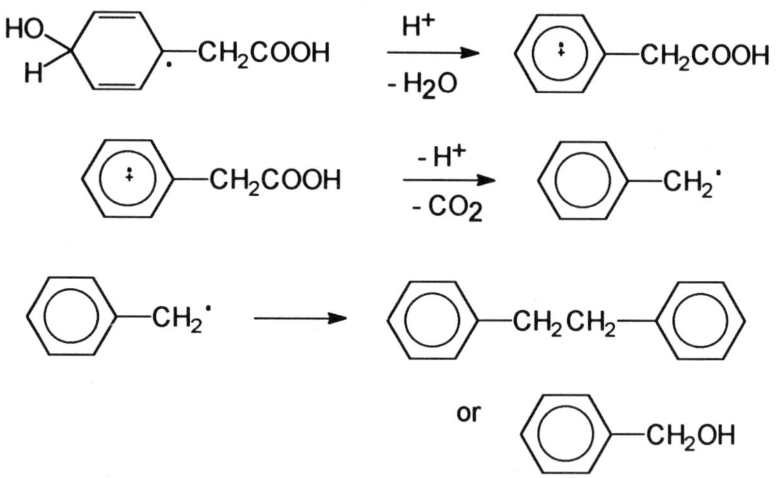

(6.47)

(6.48)

(6.49)

The benzyl radical dimerizes to bibenzyl or is oxidized to benzyl alcohol, the yield of the latter depending on the concentration of the ferric or cupric ion. Oxidation of **III** to the corresponding carbocation, followed by loss of a proton, gives the isomeric *o*-, *m*-, and *p*-hydroxyphenylacetic acids.

Other substituted benzenes behave similarly. Toluene [98,118], at low pH in the absence of added oxidants, gives bibenzyl as the main product and benzyl alcohol and cresols as minor products. At higher pH, or in the presence of added cupric ion, the cresols become the major product. The bibenzyl formed from toluene is not formed by abstraction of the alpha hydrogen, but by way of the adduct of HO· to the ring.

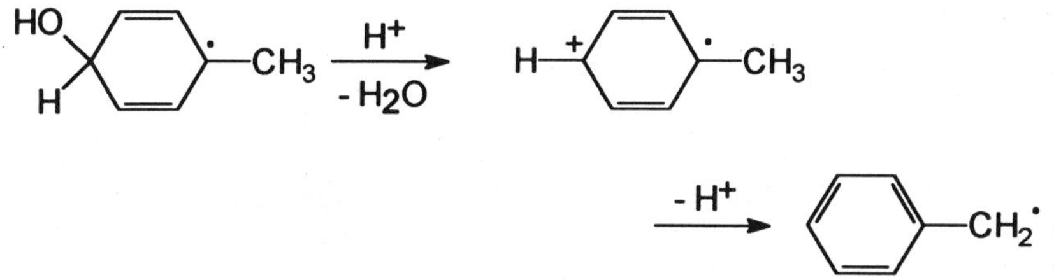

(6.50)

Phenols are oxidized to phenoxy radicals, but again the mechanism is an indirect one [104]. Although phenoxy is the only species detectable by ESR, pulse radiolysis experiments with aqueous solutions of phenol show that the phenoxy radical, absorbing light at 400 nm, is preceded by a different species absorbing at 330 nm. This species is probably the dihydroxycyclohexadienyl radical (Eq. 6.51) [119].

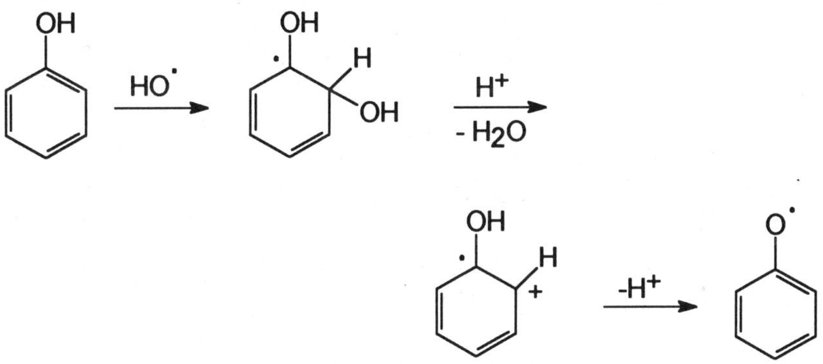

(6.51)

REFERENCES

1. Weyl, W. *Pogg. Ann.* **1864,** *123,* 350.
2. Kraus, C. A. *J. Am. Chem. Soc.* **1908,** *30,* 1323.
3. Atkins, P. W.; Symons, M. C. R. *The Structure of Inorganic Radicals,* Elsevier: Amsterdam, 1967.
4. Hart, E. J.; Boag, J. W. *J. Am. Chem. Soc.* **1962,** *84,* 4090.
5. Boag, J. W.; Hart, E. J. *Nature* **1963,** *197,* 45.
6. Hart, E. J. *Acc. Chem. Res.* **1969,** *2,* 161.
7. Kevan, L. *Acc. Chem. Res.* **1981,** *14,* 138.
8. Ichikawa, T.; Yoshida, H. *J. Phys. Chem.* **1989,** *93,* 5943.
9. Migus, A.; Gauduel, Y.; Martin, J. L.; Antonetti, A. *Phys. Rev. Lett.* **1987,** *58,* 1559.
10. Piotrowiak, P.; Miller, J. R. *J. Am. Chem. Soc.* **1991,** *113,* 5086.
11. Avery, E. C.; Remko, J. R.; Smaller, B. *J. Chem. Phys.* **1968,** *49,* 951.
12. Feng, D.-F.; Kevan, L. *Chem. Rev.* **1980,** *80,* 1.

13. Ichikawa, T.; Yoshida, H. *J. Chem. Phys.* **1980,** *73,* 1540.
14. Narayana, M.; Kevan, L. *J. Am. Chem. Soc.* **1981,** *103,* 1618.
15. Holton, D. M.; Edwards, P. P.; McFarlane, W.; Wood, B. *J. Am. Chem. Soc.* **1983,** *105,* 2104.
16. Grossweiner, L. I.; Joschek, H.-I. In *The Solvated Electron,* American Chemical Society Publications: Washington, D.C., 1965, Chapter 20.
17. Hart, E. J.; Fielden, E. M. *J. Phys. Chem.* **1968,** *72,* 577.
18. Fielden, E. M.; Hart, E. J. *Trans. Faraday Soc.* **1968,** *64,* 3158.
19. Hickel, B.; Sehested, K. *J. Phys. Chem.* **1985,** *89,* 5271.
20. Bunnett, J. F. *Acc. Chem. Res.* **1978,** *11,* 413.
21. Anbar, M. *The Solvated Electron,* American Chemical Society Publications: Washington, D. C., 1965.
22. Behar, D.; Neta, P. *J. Am. Chem. Soc.* **1981,** *103,* 2280.
23. Hayon, E.; Simic, M. *Acc. Chem. Res.* **1974,** *7,* 114.
24. Herak, J. N. *Rev. Sci. Instr.* **1967,** *38,* 1669.
25. Knox, J. H.; Dalgleish, D. G. *Int. J. Chem. Kinet.* **1969,** *1,* 69.
26. Rennert, A. M.; Wijnen, J. *Ber. Bunsenges. Phys. Chem.* **1968,** *72,* 222.
27. Currie, C. L.; Thrush, B. A. *Trans. Faraday Soc.* **1968,** *64,* 390.
28. Gonzalez, R. D.; Kokes, R. J. *J. Phys. Chem.* **1966,** *70,* 2535.
29. Gonzalez, R. D.; Lee, P.-N. *Trans. Faraday Soc.* **1970,** *66,* 1254.
30. Morgan, C. U.; White, K. J. *J. Am. Chem. Soc.* **1970,** *92,* 3309.
31. Klein, R.; Scheer, M. D.; Kelley, R. *J. Phys. Chem.* **1964,** *68,* 598.
32. Xu, Z.; Koplitz, B.; Wittig, C. *J. Chem. Phys.* **1989,** *90,* 2692.
33. Johnston, G. W.; Park, J.; Satyapal, S.; Shafer, N.; Tsukiyama, K.; Bersohn, R.; Katz, B. *Acc. Chem. Res.* **1990,** *23,* 232.
34. Chuang, M.-C.; Folz, M. F.; Moore, C. B. *J. Chem. Phys.* **1987,** *87,* 3855.
35. Pryor, W. A.; Stanley, J. P. *J. Am. Chem. Soc.* **1971,** *93,* 1412.
36. Pryor, W. A.; Lin, T. H.; Stanley, J. P.; Henderson, R. W. *J. Am. Chem. Soc.* **1973,** *95,* 6993.
37. Pryor, W. A.; Henderson, R. W. *J. Am. Chem. Soc.* **1970,** *92,* 7234.
38. Calvert, J. G.; Pitts, J. N. *Photochemistry,* Wiley: New York, 1966.
39. Brown, S. H.; Crabtree, R. H. *J. Am. Chem. Soc.* **1989,** *111,* 2935.
40. Johnson, R. L.; Setser, D. W. *Chem. Phys. Lett.* **1969,** *3,* 207.
41. Trotman-Dickenson, A. F. *Free Radicals,* Methuen: London, 1959.
42. Mortensen, E. M. *J. Chem. Phys.* **1968,** *48,* 4029.
43. Mortensen, E. M. *J. Chem. Phys.* **1968,** *49,* 3526.
44. Ingold, K. U.; Roberts, B. P. *Free Radical Substitution Reactions,* Wiley-Interscience: New York, 1971.
45. Bonacic-Koutecky, V.; Koutecky, J.; Salem, L. *J. Am. Chem. Soc.* **1977,** *99,* 842.
46. Doubleday, C.; McIver, J.; Page, M.; Zielinski, T. *J. Am. Chem. Soc.* **1985,** *107,* 5800.
47. Anbar, M.; Neta, P. *Int. J. Appl. Radiat. Isotopes* **1967,** *18,* 493.

48. Pincock, R. E. *J. Am. Chem. Soc.* **1964,** *86,* 1820.
49. Benson, S. W.; Shaw, R. *J. Am. Chem. Soc.* **1967,** *89,* 5351.
50. Dobis, O.; Benson, S. W. *J. Am. Chem. Soc.* **1990,** *112,* 1023.
51. Poutsma, M. L., J. K. Kochi, Ed., In *Free Radicals,* Wiley-Interscience: New York, 1973, Chapter 15.
52. Juneja, P. S.; Hodnett, E. M. *J. Am. Chem. Soc.* **1967,** *89,* 5685.
53. Bunce, N. J.; Ingold, K. U.; Landers, J. P.; Lusztyk, J.; Scaiano, J. C. *J. Am. Chem. Soc.* **1985,** *107,* 5464.
54. Sumiyoshi, T.; Miura, K.; Hagiwara, H.; Katayama, M. *Chem. Lett. (Jpn.)* **1987,** 1429.
55. Alfassi, Z. B.; Mosseri, S.; Neta, P. *J. Phys. Chem.* **1989,** *93,* 1380.
56. Russell, G. A. *J. Am. Chem. Soc.* **1958,** *80,* 4987.
57. Russell, G. A.; Ito, A.; Hendry, D. G. *J. Am. Chem. Soc.* **1963,** *85,* 2976.
58. Ingold, K. U., J. K. Kochi, Ed., In *Free Radicals,* Wiley-Interscience: New York; 1973, Chapter 2.
59. Leffler, J. E. *Science* **1953,** *117,* 340.
60. Hammond, G. S. *J. Am. Chem Soc.* **1955,** *77,* 334.
61. Walling, C.; Mayahi, M. F. *J. Am. Chem. Soc.* **1959,** *81,* 1485.
62. Walling, C. *J. Org. Chem.* **1988,** *53,* 305.
63. Bühler, R. E. *Helv. Chim. Acta* **1968,** *51,* 1558.
64. Chateauneuf, J. E. *J. Am. Chem. Soc.* **1990,** *112,* 442.
65. Raner, K. D.; Lusztyk, J.; Ingold, K. U. *J. Am. Chem. Soc.* **1989,** *111,* 3652.
66. Skell, P. S.; Baxter, H. N., III; Taylor, C. K. *J. Am. Chem. Soc.* **1983,** *105,* 120.
67. Ingold, K. U.; Lusztyk, J.; Raner, K. D. *Acc. Chem. Res.* **1990,** *23,* 219.
68. Breslow, R.; Brandl, M.; Hunger, J.; Adams, A. D. *J. Am. Chem. Soc.* **1987,** *109,* 3799.
69. Asmus, K.-D. *Acc. Chem. Res.* **1979,** *12,* 436.
70. Abu-Raqabah, A.; Symons, M. C. R. *J. Am. Chem. Soc.* **1990,** *112,* 8614.
71. Khanna, R. K.; Armstrong, B.; Cui, H.; Tanko, J. M. *J. Am. Chem. Soc.* **1992,** *114,* 6003.
72. Walling, C.; Jacknow, B. B. *J. Am. Chem. Soc.* **1960,** *82,* 6113.
73. Walling, C.; Miller, B. *J. Am. Chem. Soc.* **1957,** *79,* 4181.
74. Pearson, R. E.; Martin, J. C. *J. Am. Chem. Soc.* **1963,** *85,* 3142.
75. Traynham, J. G.; Lee, Y.-S. *J. Am. Chem. Soc.* **1974,** *96,* 3590.
76. Tanner, D. D.; Rowe, J. E.; Pace, T.; Kosugi, Y. *J. Am.Chem. Soc.* **1973,** *95,* 4705.
77. Skell, P. S.; Tlumak, R. L.; Seshadri, S. *J. Am. Chem. Soc.* **1983,** *105,* 5125.
78. Friedrich, S. S.; Andrews, L. J.; Keefer, R. M. *J. Org. Chem.* **1970,** *35,* 944.
79. Low, T. P.; Lee, K. H. *J. Chem. Soc. (B)* **1970,** 535.
80. Gleicher, J. G. *J. Org. Chem.* **1968,** *33,* 332.
81. Friedrich, S. S.; Friedrich, E. C.; Andrews, L. J.; Keefer, R. M. *J. Org. Chem.* **1969,** *34,* 900.

82. Huang, R. L.; Lee, K. H. *J. Chem. Soc.* **1964,** 5963.
83. Chang, E. P.; Huang, R. L.; Lee, K. H. *J. Chem. Soc. (B)* **1969,** 878.
84. Huang, R. L.; Lee, K. H. *J. Chem. Soc. (C)* **1966,** 935.
85. Martin, M. M.; Gleicher, G. J. *J. Org. Chem.* **1963,** *28,* 3266.
86. Huang, R. L.; Lee, H. H.; Malhotra, M. S. *J. Chem. Soc.* **1964,** 5947.
87. Wiberg, K. B.; Slaugh, L. H. *J. Am. Chem. Soc.* **1958,** *80,* 3033.
88. Anderson, J. G.; Toohey, D. W.; Brune, W. H. *Science* **1991,** *251,* 39.
89. Lyons, A. R.; Symons, M. C. R. *J. Am. Chem. Soc.* **1971,** *93,* 7330.
90. Everly, C. R.; Traynham, J. G. *J. Org. Chem.* **1979,** *44,* 1784.
91. Miller, B.; Walling, C. *J. Am. Chem. Soc.* **1957,** *79,* 4187.
92. Wassermann, E.; Falconer, W. E.; Yager, W. A. *Ber. Bunsenges. phys. Chem.* **1968,** *72,* 248.
93. Wolf, W.; Kharasch, N. *J. Org. Chem.* **1961,** *26,* 283.
94. Kharasch, N.; Wolf, W.; Erpelding, T. J.; Naylor, P. G.; Tokes, L. *Chem. Ind.* **1962,** 1720.
95. Nakashima, M.; Mok, C. Y.; Noyes, R. M. *J. Am. Chem. Soc.* **1969,** *91,* 7635.
96. Gazith, M.; Noyes, R. M. *J. Am. Chem. Soc.* **1955,** *77,* 6091.
97. Muizebelt, W. J.; Nivard, R. J. F. *J. Chem. Soc. (B)* **1968,** 913.
98. Jefcoate, C. R. E.; Lindsay-Smith, J. R.; Norman, R. O. C. *J. Chem. Soc. (B)* **1969,** 1013.
99. Breen, J. E.; Glass, G. P. *J. Chem. Phys.* **1970,** *52,* 1082.
100. Tezuka, T.; Narita, N. *J. Am. Chem. Soc.* **1979,** *101,* 7413.
101. Walling, C. *Acc. Chem. Res.* **1975,** *8,* 125.
102. Richter, H. W.; Waddell, W. H. *J. Am. Chem. Soc.* **1982,** *104,* 4630.
103. Fenton, H. J. H. *J. Chem. Soc.* **1894,** *65,* 899.
104. Dixon, W. T.; Norman, R. O. C. *J. Chem. Soc.* **1963,** 3119.
105. Norman, R. O. C.; Pritchett, R. J. *J. Chem. Soc. (B)* **1967,** 926.
106. Brook, M. A.; Castle, L.; Lindsay-Smith, J. R.; Higgins, R.; Morris, K. P. *J. Chem. Soc. Perkin II* **1982,** 687.
107. Ogata, Y.; Tomizawa, K.; Furuta, K. *The Chemistry of Peroxides,* Wiley: New York, 1983.
108. Saito, I.; Takayama, M.; Matsuura, T.; Matsugo, S.; Kawanishi, S. *J. Am. Chem. Soc.* **1990,** *112,* 883.
109. King, P. A.; Anderson, V. E.; Edwards, J. O.; Gustafson, G.; Plumb, R. C.; Suggs, J. W. *J. Am. Chem. Soc.* **1992,** *114,* 5430.
110. Hayon, E.; Ibata, T.; Lichtin, N. N.; Simic, M. *J. Am. Chem. Soc.* **1970,** *92,* 3898.
111. Walling, C.; El-Taliawi, G. M. *J. Am. Chem. Soc.* **1973,** *95,* 848.
112. Walling, C.; El-Taliawi, G. M.; Johnson, R. A. *J. Am. Chem. Soc.* **1974,** *96,* 133.
113. Nofre, C.; Le Roux, Y.; Gondot, L.; Cier, A. *Bull. Soc. Chim. Fr.* **1964,** 2451.
114. Eberhardt, M. K.; Colina, R. *J. Org. Chem.* **1988,** *53,* 1071.

115. Dorfman, L. M.; Taub, I. A.; Bühler, R. E. *J. Chem. Phys.* **1962,** *36,* 3051.
116. Jefcoate, C. R. E.; Norman, R. O. C. *J. Chem. Soc. (B)* **1968,** 48.
117. Snook, M. E.; Hamilton, G. A. *J. Am. Chem. Soc.* **1974,** *96,* 860.
118. Walling, C.; Johnson, R. A. *J. Am. Chem. Soc.* **1975,** *97,* 363.
119. Land, E. J.; Ebert, M. *Trans. Faraday Soc.* **1967,** *63,* 1181.

7

RADICAL SOURCES I. DIACYL PEROXIDES AND PERESTERS

7.1 THERMOLYSIS OF DIACYL PEROXIDES

The thermolysis of diacyl peroxides is a reaction that can lead to both radical and ionic intermediates.

The rate of the decomposition reaction (or reactions) is increased not only by polar solvents but also by substituents or structures able to stabilize a positive charge. Examples of structures favoring the fast polar reaction are the *p*-methoxy substituent in Eq. 7.1 (the first example of a carboxyinversion

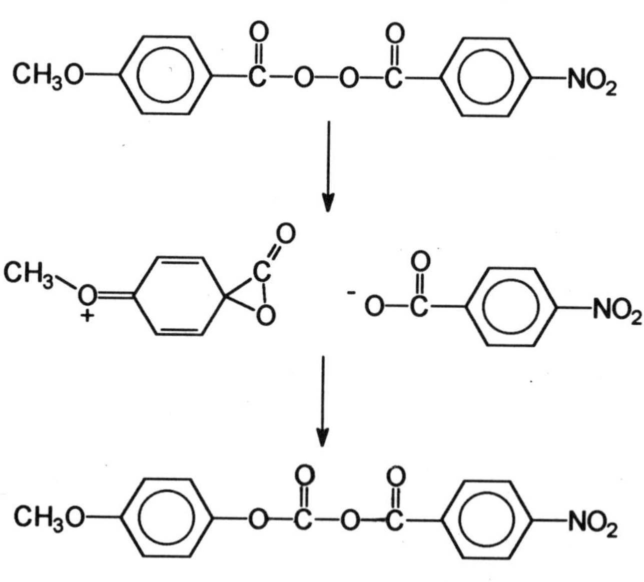

(7.1)

reaction [1]), tertiary alkyl groups as in pivaloyl peroxide [2], and secondary alkyl groups as in isobutyryl [3] and *trans*-4-*t*-butylcyclohexanecarbonyl [4] peroxides.

An example of the medium effect on the rate is the eightfold increase in rate if reaction 7.1 is carried out in nitrobenzene rather than in benzene.

In most cases the fast decomposition reactions still give radicals, but the yield of radicals is lower. For example, *p*-methoxy-*p'*-nitrobenzoyl peroxide in neat styrene gives only half as much polystyrene as the equivalent amount of benzoyl peroxide. In neat acrylonitrile, a much more polar solvent, benzoyl peroxide leads to an explosive polymerization or to complete solidification, while *p*-methoxy-*p'*-nitrobenzoyl peroxide gives only a small amount of polymer.

It now appears that almost any diacyl peroxide, with the possible exception of acetyl peroxide [5], is capable of giving a partly ionic reaction under the right circumstances. Acetyl peroxide decomposes at nearly the same rate, about 1.5 to 2×10^{-5} sec^{-1} at 70°C, in a series of solvents consisting of CCl_4, various alkanes, styrene, various aromatic hydrocarbons, and acetic acid. Various *n*-alkanoyl peroxides decompose only about 10% faster than acetyl peroxide in alkanes and styrene, but about 1.5–2 times as fast as acetyl peroxide in various aromatic solvents, and about 8 times as fast in acetic acid. It is probable that even *n*-alkanoyl peroxides decompose in part by an ionic mechanism, and do so to an increasing extent in the more polar solvents.

7.1.1 The Question of Multibond Cleavage

Although the thermolysis of peroxides gives both acyloxy radicals and alkyl or aryl radicals, the alkyl or aryl radicals may either be formed directly or indirectly in a second step in which the acyloxy radical loses carbon dioxide.

For acetyl peroxide and benzoyl peroxides, which are nonpolar and give only radicals in ordinary solvents, only the O–O bond is cleaved in the first step. The one-bond cleavage mechanism (Eq. 7.2) was established for acetyl peroxide by ^{18}O carbonyl labeling experiments [6,7]. The label was completely scrambled in the methyl acetate product and 38% scrambled in recovered acetyl peroxide.

$$\begin{aligned} &CH_3COO\text{—}OOCCH_3 \xrightleftharpoons{80^\circ} 2\,CH_3COO\cdot \\ &CH_3COO\cdot \longrightarrow CH_3\cdot + CO_2 \end{aligned} \tag{7.2}$$

When benzoyl peroxide is decomposed in CCl_4, added iodine and moisture trap 100% of the benzoyloxy groups as benzoic acid without affecting the decomposition rate [8]. This established the one-bond cleavage mechanism for that solvent.

For the decomposition of benzoyl peroxide in styrene, the one-bond mechanism was established by Bevington and Lewis's study of the end groups in the polystyrene produced [9]. It is convenient to represent the yield of benzoyloxy radicals by (1-f) and the yields of phenyl radicals by (f) as in Eq. 7.3.

$$\text{ArCOO—OOCAr} \longrightarrow 2(1\text{-}f)\ \text{ArCOO}\cdot + 2f\ \text{Ar}\cdot + 2f\ \text{CO}_2 \quad (7.3)$$

The peroxide used to initiate the polymerization of thc styrene was labeled with ^{14}C in the phenyl rings. The proportion of benzoyloxy and phenyl end groups was determined by counting the radioactive carbon in the polymer both before and after removing all of the benzoyloxy end groups by hydrolysis. The value of f varied with the styrene concentration, low concentrations giving the benzoyloxy radicals more time to decarboxylate. The fact that f became zero at high styrene concentrations means that the phenyl radicals are formed only indirectly, in the second step. As we shall see later, this is not the case for photolysis.

The thermolysis of benzoyl propanoyl and benzoyl acetyl peroxides gives the corresponding benzoate esters and alkyl benzenes [10,11]. From the CIDNP signals of those products the rate constant for the decarboxylation of the benzoyloxy radical is estimated to be about 1×10^8 sec^{-1} at 130°C. The decarboxylation of the alkanoyloxy radical was either concerted with its formation or too fast to estimate by means of CIDNP.

For peroxides such as bis(phenylacetyl) peroxide [12] or secondary or tertiary alkanoyl peroxides that give both radical pairs and ion pairs, the initial step probably breaks one of the alkyl to carbon dioxide bonds.

7.1.2 Polar Reactions in Benzoyl Peroxide Thermolysis

Although p-methoxy-p'-nitrobenzoyl peroxide gives a largely ionic reaction in polar solvents, benzoyl peroxide itself gives a good yield of radicals in ordinary solvents and decomposes at a rate that is not very sensitive to the polarity of the solvent. Adsorbed on the surface of silica, however, its behavior is quite different. The rate constant at 40°C is about 15 times that in solvents such as benzene or styrene, and the products are also quite different [13]. While *photolysis* of the peroxide on silica, a strictly radical reaction, gives an 87% yield of biphenyl, this product is found only in traces from the 40°C thermolysis on silica. Phenol, probably formed from the decomposition of the carboxyinversion compound, phenyl benzoyl carbonate, is found in up to 43% yield. Silica surfaces are an extremely polar reaction medium [14–17].

7.1.3 Alkyl Group Rearrangements

Since alkyl radicals do not rearrange, the presence of a rearranged group in a product of a peroxide decomposition indicates an alkyl cation intermediate. The decomposition of cyclobutanecarbonyl m-chlorobenzoyl peroxide is particularly interesting [18,19]. The principal products from the ionic reaction are esters and carboxyinversion compounds (Eqs. 7.4). Both the esters and carboxyinversion compounds have the original unrearranged cyclobutyl group, a rearranged cyclopropylmethyl group, and a rearranged 3-butenyl group.

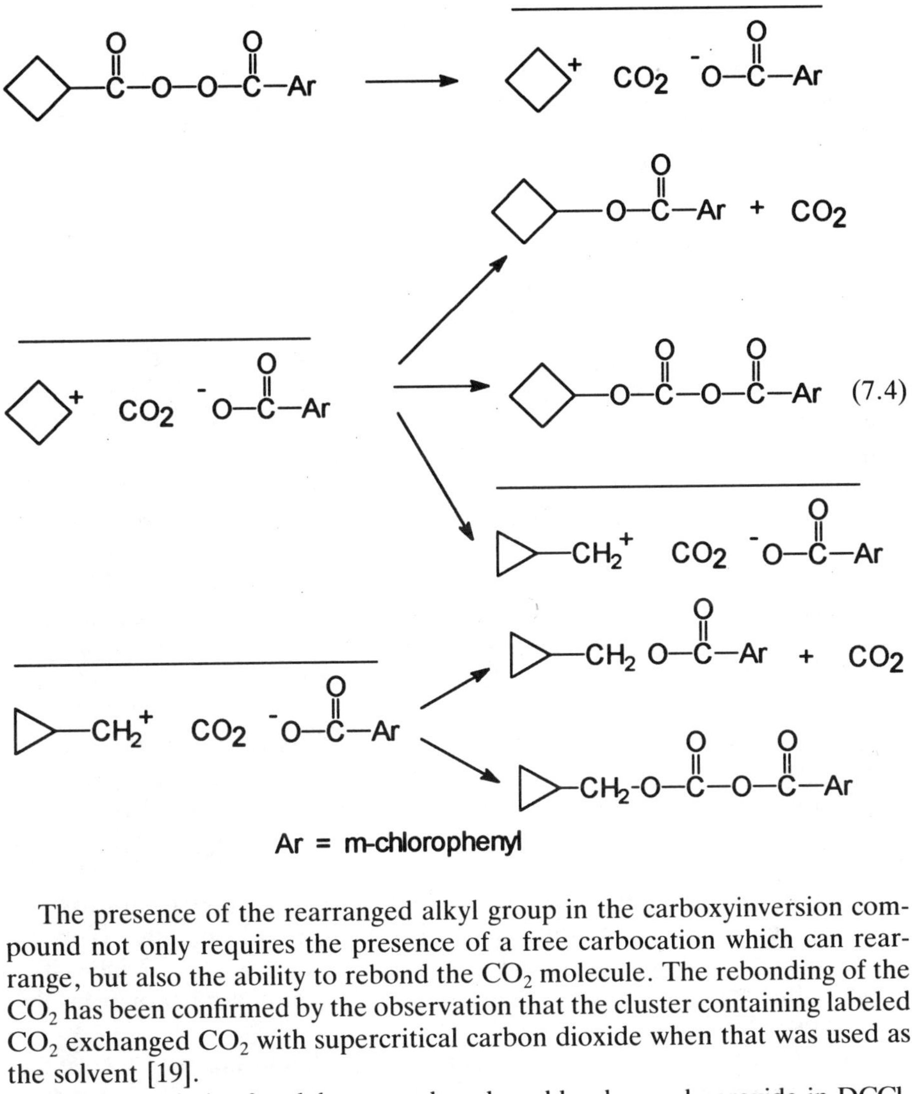

The presence of the rearranged alkyl group in the carboxyinversion compound not only requires the presence of a free carbocation which can rearrange, but also the ability to rebond the CO_2 molecule. The rebonding of the CO_2 has been confirmed by the observation that the cluster containing labeled CO_2 exchanged CO_2 with supercritical carbon dioxide when that was used as the solvent [19].

The thermolysis of cyclobutanecarbonyl *m*-chlorobenzoyl peroxide in $DCCl_3$ showed CIDNP signals corresponding to about a 5–10% of reaction by the radical pathway [18]. These signals were from CO_2 (E), cyclobutene, cyclobutyl *m*-chlorobenzoate, cyclobutyl chloride, and deuterocylobutane. There was also a very weak signal that might be due to the methine carbon of cyclopropylmethyl *m*-chlorobenzoate. That polarization may have occurred in a cyclobutyl radical, which was then converted to a cyclobutyl cation, which then rearranged before the polarization was lost.

The thermolysis of cyclopropylmethanecarbonyl *m*-chlorobenzoyl peroxide also gave ester and carboxyinversion compound with rearranged alkyl groups (cyclobutyl and 3-butenyl). Since there were no CIDNP signals, the thermolysis of this peroxide appears to be almost entirely ionic [18].

The decomposition of *t*-butylacetyl *m*-chlorobenzoyl peroxide, discussed in Chapter 5, also gave some products in which the alkyl group had undergone a carbonium ion rearrangement [20]. The products with the rearranged alkyl group also showed CIDNP. Since CIDNP requires a radical pair, and since the neopentyl radical will not rearrange, a sequence involving a radical pair and an ion pair is needed. The ion pair is formed by electron transfer in the radical pair.

7.1.4 Retention of Configuration

Although many of the polar peroxide decomposition reactions appear to involve free carbonium ions, that is, carbonium ions not covalently bonded to CO_2, the original configuration of the alkyl group is often retained in the carboxyinversion product or ester.

In the case of *trans*-4-*t*-butylcyclohexanecarbonyl peroxide in refluxing CCl_4, two thirds of the product is the carboxyinversion compound [4]. This product, which is entirely trans, is converted on further heating into the trans ester and CO_2. A radical product, 4-*t*-butylcyclohexyl chloride, consists of the cis and trans isomers in a 2.7:1 ratio. The chiral peroxides ethylmethylacetyl peroxide [4,21] and β-phenylisobutyryl peroxide [5] also gave the carboxyinversion compound and then ester with retention of configuration in the alcohol moiety.

The interpretation of reactions in which only the ester is observed is less clear cut, since the ester could be formed directly from an ion pair, indirectly from a radical pair, or from a carboxyinversion compound. The yield of *n*-hexyl heptanoate from diheptanoyl peroxide in various hydrocarbon–CCl_4 mixtures is about 10–15% [22]. Since the alkyl group is a primary one and the solvents were not very polar we may take this as an upper limit for the yield of ester from the radical reaction. When the yield of ester is much greater than this, or the rate of decomposition of the peroxide is high, it is probable that most of the ester is formed either from ion pairs or from the carboxyinversion compound.

Optically active β-phenylisobutyryl peroxide decomposing in CCl_4 gave the racemic alkyl chloride and about a 60% yield of the ester with 92% retention of configuration in the alcohol moiety [5,23].

Hydratropoyl *m*-chlorobenzoyl peroxide (methylphenylacetyl *m*-chlorobenzoyl peroxide) was too unstable to isolate, but the products formed during the attempted preparation included about a 7% yield of 2,3-diphenylbutane, 18–37% of the ester hydratropyl hydratropoate, and 5–15% of hydratropyl alcohol [24]. The alcohol moiety of the ester had a 91% retention. The free alcohol, probably formed by reaction of water with the carbonium ion, had about 51% retention. The 2,3-diphenylbutane was a mixture of the meso and *dl* forms and had a slight rotation.

7.1.5 Nucleophilic Trapping

Optically active 2-methylbutanoyl *m*-chlorobenzoyl peroxide decomposing in acetonitrile, a solvent that is nucleophilic as well as polar, gave a 36% yield of the carboxyinversion compound, with 100% net retention, and a 40% yield of *N*-acetyl-*N*-*m*-chlorobenzoyl-2-butylamine with 14.4% net inversion (Eq. 7.5) [25]. Since no CIDNP signals were detected, the decomposition of this peroxide in acetonitrile is probably entirely ionic. The carboxyinversion compound is probably formed from a tight R^+/CO_2/carboxylate cluster and the amide from a cluster loose enough to be penetrated by an acetonitrile molecule.

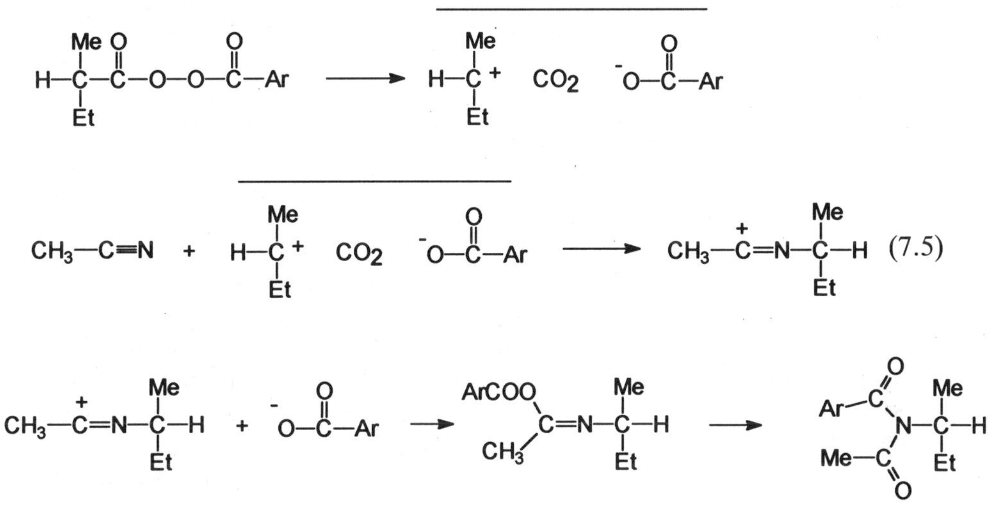

(7.5)

7.2 PHOTOLYSIS OF DIACYL PEROXIDES

7.2.1 Direct Photolysis

Dialkanoyl peroxides have a weak continuous absorption band beginning at about 2800 Å in the ultraviolet. The band is continuous because the excited state is dissociative, meaning that it has a continuous distribution of kinetic energies. In the gas phase [26] and in hexane solution [27] the photolysis quantum yield is near unity. Because decarboxylation of the acyloxy radicals is fast, photolysis gives the corresponding alkyl radicals in high yield [27].

In contrast to thermolysis, photolysis of alkanoyl peroxides has never been unequivocally shown to generate anything but radical pairs despite the theoretical possibility of subsequent electron transfer. For example, the photolyses of pivaloyl [27] and bicyclo[2.2.2]octane-1-formyl [2] peroxides probably give radicals exclusively, although the thermolysis is fast enough even at low temperatures to give some ion pair-derived product in competition with the photolysis [2]. Because of the concurrent thermolysis, ester yields reported from the photolysis of secondary or tertiary alkanoyl peroxides cannot be attributed entirely to the photolysis.

The yields of carbon dioxide in the photolysis of dialkanoyl peroxides approach two moles per mole of peroxide. In the photolysis of propionyl, butanoyl, and pentanoyl peroxides in hydrocarbon solvents at 30°C the yields of esters were less than 1% and no acids were found [27]. However, photolysis of acetyl peroxide in acetonitrile gave a 3% yield of methyl acetate. The methyl acetate could not have been formed by way of a carboxyinversion compound since that ester anhydride is too stable under the reaction conditions.

The photolysis of propionyl, butanoyl, and pentanoyl peroxides has been carried out in argon matrices. Peroxide vapor (about 1 mm), and then argon (about 600 mm), are leaked into an evacuated mixing chamber, mixed, and then leaked into a cold infrared cell where the matrix is deposited. Photolysis of the peroxide in the argon matrix gives a cluster consisting of the two alkyl radicals and two molecules of CO_2, a reaction like that in hydrocarbon solutions.

ESR signals from the radicals produced by the photolysis of benzoyl peroxide in CCl_4 have been recorded using the time-resolved technique and laser excitation [28]. The decay of the signal from the benzoyloxy radical (g = 2.0123) matched the rise in the signal from the trichloromethyl radical (g = 2.0096, a_{Cl} 6.2 G). A phenyl radical signal was not seen because its reaction with the CCl_4 solvent is too fast. Laser excitation combined with time-resolved visible/ultraviolet spectroscopy has been used to prepare and study a number of substituted benzoyloxy radicals in solution [29,30].

Although many radicals are hard to detect spectroscopically in an ordinary solvent, they can be detected if steps are taken to prevent reactions that destroy the radical. One method is to photolyze a crystalline diacyl peroxide [31–33]. The crystal lattice immobilizes the radical pair and immediate reaction within the pair is prevented by the one or two intervening carbon dioxide molecules. A similar approach is to photolyze the peroxide in an argon matrix, a technique which has been used to obtain infrared spectra of alkyl radicals [34–37].

Photolysis of crystalline acetyl benzoyl peroxide gives the cage products methyl benzoate from the loss of one carbon dioxide and toluene from the loss of both [33]. Very brief photolysis allows the ESR spectrum of the methyl benzoyloxy radical pair to be seen, and continued photolysis gives the ESR spectrum of the methyl phenyl pair.

The Decarboxylation Step in Photolysis. Bevington and Lewis applied their labeled polymer technique, already discussed, to the photolysis of benzoyl peroxides as well as thermolysis [9]. Although thermally induced homolysis of benzoyl peroxides gives CO_2 only indirectly by decarboxylation of the benzoyloxy radicals, the photolysis gives some CO_2 directly. Photolysis of benzoyl peroxide at 25°C showed that the fraction f of initially decarboxylated benzoyloxy radicals was 0.71, rather than zero. Photolysis of anisoyl peroxide gave an even higher fraction, about 0.93. Since these fractions were independent of the temperature at which the photolyses were carried out, either the photolysis gives some CO_2 directly or it produces aroyloxy radicals in a state that leads to very fast decarboxylation.

The photolysis of benzoyl chloroacetyl peroxide has been studied by means of CIDNP [38]. Decarboxylation of the chloroacetoxy radical in CCl_4 occurs either in the initial step or so quickly as to be instantaneous on the CIDNP time scale. The fraction f of benzoyloxy radicals losing CO_2 depends on how much energy is put into the peroxide molecule, and presumably, passed on in part to the benzoyloxy radical. An f value of 0.75, obtained for direct photolysis, was decreased when shorter wavelengths were excluded by a filter. For singlet-sensitized photolysis using anthracene (singlet energy 0.76 kcal mole^{-1}) f was 0.25, and for benzophenone triplet-sensitized photolysis (energy 69 kcal mole^{-1}) it was only 0.07. As expected the signs of the CIDNP effects were reversed for the singlet and triplet radical pairs. Also as expected, the triplet radical pairs gave lower yields of geminate reaction products. The rate constant for the decarboxylation of thermally generated benzoyloxy radicals at 130°C is about 1×10^8 sec^{-1} [10,11], while that for the photochemically generated radicals can be as high as 10^{10} sec^{-1} even at room temperature [38].

Direct photolysis and singlet-sensitized photolysis of benzoyl peroxide gives higher yields (0.05–0.16 M/M) of phenyl benzoate than thermolysis (0.004–0.011 M/M), because the singlet-sensitized and direct photolyses produce caged phenyl–benzoyloxy pairs directly [39]. Styrene has no effect on the yield because phenyl benzoate is entirely a geminate product.

7.2.2 Characteristics of Sensitized Photolysis

The Energy of the Sensitizer. When the energy of the sensitizer is less than the bond dissociation energy of the peroxide, sensitized photolysis can still occur (although with lower efficiency), if the peroxide is in a vibrationally excited state [40,41]. This is illustrated by the behavior of di-t-butyl peroxide with various sensitizers [40]. Quenching of the triplet states of the sensitizers was observed directly by monitoring the triplet–triplet absorptions of the sensitizers. The pseudo first order rate constant k_d for the decay of the sensitizer is given by Eq. 7.6, in which k_q is the rate constant for the transfer of the excitation energy of the sensitizer triplet to the peroxide.

$$k_d = k_d^0 + k_q[\text{BuO–OBu}] \tag{7.6}$$

The values of k_q increase with increasing E_T of the sensitizer. For p-methoxypropiophenone (E_T = 72.5 kcal mole^{-1}) as the sensitizer, k_q is about 7.9 × 10^6 M^{-1} sec^{-1} at 25°C while for fluorenone (E_T = 51) it is only 2.4 × 10^5. The activation energy for sensitization by *p*-methoxypropiophenone is zero, because the triplet energy is sufficient to excite the peroxide to its dissociative state directly from the ground state. For fluorenone an activation energy of 3.94 kcal/mole is needed because the peroxide has to be in a higher vibrational state.

The k_q values for *t*-butyl perbenzoate [41] and a number of diaroyl peroxides [42] also tend to increase as the E_T of the triplet sensitizer increases.

Internal Chromophores or Sensitizers. Peroxides with conjugated naphthyl or biphenylyl groups quench benzophenone triplets with k_q values 10^3–10^4 times that for benzoyl peroxide [42]. The corresponding hydrocarbons, naphthalene and biphenyl, are themselves good quenchers, suggesting that energy is transferred to the peroxide bond indirectly by way of the naphthyl or biphenylyl group, and then to the peroxide bond.

The benzophenone-sensitized photolysis of benzoyl peroxide has a quantum yield of about 0.25 in the limit of high peroxide concentrations [43]. This relatively low quantum yield is due to two factors, a rate of energy transfer from the benzophenone triplet to the peroxide orders of magnitude below the diffusion-controlled rate and an inherently low efficiency of utilization of the energy even after it has been transferred. When the peroxide has a benzophenone-like sensitizer as part of the same molecule, the results are quite different [44]. Energy transfer from external benzophenone sensitizers becomes more efficient and sensitization by the internal sensitizer is even better. For example, the quantum yield for the direct photolysis in CCl_4 of benzoyl *m*-benzoylbenzoyl peroxide (*m*BBP, Eq. 7.7), irradiated on the long-wavelength side of the $n\pi^*$ ketone band, is 0.91. The yields of chlorobenzene and chlorobenzophenone are 89–91%. The chlorobenzene shows CIDNP with enhanced absorption as expected for a process generating triplet radical pairs.

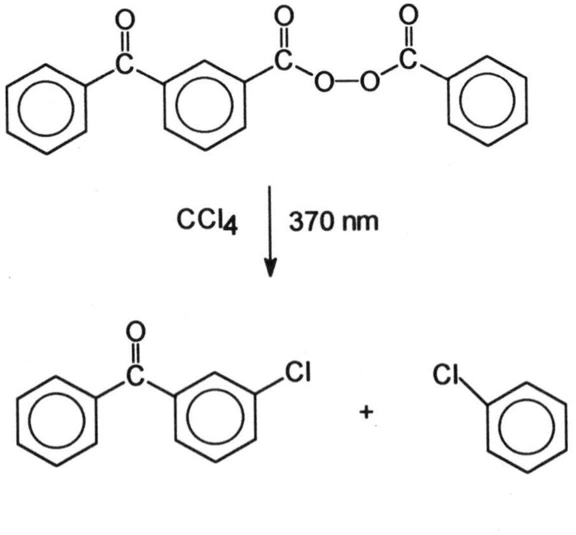

(7.7)

Thermolysis of *m*BBP gives a chlorobenzene with an enhanced emission signal.

The quantum yield of the photolysis and the yields of chlorobenzophenone and chlorobenzene were also high for the para isomer, *p*BBP.

Even the keto peroxide *p*BEP, in which the chromophore is insulated, has a quantum yield of 0.57, more than twice that of the benzoyl peroxide plus benzophenone system.

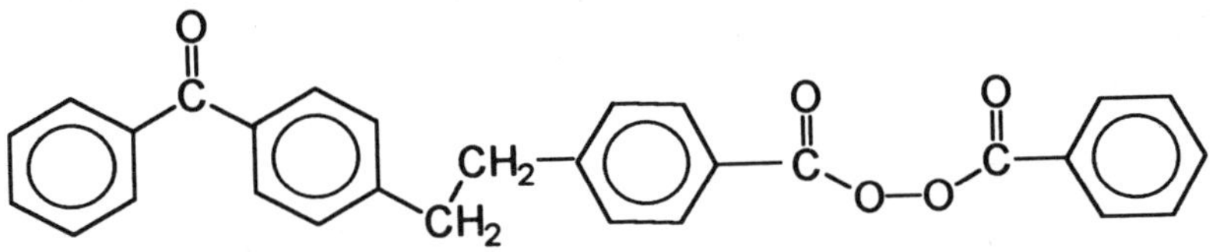

The diperoxide *m,p*-DP has a central benzophenone chromophore with a benzoyl peroxide at a meta position on one ring and at the para position on the other.

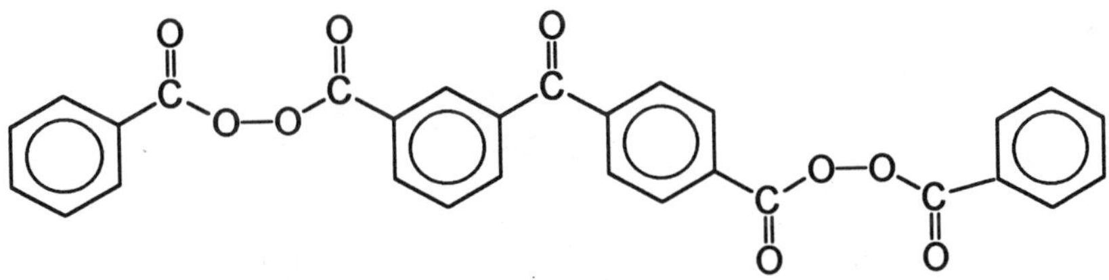

Photolysis of *m,p*-DP (Eqs. 7.8 and 7.9) can cleave either peroxide function, but not both in one photolysis step, even though the energy should suffice for both if the excess were not dissipated. There is also an 18:1 preference for activating the peroxide group that is conjugated, that is, the one in the para position.

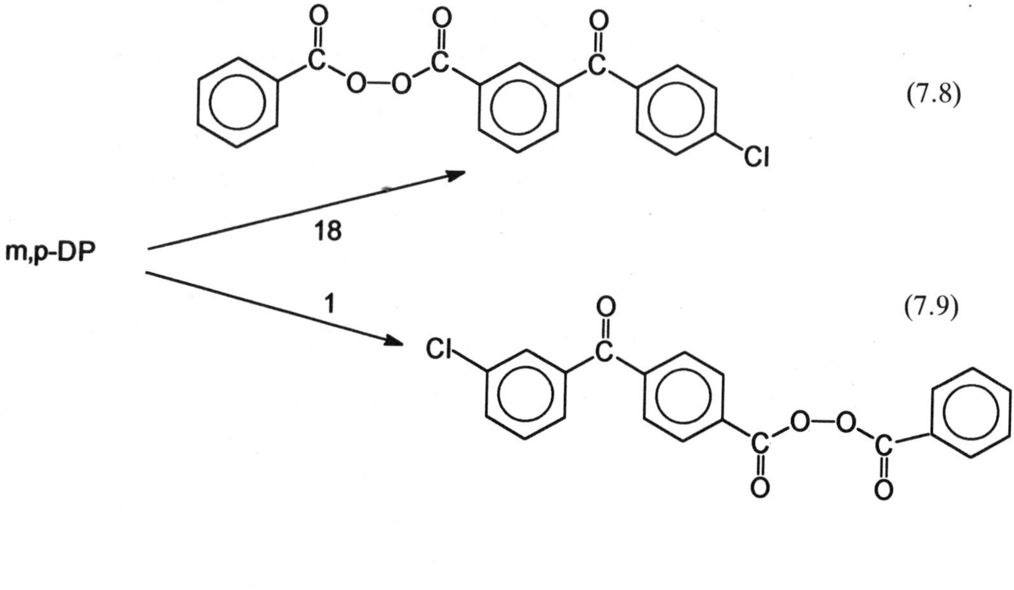

(7.8)

(7.9)

7.3 THERMOLYSIS OF PERESTERS

Because peresters of secondary and primary alcohols are subject to a fast cyclic or base-catalyzed decomposition into the corresponding carboxylic acid and ketone or aldehyde [46,47], they are not effective radical sources in solution and will not be discussed further.

Peresters of tertiary alcohols [45] are similar in many ways to diacyl peroxides. Some decompose only to ion pairs, some only to radical pairs, and some to give both ion and radical pairs.

7.3.1 Heterolysis

An interesting example of the purely ionic reaction is the Criegee rearrangement of *trans*-9-decalyl perbenzoate (Eq. 7.10) [48]. The reaction is accelerated both by polar solvents and by electron-withdrawing substituents on the acyloxy group. In fact, the *trans*-9-decalyl perester of trichloroacetic acid is not isolable, and the perester of *p*-nitrobenzoic acid requires mild conditions.

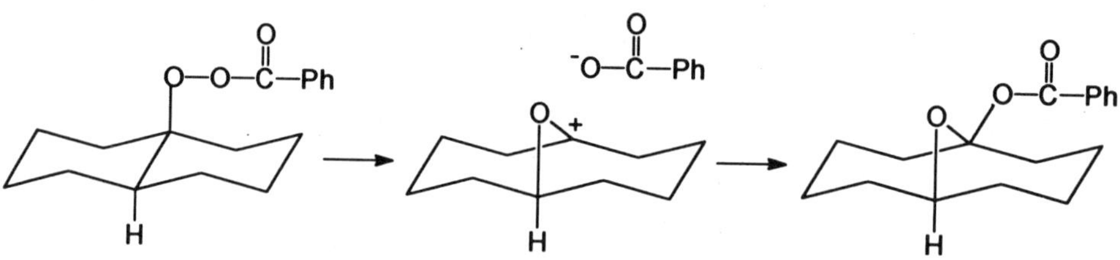

(7.10 and 7.11)

The ion pair is an intimate one not easily diverted from collapse to the hemiketal benzoate. The product is still largely the benzoate, and not the methyl ether, even in methanol solution [49]. There is also no exchange with benzoate ions [50], and no scrambling of labeled oxygen in the benzoyloxy group [51].

The decomposition of *trans*-9-decalyl perphenylacetate gives both ions and radicals, the balance between the two modes of reaction depending on the polarity of the solvent. In acetonitrile the reaction is still fast and mostly ionic, with Criegee rearrangement products accounting for 55% of the product. In ethylbenzene the polar reaction accounts for only about 3% of the reaction [45].

Although most *t*-butyl peresters decompose only to radical pairs [52–54], *t*-butyl trifluoroperacetate [55] and *t*-butyl arylpersulfonates [56] decompose to ion pairs. This mechanism is suggested by strong medium and substituent effects on the rate and confirmed by the quantitative yield of acetone and benzenesulfonic acid in methanol (Eq. 7.12).

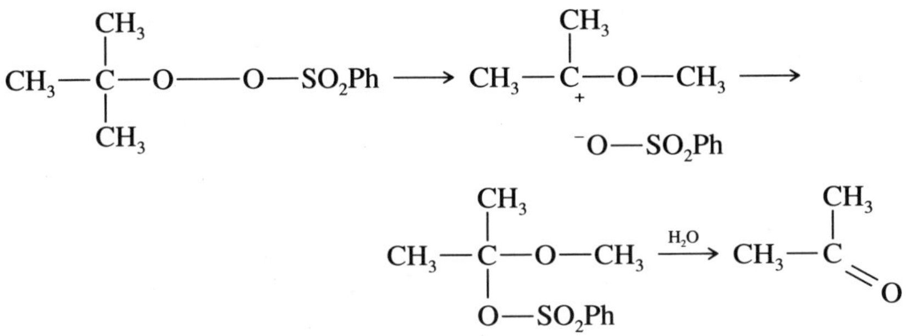

(7.12)

Cumyl peracetate decomposes by a heterolytic, homolytic, or mixed mechanism depending on the reaction medium [57,58]. In nonpolar media the reaction is slow at 90°C and gives acetophenone as a major product. In polar media the reaction is faster and ionic and takes place with migration of phenyl to give 2-phenoxypropene or 2-phenoxy-2-acetoxypropane as a major product. In silica slurries the reaction is fast and ionic even at room temperature [15].

With appropriate structures, and assisted by a catalyzing acid, heterolysis of a perester in the opposite sense is possible [59]. In Eq. 7.13 the *t*-butoxy moiety has been protonated by perchloric acid and the alkyl group of the acyloxy moiety gives tropylium ion, an unusually stable carbenium ion.

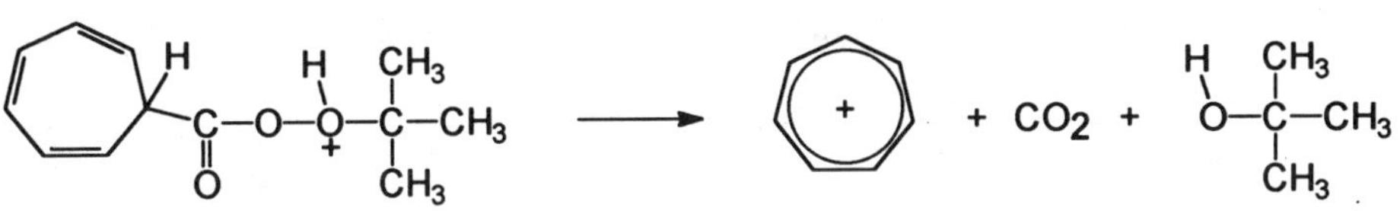

(7.13)

7.3.2 Homolysis

Di-*t*-butyl peroxide, *t*-butyl perbenzoate, and dibenzoyl peroxide are three peroxides that thermolyze exclusively to radical pairs in ordinary solvents. The rate constant for the homolysis of *t*-butyl perbenzoate is approximately 10^{-9} sec^{-1} at 60°C, which is about a factor of 40–50 faster than the rate constant for di-*t*-butyl peroxide and about a factor of 40–50 slower than that for dibenzoyl peroxide [45]. Most *t*-butyl peresters are thermally more stable than the corresponding peroxides and, as already noted, their thermolysis [52–54] is usually free from competing ionic reactions.

One-Bond versus Two-Bond Cleavage. On the basis of scrambling of ^{18}O label in the recovered perester as well as viscosity effects on the decomposition rate, the decomposition of *t*-butyl perbenzoate probably involves the cleavage of only one bond in the first step (Eq. 7.14) [60,61].

$$t\text{-Bu—O—OOC—}C_6H_5 \rightleftharpoons t\text{-BuO}\cdot + C_6H_5\text{—COO}\cdot \quad (7.14)$$

$$C_6H_5\text{—COO}\cdot \longrightarrow C_6H_5\cdot + CO_2 \quad (7.15)$$

The slower decomposition rate in more viscous solvents is due to geminate recombination of the *t*-butoxy and under carboxylated benzoyloxy radicals, a process that regenerates the perester and decreases the net rate. Geminate reassembly of a phenyl radical, a CO_2 molecule, and a benzoyloxy radical is much less likely.

The decomposition of *t*-butyl peracetate is also believed to cleave only the O–O bond in the first step. The evidence in this case is scrambling of ^{18}O carbonyl label in recovered perester [62,63], the small magnitude of the secondary deuterium isotope effect [64], and a solvent viscosity effect [60,64].

In the series of *t*-butyl cycloalkanepercarboxylates, solvent viscosity and deuterium isotope effects indicate one-bond cleavage for the cyclopropanepercarboxylate, concurrent one- and two-bond cleavage for the 4- and 5-ring peresters, and two-bond cleavage for the 6- and 7-ring peresters [65]. The decomposition rates increase with ring size.

In the *t*-butyl peresters to be discussed next (Table 7.1) [55] the methyl of the acetoxy moiety has been replaced by a series of groups with increasing amounts of conjugation or alpha branching. These changes are accompanied by increasing decomposition rates.

In the series where the alkyl groups are methyl, benzyl, and diphenylmethyl the rates span a factor of 10^4, and in the series of tertiary radicals *t*-butyl, cumyl, and α,α-diphenylethyl the rates encompass a factor of 50. Both the corresponding radicals and the carbenium ions are increasingly more stable.

The effect of substituents in the benzene ring of *t*-butyl phenylperacetate on the decomposition rate obeys a Hammett $\rho\sigma^+$ relationship. The value of

TABLE 7.1 Decomposition of *t*-Butyl Peresters $(CH_3)_3CO—OOC—R$ [55][a]

R	Half Life at 60° (min)	$\Delta H^\ddagger$ (kcal mole^{-1})	$\Delta S^\ddagger$ (cal mol^{-1} deg^{-1})
CH_3	500,000	38	+17
$C_6H_5CH_2$	1700	28.7	+3.9
$(CH_3)_3C$	300	30.6	+13
$C_6H_5CH{=}CHCH_2$	100	23.5	−5.9
$(C_6H_5)_2CH$	26	24.3	−1.0
$C_6H_5(CH_3)_2C$	12	26	5.8
$(C_6H_5)_2CH_3C$	6	24.7	3.3
$C_6H_5(CH_2{=}CH)CH$	4	23.0	−1.1

[a]Reprinted with permission from P. D. Bartlett and R. R. Hiatt, *J. Am. Chem. Soc.* **1958,** *80,* 1399. Copyright 1958 American Chemical Society.

ρ is -1.09 at 91°C, which means that the alkyl group has some carbenium ion character in the transition state although the products are still radicals [66]. The fact that *t*-butyl phenylperacetate decomposes 50% faster in chlorobenzene than in cumene also indicates a polar transition state (Eq. 7.16) [67].

$$PhCH_2{-}CO_2{-}O{-}C(CH_3)_3 \longrightarrow \overset{\delta+}{PhCH_2}...CO_2...O{-}\overset{\delta-}{C}(CH_3)_3 \quad (7.16)$$

The reason that a polar transition state can still lead to nonpolar products may have to do with the relative positions of the covalent and ionic potential energy surfaces [68]. In peresters the covalent surface is lower than the ionic surface and comes close enough to acquire some ionic character only at its highest point, that is, the transition state. In some diacyl peroxides the two surfaces are close enough to mix at the product end of the reaction coordinate as well as at the transition state.

Although the alkyl-to-CO_2 bond is evidently weakened and polarized in the transition state, this need not increase to the extent of complete cleavage in the product. A more direct test for two-bond cleavage is the effect of viscosity on the decomposition rate. The rates for *t*-butyl α-phenylperisobutyrate [53], *t*-butyl perpivalate [60], *t*-butyl *p*-methylphenylperacetate, and *p*-methoxyphenylperacetate are independent of the viscosity, confirming the two-bond cleavage [69]. The rates for *t*-butyl phenylperacetate and *t*-butyl *p*-nitrophenylperacetate, on the other hand, were found to be viscosity dependent, indicating one-bond cleavage.

Peresters in which the alkyl-to-CO_2 bond is weaker tend to have lower enthalpies of activation and much less positive or even negative entropies of activation, as shown in Table 7.1 [55]. These enthalpies and entropies are roughly correlated, forming an approximate isokinetic relationship. A series of substituted *t*-butyl phenylperacetates could also be included in the same correlation [66].

One explanation for the trend in $\Delta H^\ddagger$ and $\Delta S^\ddagger$ is in terms of bond rotations that might be frozen or partially frozen in the transition state [55]. In peresters in which the alkyl group corresponds to a radical or carbenium ion more stable than methyl, the transition state might have a contributing structure in which both of the sigma bonds to the CO_2 moiety are broken. If so, the energy of the transition state is lowered by some fraction of the heat of formation of a CO_2 molecule. But to make something like a CO_2 molecule, the two sigma bonds need to be approximately antiperiplanar. This results in the loss of rotational freedom in the carbonyl to oxygen bond, and hence in a lower entropy. If resonance stabilization of the developing alkyl fragment requires a coplanar structure as in the case of benzyl and diphenylmethyl, an additional one or two bond rotations may be lost. For *t*-butyl peracetate the transition state has an intact *t*-butyl-to-CO_2 bond, and no bond rotations are frozen. For *t*-butyl diphenylperacetate (Fig. 7.1), three bonds are frozen, one for the CO_2 and one each for the two conjugated benzene rings.

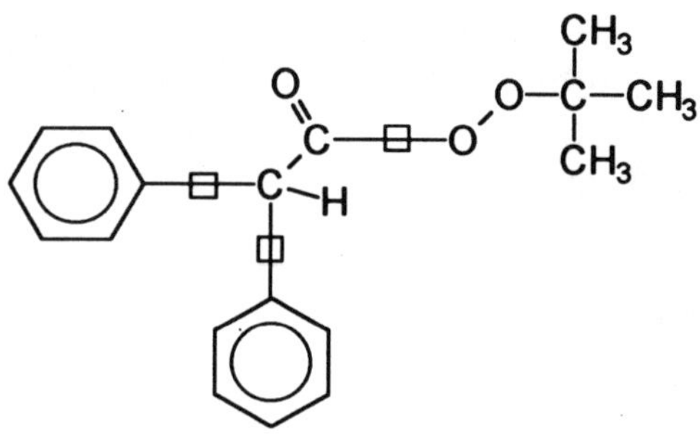

FIG. 7.1 Rectangles indicate frozen bond rotations.

An alternative, or perhaps an additional, explanation for the enthalpy and entropy effects is that the transition states in this series of peresters are increasingly more polar. The negative ρ for the series of substituted *t*-butyl perphenylacetates shows the importance of polar contributions in that series. A transition state in which a carbenium ion structure is important will be more strongly solvated. In nonhydroxylic solvents solvation involves a net loss of freedom of motion of solvent molecules, and therefore a decrease in entropy.

In *t*-butyl perpivalate the alkyl group is *t*-butyl and the homolysis rate, about 10^4 times that of *t*-butyl peracetate at 80°C, is associated with the weakening of the alkyl-to-carboxyl bond. Replacing the *t*-butyl group with a 1-adamantyl group has very little further effect on the homolysis rate [70,71]. The 1-bicyclo[2.2.2]octyl group decreases the rate by about a factor of 10, and the 1-norbornyl group decreases the rate by about a factor of 10^3 [71]. This sequence of decreasing rates is the same as the deviations from coplanarity in the corresponding radicals or carbenium ions. The differences are minute in comparison to the relative rates for the same series of alkyl groups in the solvolysis of alkyl bromides at 25°C, which are *t*-butyl 1, adamantyl 1.2×10^{-2}, bicyclooctyl 2.4×10^{-7}, and norbornyl 2.0×10^{-12} [72].

7.3.3 Effects of Anionic or Radical Substituents

Carbon Radical Substituents. The homolysis of di-*t*-butylperoxy diphenylmalonate (Eq. 7.17) in cumene at 70°C gives a perester with a benzhydryl radical substituent [73]. Because a radical of this type will neither dimerize

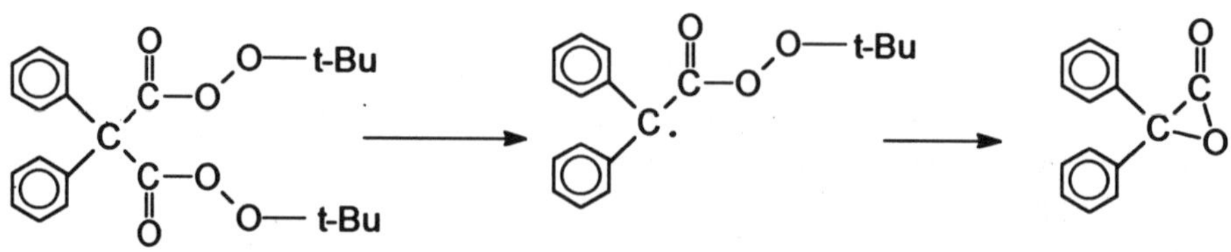

(7.17)

nor abstract hydrogen from the cumene solvent, the next step is formation of an α-lactone with expulsion of a *t*-butoxy radical. The α-lactone is not isolable but polymerizes to the polyester of benzylic acid.

Similar perester radicals are generated when the decomposition of a perester is induced by the abstraction of an alpha hydrogen in the acyloxy moiety. For example, the thermolysis of *t*-butylperoxy isobutyrate neat or in chlorobenzene gives the polyester of α-hydroxyisobutyric acid via dimethylacetolactone.

In the thermolysis of di-*t*-butylperoxy homoterephthalate (Eq. 7.18) the peracetate infrared band disappears completely while the perbenzoate band is unchanged [74]. Formation of the α-lactone is unable to compete with other reactions of the benzyl radical such as dimerization and radical combination.

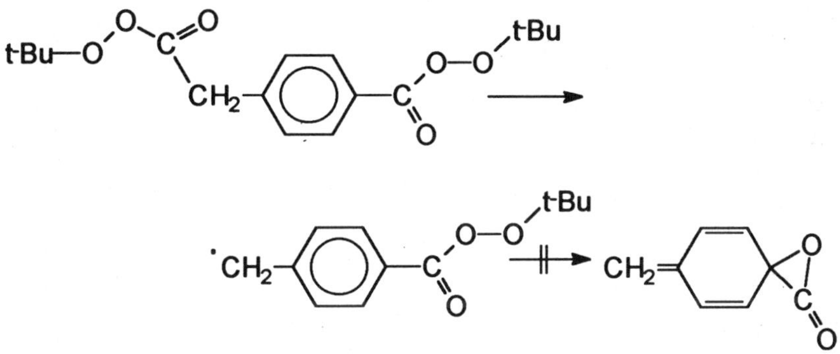

(7.18)

Oxygen Radical or Anionic Substituents. Conversion of the hydroxyl group in the phenolic perester *t*-butyl 3,5-di-*t*-butyl-4-hydroxyperbenzoate either to a phenoxy radical or to a phenoxide anion induces a rapid *radical* chain decomposition of the perester [75]. The usual bimolecular reaction between phenols and acyl peroxides is prevented by the hindering *t*-butyl groups.

The rate constant for the decomposition of this perester in very dilute solution in degassed benzene at 100°C is 3.18×10^{-5} sec^{-1}, which is very close to the value predicted from the ρ value and activation energies for substituted *t*-butyl perbenzoates [76]. Hence there is no unusual interaction between the *unmodified* phenolic hydroxyl and the perester group. The products from the decomposition in the dilute benzene solution are mainly 3,5-di-*t*-butyl-4-hydroxybenzoic acid, 4-phenyl-2,6-di-*t*-butylphenol, acetone, and *t*-butyl alcohol, as expected from a simple homolysis (Eq. 7.19). In the presence of di-*t*-butylcresol as a radical trap, the carboxylic acid and *t*-butyl alcohol are formed quantitatively.

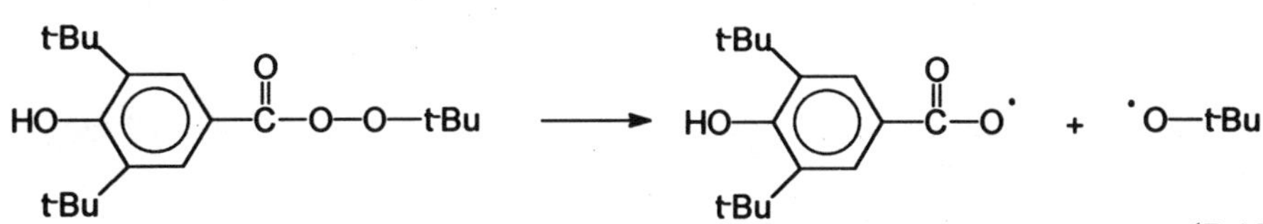

(7.19)

At higher concentrations of the perester, however, it becomes evident from the increased rate that this perester is unusually sensitive to induced decomposition. The yields of the carboxylic acid, acetone, and 3,5-di-*t*-butyl-4-hydroxybiphenyl are lower, whereas that of *t*-butyl alcohol is higher. The decomposition of *t*-butyl 3,5-di-*t*-butyl-4-*deuter*oxyperbenzoate does not show these effects of initial concentration on the rate and products, which supports a mechanism in which the attacking radicals, mainly *t*-butoxy, abstract hydrogen from the phenolic hydroxyl group. Manganese dioxide, a commonly used reagent for converting hindered phenols to phenoxy radicals, decomposes the perester in benzene solution rapidly at room temperature, giving 2,6-di-*t*-butylbenzoquinone and 3,5,3′,5′-tetra-*t*-butyldiphenoquinone as major products. None of the phenolic acid was found, but that compound is oxidized to the diphenoquinone by manganese dioxide. An ESR spectrum resembling that of the 2,6-di-*t*-butyl-4-*t*-butoxyphenoxy radical was observed in solutions of the perester freshly exposed to MnO_2.

The perester is extremely sensitive to traces of base, especially in solvents, such as alcohols or ethers, that are capable of radical chain transfer. For example, the decomposition is fast even at −20°C in tetrahydrofuran containing a few drops of pyridine. CIDNP signals are observed from the vinyl protons of 2,3-dihydrofuran and from the ortho protons of the 3,5-di-*t*-butyl-4-hydroxybenzoic acid.

Both the radical- and base-initiated reactions seem to be best explained in terms of an α-lactone with some diradical character or in equilibrium with a diradical (Eq. 7.20).

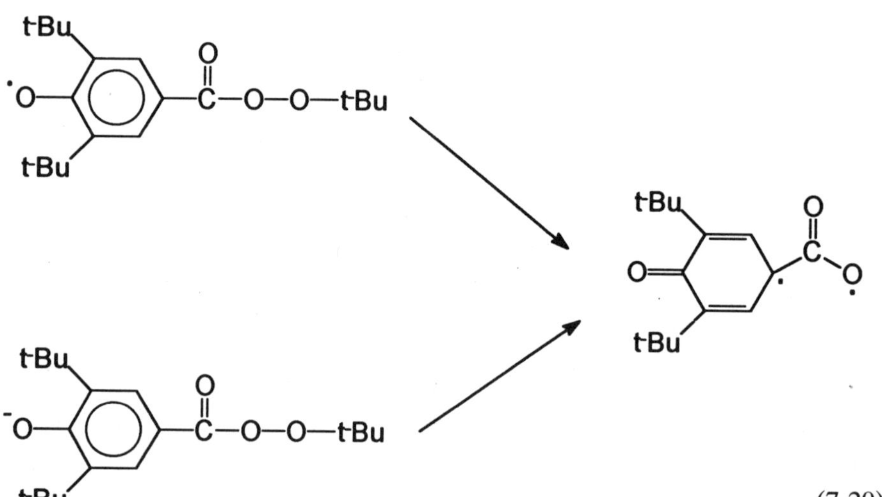

(7.20)

The strongly enhanced absorption of the ortho protons in the CIDNP signal of the acid is significant, since carboxylic acids from ordinary acyloxy radicals, as a result of the very small hyperfine coupling, never exhibit such signals. The polarization probably occurs in the radical pair shown in Eq. 7.21.

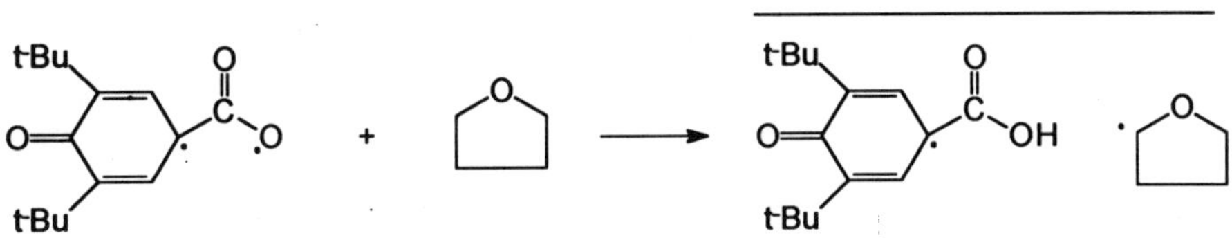

(7.21)

In degassed benzene at 26°C the products from the pyridine-initiated decomposition include some that are also formed in the uncatalyzed decomposition at 100°C, for example, the carboxylic acid and the quinones. This degree of similarity is to be expected if much of the reaction is still a radical chain reaction. The products do differ in several ways, however. There is less of the carboxylic acid, and no 4-phenyl-2,6-di-*t*-butylphenol, but there is more *t*-butyl alcohol. Added di-*t*-butylcresol forms a para adduct (Eq. 7.22) at the expense of the 3,5,3′,5′-tetra-*t*-butyldiphenoquinone.

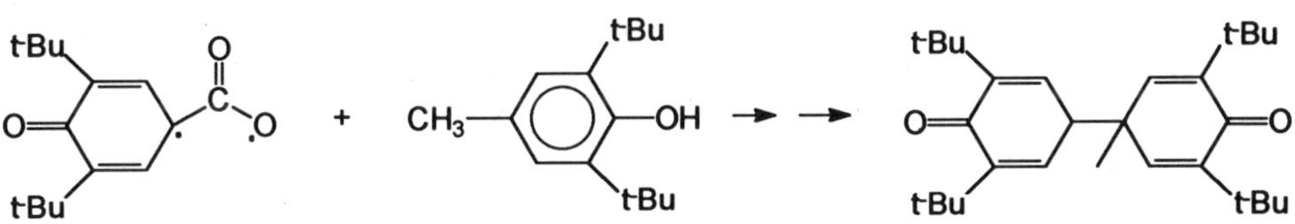

(7.22)

An isomeric phenolic perester in which the hydroxyl is ortho to the perester function decomposes rapidly, but with no sign of the radical- or base-initiated processes [77]. This reaction probably involves anchimeric participation [78] of the hydrogen-bonded hydroxyl.

Peresters of 3,5-di-*t*-butyl-4-hydroxyperbenzoic acid in which the alcohol moiety is a 1-alkyl-3,5-di-*t*-butyl-4-oxo-2,5-cyclohexadienol rather than *t*-butyl alcohol are also extremely sensitive to radical- and base-induced decomposition [79].

7.4 PHOTOLYSIS OF PERESTERS

The photolysis of the *t*-butyl peresters of *aliphatic* acids gives quantitative yields of CO_2. The lack of any viscosity effects on the rate [80], and the failure to trap the acyloxy radical as the carboxylic acid or by addition to alkenes [80,81], mean that decarboxylation is very fast. However, as we shall see, it is nevertheless not concerted with the breaking of the peroxide bond.

Photolysis of peresters can be carried out directly [80,82,83], with external singlet or triplet sensitizers [41,84], or with built-in ketonic or aromatic sensitizers [85].

Photolysis of peresters like α-phenylethyl peracetate (**I**), in which the alkoxy moiety has an α hydrogen give the corresponding carboxylic acid and ketone. The ketone is formed in an excited state, which then emits [46].

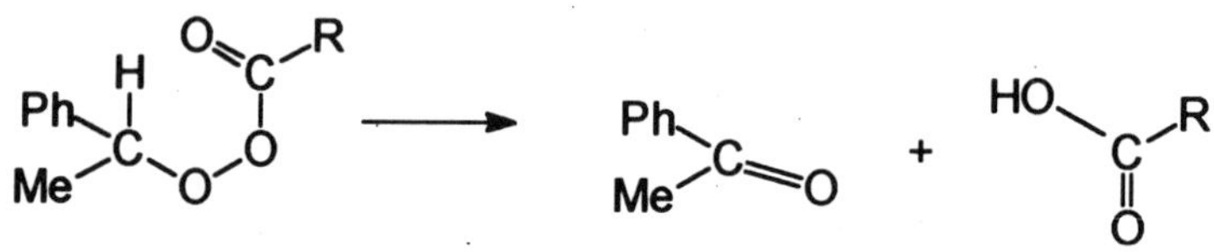

I

The decarboxylation of acyloxy radicals from the photolysis of peresters is very fast, but not concerted. This was shown by a laser flash study of the 9-methylfluorenyl perester in Eq. 7.23 [86]. On a nanosecond time scale the absorption spectrum of the 9-methylfluorenyl radical is seen immediately. Using a picosecond flash, however, the spectrum of the 9-methylfluorenyl radical is not developed instantaneously but is seen growing in at the expense of some precursor. The precursor, which is almost certainly the decarboxylating acyloxy radical, has a 55-ps half-life and decays with a rate constant of about $1.8 \times 10^{10}\ sec^{-1}$ at room temperature.

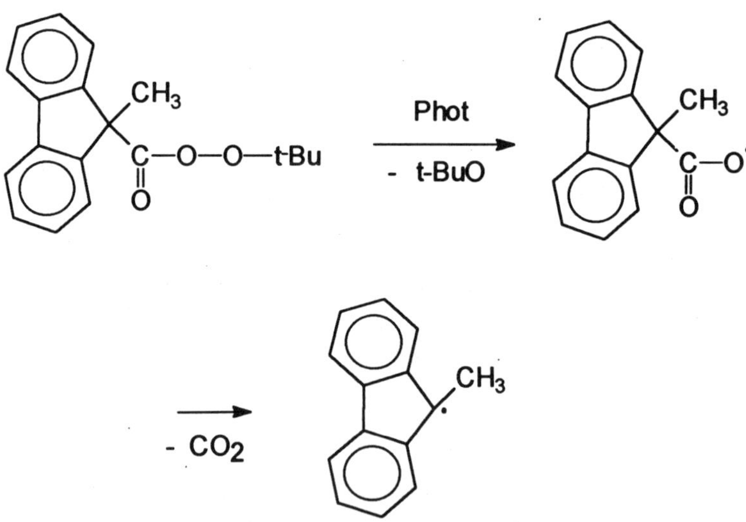

(7.23)

Since the 9-methylfluorenyl radical is very stable, the decarboxylation of acyloxy radicals corresponding to less stable aliphatic radicals should be somewhat slower. The rate of decarboxylation of acetoxy radicals, formed by the thermolysis of acetyl peroxide [87], is estimated to be about $1.6 \times 10^{9}\ sec^{-1}$.

Benzoyloxy radicals from the photolysis of *t*-butyl perbenzoate in the presence of alkenes give the ESR signal of the benzoyloxy adduct [81]. The photolysis of *t*-butyl perbenzoate in hexane gives 11% benzoic acid, and about 4% scrambling of ^{18}O carbonyl label in recovered perester [80]. A rate constant of 2×10^6 sec^{-1} at 24°C in CCl_4 has been obtained for the decarboxylation of benzoyloxy radical from benzoyl peroxide in a laser flash study [29,30].

7.5 PHOTOLYSIS OF DIALKYL PEROXIDES

Laser flash photolysis permits the generation of alkoxy radicals and the observation of their reactions on the nanosecond time scale [88,89]. In the case of di-*t*-butyl peroxide a competitive probe technique had to be used [89] because the *t*-butoxy radical itself has only a weak absorption tail in the ultraviolet. Benzyloxy radical, however, absorbs strongly in the ultraviolet at 460 nm and can be observed directly [90]. The absorption band from benzyloxy and other arylcarbinyloxy radicals may be due to internal charge transfer from the aryl group to the oxygen. Substituents shift the absorption into the visible, so that cumyloxy is orange (485 nm) and 4-methoxycumyloxy (590 nm) is blue.

REFERENCES

1. Leffler, J. E. *J. Am. Chem. Soc.* **1950,** *72,* 67.
2. Leffler, J. E.; More, A. A. *J.Am. Chem. Soc.* **1972,** *94,* 2483.
3. Walling, C.; Waits, H. P.; Milovanovic, J.; Pappiaonnou, C. G. *J. Am. Chem. Soc.* **1970,** *92,* 4927.
4. Greene, F. D.; Stein, H. P.; Chu, C.-C.; Vane, F. M. *J. Am. Chem. Soc.* **1964,** *86,* 2080.
5. Kashiwagi, T.; Kozuka, S.; Oae, S. *Tetrahedron* **1970,** *26,* 3619.
6. Taylor, J. W.; Martin, J. C. *J. Am. Chem. Soc.* **1966,** *88,* 3650.
7. Taylor, J. W.; Martin, J. C. *J. Am. Chem. Soc.* **1967,** *89,* 6904.
8. Hammond, G. S.; Soffer, L. M. *J. Am. Chem. Soc.* **1950,** *72,* 4711.
9. Bevington, J. C.; Lewis, T. D. *Trans. Faraday Soc.* **1958,** *54,* 1340.
10. Schwerzel, R. E.; Lawler, R. G.; Evans, G. T. *Chem. Phys. Lett.* **1974,** *29,* 106.
11. den Hollander, J. A. *Chem. Phys.* **1975,** *10,* 167.
12. Bartlett, P. D.; Leffler, J. E. *J. Am. Chem. Soc.* **1950,** *72,* 3030.
13. Flowers, G. C.; Leffler, J. E. *J. Org. Chem.* **1989,** *54,* 3995.
14. Leffler, J. E.; Barbas, J. T.; Flowers, G. C. *J. Org. Chem.* **1982,** *47,* 2286.
15. Barbas, J. T.; Leffler, J. E. *J. Org. Chem.* **1985,** *50,* 705.
16. Flowers, G. C.; Leffler, J. E. *J. Org. Chem.* **1985,** *50,* 4406.
17. Kropp, P. J.; Daus, K. A.; Crawford, S. D.; Tubergen, M. W.; Kepler, K. D.; Craig, S. L.; Wilson, V. P. *J. Am. Chem. Soc.* **1990,** *112,* 7433.

18. Taylor, K. G.; Govindan, C. K.; Kaelin, M. S. *J. Am. Chem. Soc.* **1979,** *101,* 2091.
19. Sigman, M. E.; Barbas, J. T.; Leffler, J. E. *J. Org. Chem.* **1987,** *52,* 1754.
20. Lawler, R. G.; Barbara, P. F.; Jacobs, D. *J. Am. Chem. Soc.* **1978,** *100,* 4912.
21. Kharasch, M. S.; Kuderna, J.; Nudenberg, W. *J. Org. Chem.* **1954,** *19,* 1283.
22. DeTar, D. F.; Wells, D. V. *J. Am. Chem. Soc.* **1960,** *82,* 5839.
23. DeTar, D. F.; Weis, C. *J. Am. Chem. Soc.* **1957,** *79,* 3045.
24. Greene, F. D. *J. Am. Chem. Soc.* **1955,** *77,* 4869.
25. Walling, C.; Sloan, J. P. *J. Am. Chem. Soc.* **1979,** *101,* 7679.
26. Calvert, J. G.; Pitts, J. N. *Photochemistry,* Wiley: New York, 1966.
27. Sheldon, R. A.; Kochi, J. K. *J. Am. Chem. Soc.* **1970,** *92,* 4395.
28. Yamauchi, S.; Hirota, N.; Takahara, S.; Sakuragi, H.; Tokumaru, K. *J. Am. Chem. Soc.* **1985,** *107,* 5021.
29. Chateauneuf, J.; Lusztyk, J.; Ingold, K. U. *J. Am. Chem. Soc.* **1988,** *110,* 2886.
30. Chateauneuf, J.; Lusztyk, J.; Ingold, K. U. *J. Am. Chem. Soc.* **1988,** *110,* 2877.
31. Box, H. C.; Budzinski, E. E.; Freund, H. G. *J. Am. Chem. Soc.* **1970,** *92,* 5305.
32. Walter, D. W.; McBride, J. M. *J. Am. Chem. Soc.* **1981,** *103,* 7069.
33. McBride, J. M. *Acc. Chem. Res.* **1983,** *16,* 304.
34. Pacansky, J.; Horne, D. E.; Gardini, G. P.; Bargon, J. *J. Phys. Chem.* **1977,** *81,* 2149.
35. Pacansky, J.; Gutierrez, A. *J. Phys. Chem.* **1983,** *87,* 3074.
36. Pacansky, J.; Bargon, J. *J. Am. Chem. Soc.* **1975,** *97, 6896.*
37. Pacansky, J.; Coufal, H. *J. Chem. Phys.* **1980,** *72,* 3298.
38. den Hollander, J. A.; van der Ploeg, J. P. M. *Tetrahedron* **1976,** *32,* 2433.
39. Kitamura, A.; Sakuragi, H.; Yoshida, M.; Tokumaru, K. *Bull. Chem. Soc. Jpn.* **1980,** *53,* 1393.
40. Scaiano, J. C.; Wubbels, G. G. *J. Am. Chem. Soc.* **1981,** *103,* 640.
41. Engel, P. S.; Woods, T. L.; Page, M. A. *J. Phys. Chem.* **1983,** *87,* 10.
42. Takahara, S.; Sakuragi, H.; Tokumaru, K.; Kitamura, A.; Shiga, A. *J. Photochem.* **1985,** *31,* 239.
43. Walling, C.; Gibian, M. J. *J. Am. Chem. Soc.* **1965,** *87,* 3413.
44. Leffler, J. E.; Miley, J. W. *J. Am. Chem. Soc.* **1971,** *93,* 7005.
45. Rüchardt, C. *Fortschr. chem. Forsch.* **1966,** *6,* 251.
46. Dixon, B. G.; Schuster, G. B. *J. Am. Chem. Soc.* **1979,** *101,* 3116.
47. Hiatt, R. R.; Glover, L. C.; Mosher, H. S. *J. Am. Chem. Soc.* **1975,** *97,* 1556.
48. Criegee, R. *Ber. Deutsch. Chem. Ges.* **1944,** *77,* 722.
49. Goering, H. L.; Olson, A. C. *J. Am. Chem. Soc.* **1953,** *75,* 5853.
50. Bartlett, P. D.; Kice, J. L. *J. Am. Chem. Soc.* **1953,** *75,* 5591.
51. Denney, D. B.; Denney, D. G. *J. Am. Chem. Soc.* **1957,** *79,* 4806.
52. Ernst, J. A.; Thankachan, C.; Tidwell, T. T. *J. Org. Chem.* **1974,** *39,* 3614.
53. Herkes, F. E.; Friedman, J.; Bartlett, P. D. *Int. J. Chem. Kinet.* **1969,** *1,* 193.
54. Koenig, T., In *Free Radicals,* J. K. Kochi, Ed. Wiley-Interscience: New York, 1973, Chapter 3.
55. Bartlett, P. D.; Hiatt, R. R. *J. Am. Chem. Soc.* **1958,** *80,* 1398.

56. Bartlett, P. D.; Storey, B. T. *J. Am. Chem. Soc.* **1958,** *80,* 4954.
57. Yablokov, V. A.; Shushunov, V. A.; Kolyaskina, L. V. *Zh. Obshch. Khim.* **1962,** *32,* 2714.
58. Leffler, J. E.; Scrivener, F. E. J. *J. Org. Chem.* **1972,** *37,* 1794.
59. Rüchardt, C.; Scwarzer, H. *Chem. Ber.* **1966,** *99,* 1878.
60. Pryor, W. A.; Morkved, E. H.; Bickley, H. T. *J. Org. Chem.* **1972,** *37,* 1999.
61. Koenig, T.; Deinzer, M.; Hoobler, J. A. *J. Am. Chem. Soc.* **1971,** *93,* 938.
62. Goldstein, M. J.; Judson, H. A. *J. Am. Chem. Soc.* **1970,** *92,* 4120.
63. Koenig, T.; Deinzer, M. *J. Am. Chem. Soc.* **1968,** *90,* 7014.
64. Koenig, T.; Huntington, J.; Cruthoff, R. *J. Am. Chem. Soc.* **1970,** *92,* 5413.
65. Wolf, R. A.; Migliore, M. J.; Fuery, P. H.; Gagnier, P. R.; Sabeta, I. C.; Trocino, R. J. *J. Am. Chem. Soc.* **1978,** *100,* 7967.
66. Bartlett, P. D.; Rüchardt, C. *J. Am. Chem. Soc.* **1960,** *82,* 1756.
67. Neuman, R. C., Jr.; Behar, J. V. *J. Am. Chem. Soc.* **1969,** *91,* 6024.
68. Leffler, J. E. Mixing of Ionic and Covalent Potential Energy Surfaces in Acyl Peroxide Thermolysis. Abstract, Session I, 26th IUPAC Conference, Tokyo, 1977, p. 47.
69. Pryor, W. A.; Smith, K. *J. Am. Chem. Soc.* **1970,** *92,* 5403.
70. Lorand, J. P.; Chodroff, S. D.; Wallace, R. W. *J. Am. Chem. Soc.* **1968,** *90,* 5266.
71. Humphrey, L. B.; Hodgson, B.; Pincock, R. E. *Can. J. Chem.* **1968,** *46,* 3099.
72. Gleicher, G. J.; Schleyer, P. von R. *J. Am. Chem. Soc.* **1967,** *89,* 582.
73. Bartlett, P. D.; Gortler, L. B. *J. Am. Chem. Soc.* **1963,** *85,* 1864.
74. Dalton, A. I.; Tidwell, T. T. *J. Org. Chem.* **1972,** *37,* 1504.
75. Huček, A. M.; Barbas, J. T.; Leffler, J. E. *J. Am. Chem. Soc.* **1973,** *95,* 4698.
76. Blomquist, A. T.; Berstein, I. A. *J. Am. Chem. Soc.* **1951,** *73,* 5546.
77. Barbas, J. T.; Leffler, J. E. *J. Am. Chem. Soc.* **1975,** *97,* 7270.
78. Bentrude, W. G.; Martin, J. C. *J. Am. Chem. Soc.* **1962,** *84,* 1561.
79. Nishinaga, A.; Nakamura, K.; Matsuura, T. *J. Org. Chem.* **1983,** *48,* 3700.
80. Sheldon, R. A.; Kochi, J. K. *J. Am. Chem. Soc.* **1970,** *92,* 5175.
81. Edge, D. J.; Kochi, J. K. *J. Am. Chem. Soc.* **1973,** *95,* 2635.
82. Krusic, P. J.; Rettig, T. A.; Schleyer, P. von R. *J. Am. Chem. Soc.* **1972,** *94,* 995.
83. Kochi, J. K.; Krucic, P. J. *J. Am. Chem. Soc.* **1969,** *91,* 3940.
84. Koenig, T.; Hoobler, J. A. *Tetrahedron Lett.* **1972,** 1803.
85. Morlino, E. A.; Bohorquez, M. D.; Neckers, D. C.; Rodgers, M. A. J. *J. Am. Chem. Soc.* **1991,** *113,* 3599.
86. Falvey, D. E.; Schuster, G. B. *J. Am. Chem. Soc.* **1986,** *108,* 7419.
87. Braun, W.; Rajbenbach, L.; Eirich, F. R. *J. Phys. Chem.* **1962,** *66,* 1591.
88. Small, R. D. J.; Scaiano, J. C. *J. Am. Chem. Soc.* **1978,** *100,* 296.
89. Paul, H.; Small, R. D. J.; Scaiano, J. C. *J. Am. Chem. Soc.* **1978,** *100,* 4520.
90. Avila, D. V.; Lusztyk, J.; Ingold, K. U. *J. Am. Chem. Soc.* **1992,** *114,* 6576.

8

RADICAL SOURCES II

8.1 FORMATION OF RADICALS FROM AZO COMPOUNDS

Azoalkanes can be used to generate the corresponding alkyl radicals either thermally or photochemically [1–6].

8.1.1 Thermolysis

The thermolysis rate of substituted 1,2-diazenes depends strongly on the structure of the alkyl groups. At the extremes, *trans*-azobenzene is stable even at 600°C, while azotriphenylmethane has only a transient existence even at −40°C [7,8]. For less extreme cases, with groups such as methyl, *t*-butyl, benzyl, and allyl, the energy of activation of the symmetrically substituted diazene is an approximately linear function of the bond dissociation energy for R–H, with about unit slope [2,9]. In addition to effects associated with the stability of the radicals, strain in the azo compound can greatly increase the decomposition rate. For example, reaction 8.2 is 1300 times as fast as reaction 8.1, even though both radicals are tertiary and should have about the same R–H bond dissociation energy [10].

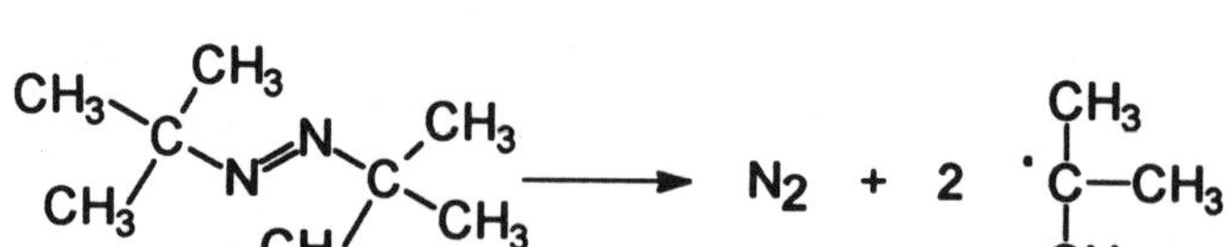

(8.1)

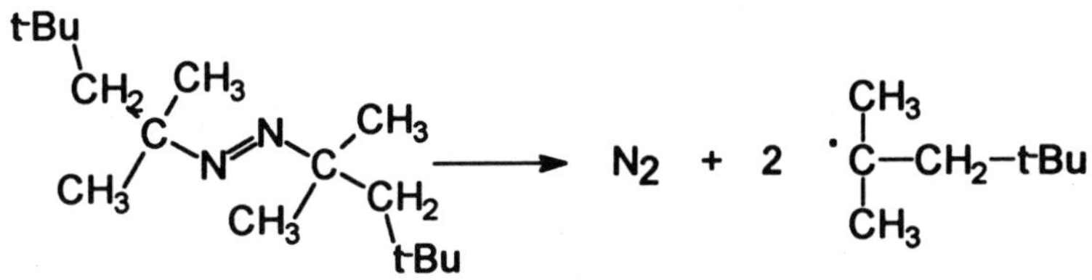

(8.2)

Poor yields of radicals are often obtained when the azo compound has an alpha hydrogen, which allows isomerization to an azine, as in Eq. 8.3. This is more of a problem with cis azo compounds than it is with trans.

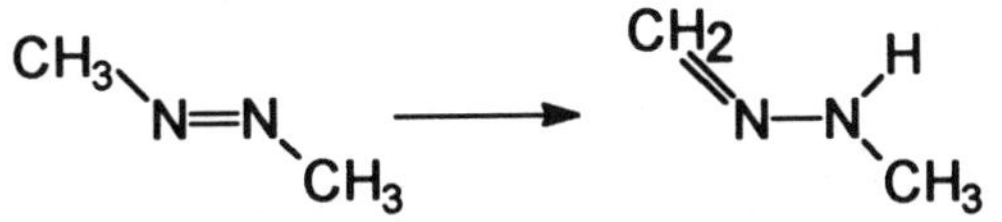

(8.3)

Another side reaction that sometimes leads to poor yields of radicals is the partial reduction of the azo compound to the hydrazo compound. In the decomposition of 2,4-dinitrophenylazo-tris(p-anisyl)methane in benzene, the azo compound is probably reduced by the dinitrophenylcyclohexadienyl radical [11]. A number of azo compounds react with diphenylhydroxymethyl radicals to give the hydrazo compound and benzophenone [12].

One-Bond versus Two-Bond Cleavage. Although the thermolysis of azoalkanes ultimately gives two alkyl radicals, the mechanism can range from completely concerted (Eq. 8.4) to stepwise (Eq. 8.5). In the stepwise decomposition only the bond to the more stable radical (R_s) is broken in the first step.

$$\mathrm{R{-}N{=}N{-}R_s \longrightarrow N_2 + R\cdot + R_s\cdot} \qquad (8.4)$$

$$\mathrm{R{-}N{=}N{-}R_s \longrightarrow R_s\cdot + R{-}N{=}N\cdot}$$

$$\mathrm{R{-}N{=}N\cdot \longrightarrow N_2 + R\cdot} \qquad (8.5)$$

If Eq. 8.4 is the correct mechanism, the decomposition of the unsymmetrical azo compound R—N=N—R_s benefits from the formation of only one stable radical R_s and its rate should fall between those of the two symmetrical azo compounds R—N=N—R (slow) and R_s—N=N—R_s (fast). If the stepwise mechanism is correct, the decomposition should not undergo any further acceleration when a second R_s group is present, and the rates of R—N=N—R_s and R_s—N=N—R_s should be about the same. Usually $\delta\Delta G_2^\ddagger$, the difference in the free energies of activation for R—N=N—R_s versus R_s—N=N—R_s, is smaller than $\delta\Delta G_1^\ddagger$, the difference for R—N=N—R_s versus R—N=N—R.

This supports a mechanism that resembles Eq. 8.5 more than Eq. 8.4 [2]. It is probable that symmetrical azoalkanes decompose by the concerted mechanism, with equal stretching of both bonds. However, for unsymmetrically substituted diazenes, $\delta\Delta G_2^\ddagger$ is rarely zero. This means that both bonds are weakened to some extent in the transition state even though one is more nearly broken than the other. The decomposition of arylazoalkanes is an extreme case in which the Ar—N=N· is a definite intermediate [13].

Two methyl radicals are probably produced in one step by the thermolysis or photolysis of azomethane [2], but thermolysis of unsymmetrical chiral azo compounds gives partly racemized starting material, indicating one-bond thermolysis, as in Eq. 8.6 [14,15].

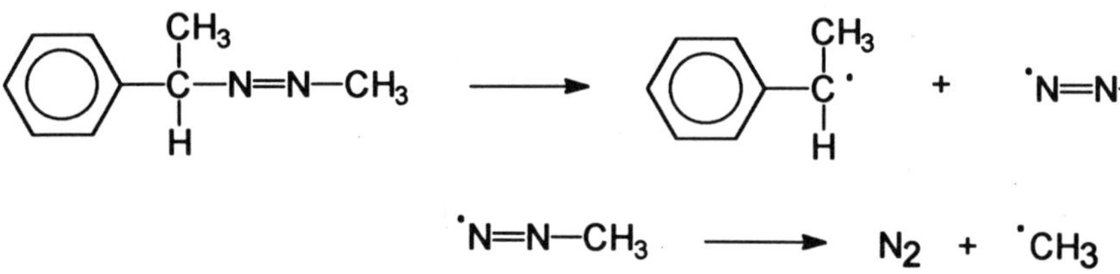

(8.6)

Observations of CIDNP signals from ^{15}N are evidence for norbornyldiazenyl radicals in the thermolysis of 1-norbornylazocumene and evidence for phenyldiazenyl radicals in the thermolysis of phenylazotriphenylmethane, phenylazocumene, and phenylazoisobutyronitrile [13]. The phenyldiazenyl radical was also trapped by triphenylmethyl radicals.

The thermolysis of cyclic azo compounds, and also of cyclic ketones, simplifies matters since only two fragments rather than three are produced. The thermolyses of several bicyclic compounds (Eq. 8.7), initiated in a shock tube, have been studied by means of a continuous wave CO laser [16].

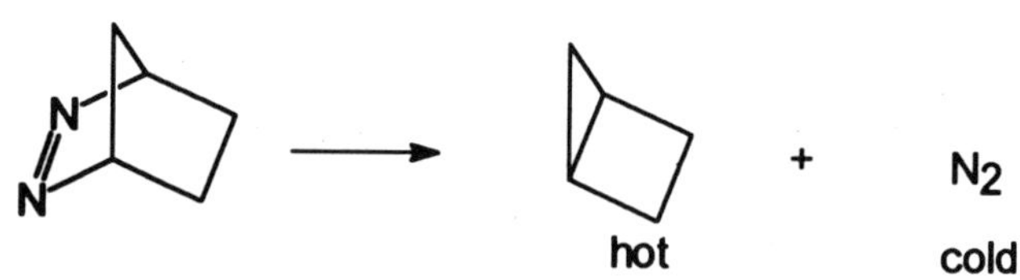

(8.7a)

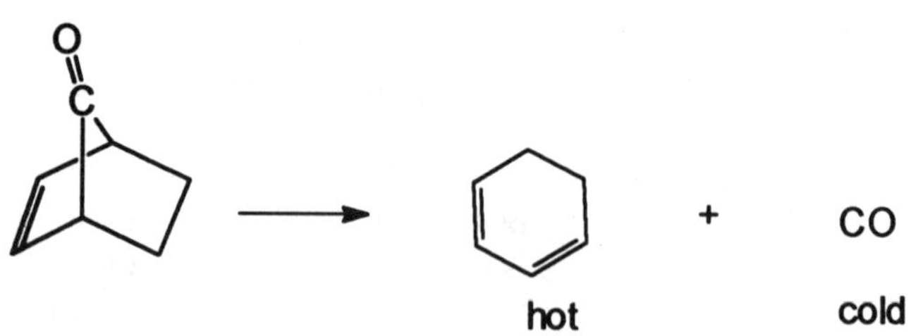

(8.7b)

If the formation of N_2 in reaction (8.7a) were concerted, breaking both bonds at once, there would have been an abrupt change in the N-to-N bond length from about 1.246 Å in the azo compound, to 1.098 Å in the N_2 molecule. This should have formed the N_2 in a vibrationally excited state (i.e., hot). Instead, the small molecules were formed cold and the "housane" (Eq. 8.7a) and cyclohexadiene (Eq. 8.7b) had the vibrational energy.

Cis and Trans Azo Compounds. Another complication in the chemistry of azo compounds is the existence of cis isomers. The cis azo compounds, which are more reactive than the trans, can be prepared from the usual trans isomer by irradiation at a low temperature [2,17]. The isomerization can be reversed by using light of a longer wavelength (Eq. 8.8).

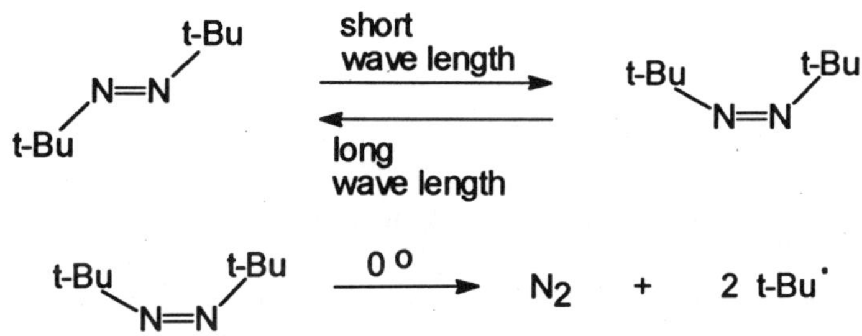

(8.8)

There is a continuum of possible reactions for cis azo compounds, ranging from exclusive homolysis at one extreme (Eq. 8.8) to exclusive isomerization to the trans isomer at the other.

The cis compound azo-1-norbornane merely isomerizes instead of decomposing (Eq. 8.9) [18]. There are two reasons for this. First, in the bridgehead structure of the norbornyl group the alpha substituents are tied back, reducing the steric strain. Second, the 1-norbornyl radical, unlike ordinary tertiary radicals, is highly reactive and affords very little driving force for homolysis.

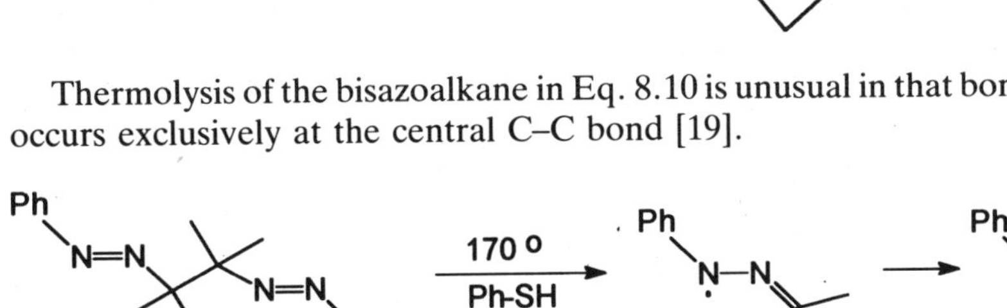

(8.9)

Thermolysis of the bisazoalkane in Eq. 8.10 is unusual in that bond cleavage occurs exclusively at the central C–C bond [19].

(8.10)

In contrast, the bisazoalkane of Eq. 8.11 cleaves only C–N bonds, probably losing one nitrogen molecule at a time.

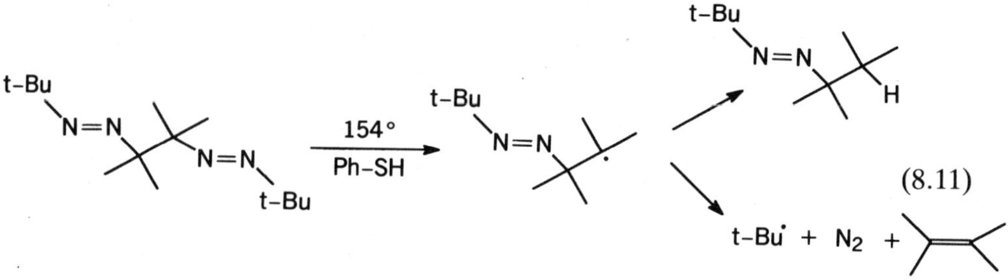

(8.11)

8.1.2 Photolysis of Azo Compounds

The characteristic $n\pi^*$ absorption band of azo compounds is found at 320–380 nm [1,20].

The photolysis of trans azo compounds usually involves isomerization to the cis isomer, which then thermolizes [15]. The photolysis of *trans*-azo-1-adamantane and *trans*-azo-1-bicyclo[2.2.2]octane at 65°C, a clean source of the corresponding radicals, are examples [21]. Although the cis isomers of these azo compounds are stable at lower temperatures, they isomerize readily at 65°C.

Quantum yields for the photolysis in solution are rarely greater than 0.5, probably because the excited state has an equal probability of returning to the trans configuration rather than the cis. In the gas phase at low enough pressures, quantum yields near 1 are obtained. A possible explanation is that the vibrationally excited cis isomer is not deactivated by collisions under these conditions [1].

The photolysis of cyclic azo compounds with 5- or 6-membered rings gives diradicals. If the diradical is a triplet, produced by means of a triplet sensitizer, closure to the singlet product may be slowed by the requirement for intersystem crossing to the singlet diradical, allowing time for rotation about C–C single bonds. For example, the cyclic azo compound in Eq. 8.12 has been thermolized and photolyzed both directly and with a thioxanthone triplet sensitizer [22]. The cyclobutanes produced via the singlet diradical, from thermolysis or direct photolysis, retained the original configuration to the extent of 95% or more. The cyclobutanes produced via the triplet diradical retained the original configuration only to the extent of 61–65%.

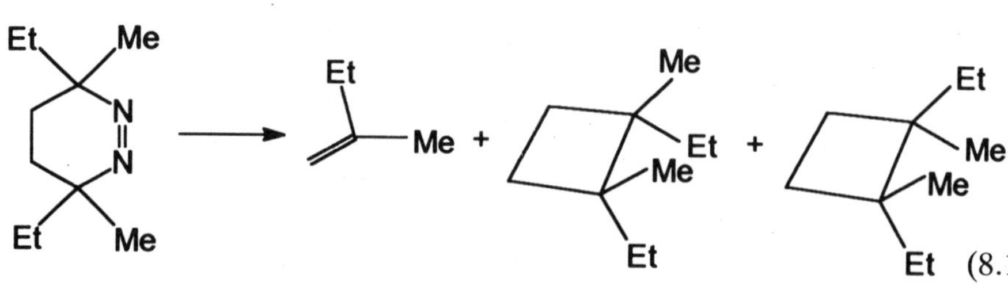

(8.12)

8.2 BETA CLEAVAGE OF OXYGEN RADICALS

8.2.1 From Naphthylmethyl Esters

The study of acyloxy radicals from the photolysis of diacyl peroxides and peresters has been supplemented by the laser flash photolysis of 1-naphthylmethyl alkanoates (Eq. 8.13). This initiates a rather complex process, but, with certain assumptions, it is possible to estimate the decarboxylation rates of the acyloxy radicals [23].

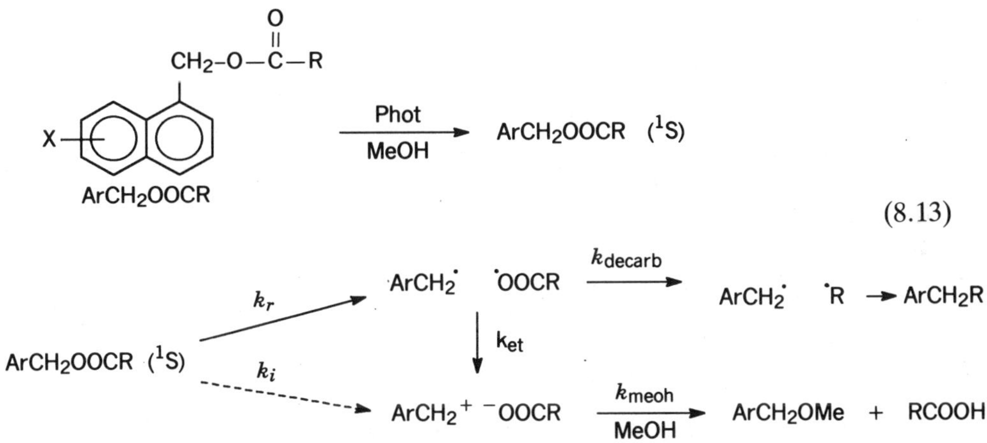

(8.13)

For the acetoxy radical the rate of decarboxylation is less than 1.3×10^9 sec^{-1}. For propionoxy it is $2.0 \times 10^9\ sec^{-1}$, for isobutyryloxy $6.5 \times 10^9\ sec^{-1}$, and for pivaloyloxy $11 \times 10^9\ sec^{-1}$. The rate approximately doubles for each α methyl substituent introduced [23].

8.2.2 The Kolbe Reaction

An early example of the formation of methyl and other alkyl radicals from acyloxy radicals is Kolbe's electrolytic oxidation of alkanoate ions in 1849 [24]. At that time the methyl group was believed to have the empirical formula C_2H_3, and Kolbe's claim to have prepared "freiem Methyl" and other free radicals was not believed until many years later.

8.2.3 Alkoxy Radicals

Methyl radical is formed along with acetone by the β cleavage of the *t*-butoxy radical [25–27]. Similarly, methyl radical and acetophenone are formed from the cumyloxy radical (Eq. 8.14), ethyl radical and acetone from 1,1-dimethylpropyloxy [26], ethyl radical and formaldehyde from propyloxy [28].

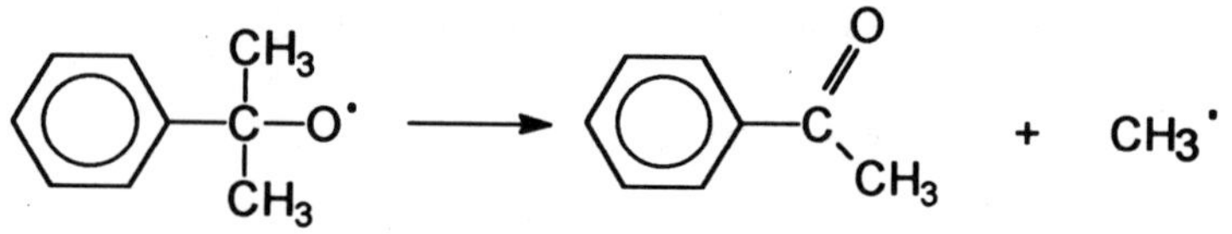

(8.14)

It should be noted that rearrangement does not compete with β cleavage unless an unsaturated group such as phenyl is present. The triphenylmethoxy radical rearranges to diphenylphenoxymethyl radical rather than cleave to phenyl radical and benzophenone.

Alkoxy radicals as precursors for the generation of alkyl radicals by β cleavage can be made not only by the thermolysis or photolysis of peresters and dialkyl peroxides, but also by the photolysis of an alkyl nitrite [28] or a hypochlorite [26] (Eqs. 8.15–8.17, R = isopropyl).

$$RCH_2ONO \longrightarrow RCH_2O\cdot + \cdot NO \qquad (8.15)$$

$$RCH_2O\cdot \longrightarrow R\cdot + CH_2{=}O$$

$$R(CH_3)_2COCl \longrightarrow R(CH_3)_2CO\cdot + \cdot Cl \qquad (8.16)$$

$$R(CH_3)_2CO\cdot \longrightarrow R\cdot + (CH_3)_2C{=}O \qquad (8.17)$$

8.3 PHOTOLYSIS OF CARBONYL COMPOUNDS

The photolysis of a ketone or an aldehyde via its n,π^* triplet is analogous to the β cleavage of an alkoxy radical. The cleavage of the ketone or aldehyde triplet diradical (Eq. 8.18), like the cleavage of an alkoxy radical, gives an alkyl radical and a fragment containing a carbonyl group [29–32].

$$CH_3{-}\overset{\overset{\dot{O}}{|}}{\underset{\cdot}{C}}{-}CH_3 \longrightarrow CH_3{-}\overset{\overset{O}{\|}}{C}\cdot + CH_3\cdot \qquad (8.18)$$

The CIDNP spectra of the molecular products from the photolysis of acetone [32] show enhanced emission from the methyl protons of rebonded acetone and enhanced absorption from methane.

The photolysis of *t*-butyl pivalate (Eq. 8.19) gives *t*-butyl radicals and a *t*-butoxyacyl radical [33]. Beta cleavage of the latter gives another *t*-butyl radical and CO_2.

$$t\text{-Bu}{-}\overset{\overset{O}{\|}}{C}{-}O{-}t\text{-Bu} \xrightarrow{\text{phot}} t\text{-Bu}\cdot \quad \cdot\overset{\overset{O}{\|}}{C}{-}O{-}t\text{-Bu} \qquad (8.19)$$

$$\cdot\overset{\overset{O}{\|}}{C}{-}O{-}t\text{-Bu} \longrightarrow CO_2 + t\text{-Bu}\cdot$$

8.4 ORGANOMETALLIC AND ORGANOMETALLOID SOURCES

The pyrolysis of organometallic compounds such as dimethylzinc [34], dimethylmercury [35], alkylmercuric chlorides [36], or tetramethyllead breaks the bond between the alkyl group and the metal.

8.4.1 Paneth

The pyrolysis of tetramethyllead was used by Paneth in his famous lead mirror experiments of 1929 [37]. The overall reaction in the dissociation of lead tetramethyl (Eq. 8.20) is reversible, favoring methyl radicals at high temperatures and lead tetramethyl at low temperatures.

$$Pb(CH_3)_4 \rightleftharpoons Pb + 4\ CH_3\cdot \qquad (8.20)$$

When a stream of carrier gas containing lead tetramethyl is passed over a spot heated by a Bunsen burner, lead from the decomposing lead tetramethyl forms a mirror at that point. If there is a second lead mirror not too far downstream, in a cool part of the tube, that mirror disappears. Since only lead tetramethyl, carrier gas, and inert hydrocarbons emerge from the end of the tube, it is clear that some transient intermediate is being generated. The lifetime of the methyl radicals, estimated from the flow rate and the maximum distance at which the second mirror will still react, is about 6×10^{-3} sec.

8.4.2 Arsenic

Photolysis of *t*-butyl peroxide in the presence of trimethylarsine gives an organoarsenic radical by addition of the *t*-butoxy radical to the arsine and expansion of the valence shell (Eq. 8.21). Subsequent cleavage of a methyl–arsenic bond (Eq. 8.22) produces a methyl radical. The overall process of Eqs. 8.21 and 8.22 amounts to an S_H2 displacement of methyl by *t*-butoxy [38].

$$t\text{-Bu—O}\cdot + (CH_3)_3As \longrightarrow (CH_3)_3As\cdot\text{—O—}t\text{-Bu} \qquad (8.21)$$

$$(CH_3)_3As\cdot\text{—O—}t\text{-Bu} \longrightarrow (CH_3)_2As\text{—O—}t\text{-Bu} + CH_3\cdot \qquad (8.22)$$

8.4.3 Phosphorus

n-Pentyl radicals have been made in solution by the thiol plus phosphite chain reaction (Eq. 8.23 and 8.24, R = *n*-pentyl) [39].

$$R\cdot + RSH \longrightarrow RH + RS\cdot \qquad (8.23)$$

$$RS\cdot + P(OEt)_3 \longrightarrow R\cdot + SP(OEt)_3 \qquad (8.24)$$

8.5 THERMOLYSIS OF STRAINED MOLECULES

The pyrolysis of highly strained compounds such as tri-*t*-butylmethane [40] (Eq. 8.25) or tri-*t*-butylcarbinol [41,42] (Eq. 8.26) gives *t*-butyl radicals. The fast homolysis means that much of the strain is relieved in the transition state, which resembles the radicals more than it does the molecule. As calculated by a molecular mechanics program [43], both the bond angles and bond lengths of tri-*t*-butylmethane are abnormal. Its heat of formation (-53.69 kcal mole^{-1}) is also unusually large. The rate constant for reaction 8.25 in decalin at 239.5°C is 2.33×10^{-4} sec^{-1}, the enthalpy of activation is 47.3 kcal mole^{-1}, and the entropy of activation is $+16.2$ cal mol^{-1} deg^{-1}. The enthalpy of activation for the thermolysis of the carbinol is 43.0 kcal mole^{-1}.

$$(t\text{-Bu})_3\text{C—H} \longrightarrow (t\text{-Bu})_2\dot{\text{C}}\text{—H} + t\text{-Bu}\cdot \tag{8.25}$$

$$(t\text{-Bu})_3\text{C—OH} \longrightarrow (t\text{-Bu})_2\dot{\text{C}}\text{—OH} + t\text{-Bu}\cdot \tag{8.26}$$

For the loss of *t*-butyl radical from a series of carbinols RR′BuCOH the free energy of activation is fairly well correlated both with the Taft–Ingold steric parameters $E_s(\text{R} + \text{R}')$ and with the strain energy of the carbinol as calculated by molecular mechanics [41].

A particularly interesting example is the aromatic diradical formed by relief of strain in the naturally occurring antibiotic dynamicin A (Eq. 8.27) [44].

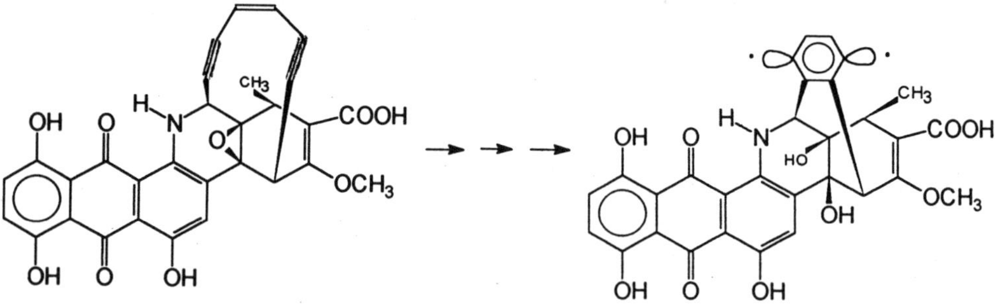

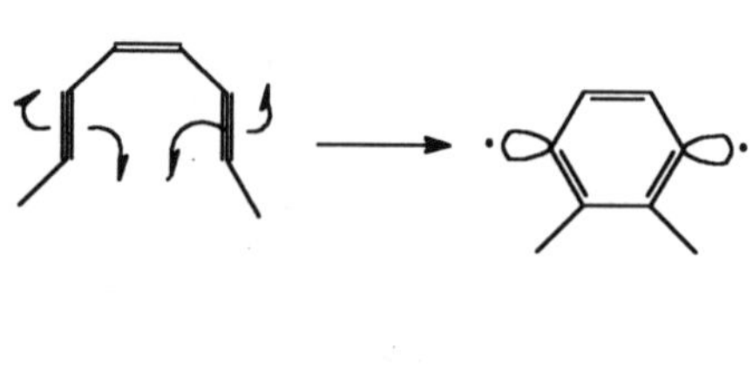

(8.27)

Dynamicin A is believed to act by causing DNA strand breaks. After intercalation of the anthraquinone core, the diyne changes to the diradical and the diradical abstracts hydrogen from the DNA.

8.6 ATOM ABSTRACTION REACTIONS

A frequently used method of making hydrocarbon radicals is photolysis of di-*t*-butyl peroxide in the presence of the hydrocarbon. The *t*-butoxy radicals will abstract hydrogen and can give the desired radical in good yield in certain cases. The cyclopropyl radical has been generated in this way by photolyzing di-*t*-butyl peroxide in cyclopropane solution. Another example is the preparation of the di-*t*-butylmethyl radical (Eq. 8.28) [45].

$$t\text{-BuO}\cdot + Me_3C{-}CH_2{-}CMe_3 \longrightarrow t\text{-BuOH} + Me_3C{-}C\cdot H{-}CMe_3 \tag{8.28}$$

Alkyltin radicals are useful for selectively abstracting a chlorine atom. An example is the photolysis of hexamethylditin in the presence of di-*t*-butylmethyl chloride (Eq. 8.29) [46].

$$Me_3Sn\cdot + Me_3C{-}CHCl{-}CMe_3 \longrightarrow Me_3SnCl + Me_3C{-}C\cdot H{-}CMe_3 \tag{8.29}$$

REFERENCES

1. Engel, P. S.; Steel, C. *Acc. Chem. Res.* **1973,** *6,* 275.
2. Engel, P. S. *Chem. Rev.* **1980,** *80,* 99.
3. Dixon, P. S.; Stefani, A. P.; Szwarc, M. *J. Am. Chem. Soc.* **1963,** *85,* 2551.
4. Davis, W. H., Jr.; Pryor, W. A. *J. Am. Chem. Soc.* **1977,** *99,* 6365.
5. Tanner, D. D.; Rahimi, P. M. *J. Am. Chem. Soc.* **1982,** *104,* 225.
6. Pryor, W. A.; Tang, F. Y.; Tang, R. H.; Church, D. F. *J. Am. Chem. Soc.* **1982,** *104,* 2885.
7. Wieland, H.; vom Hove, H.; Börner, K. *Liebigs Ann. Chem.* **1926,** *446,* 31.
8. Barton, D. H. R.; Haynes, R. K.; Leclerc, G.; Magnus, P. D.; Menzies, I. D. *J. Chem. Soc. Perkin I* **1975,** 2055.
9. Crawford, R. J.; Takagi, K. *J. Am. Chem. Soc.* **1972,** *94,* 7406.
10. Garner, A. W.; Timberlake, J. W.; Engel, P. S.; Melaugh, R. A. *J. Am. Chem. Soc.* **1975,** *97,* 7377.
11. Leffler, J. E.; Barbato, L. M. *J. Am. Chem. Soc.* **1955,** *77,* 1690.
12. Engel, P. S.; Wu, W.-X. *J. Am. Chem. Soc.* **1989,** *111,* 1830.
13. Porter, N. A.; Dubay, G. R.; Green, J. G. *J. Am. Chem. Soc.* **1978,** *100,* 920.

14. Tsolis, A.; Mylonakis, S. G.; Nieh, M. T.; Seltzer, S. *J. Am. Chem. Soc.* **1972,** *94,* 829.
15. Porter, N. A.; Marnett, L. J. *J. Am. Chem. Soc.* **1973,** *95,* 4361.
16. Simpson, C. J. S. M.; Wilson, G. J.; Adam, W. *J. Am. Chem. Soc.* **1991,** *113,* 4728.
17. Mill, T.; Stringham, R. S. *Tetrahedron Lett.* **1969,** 1853.
18. Engel, P. S.; Melaugh, R. A.; Page, M. A.; Szilagyi, S.; Timberlake, J. W. *J. Am. Chem. Soc.* **1976,** *98,* 1971.
19. Engel, P. S.; Chen, Y.; Wang, C. *J. Am. Chem. Soc.* **1991,** *113,* 4355.
20. Meier, H.; Zeller, K.-P. *Angew. Chem. Int. Ed.* **1977,** *16,* 815.
21. Engel, P. S.; Chae, W.-K.; Baughman, S. A.; Marschke, G. E.; Lewis, E. S.; Timberlake, J. W.; Luedtke, A. E. *J. Am. Chem. Soc.* **1983,** *105,* 5030.
22. Bartlett, P. D.; Porter, N. A. *J. Am. Chem. Soc.* **1968,** *90,* 5317.
23. Hilborn, J. W.; Pincock, J. A. *J. Am. Chem. Soc.* **1991,** *113,* 2683.
24. Kolbe, H. *Liebigs Ann. Chem.* **1849,** *69,* 257.
25. Zavitsas, A. A.; Seltzer, S. *J. Am. Chem. Soc.* **1964,** *86,* 1265.
26. Zavitsas, A. A.; Ehrenson, S. *J. Am. Chem. Soc.* **1965,** *87,* 2841.
27. Tomkinson, D. M.; Pritchard, H. O. *J. Phys. Chem.* **1966,** *70,* 1579.
28. Houle, F. A.; Beauchamp, J. L. *J. Am. Chem. Soc.* **1979,** *101,* 4067.
29. Neuman, R. C., Jr.; Frink, M. E. *J. Org. Chem.* **1983,** *48,* 2430.
30. Maillard, B.; Ingold, K. U.; Scaiano, J. C. *J. Am. Chem. Soc.* **1983,** *105,* 5095.
31. Chen, H. E.; Vaish, S. P.; Cocivera, M. *J. Am. Chem. Soc.* **1973,** *95,* 7586.
32. Vaish, S. P.; McAlpine, R. D.; Cocivera, M. *J. Am. Chem. Soc.* **1974,** *96,* 1683.
33. Fischer, H.; Paul, H. *Acc. Chem. Res.* **1987,** *20,* 200.
34. Price, S. J. W.; Trotman-Dickenson, A. F. *Trans. Faraday Soc.* **1957,** *53,* 1208.
35. Weston, R. E., Jr.; Seltzer, S. *J. Phys. Chem.* **1962,** *66,* 2192.
36. Russell, G. A.; Khanna, R. K. *J. Am. Chem. Soc.* **1985,** *107,* 1450.
37. Paneth, F.; Hofeditz, W. *Ber. Deutsch. Chem. Ges.* **1929,** *62,* 1335.
38. Chatgilialoglu, C.; Ingold, K. U.; Scaiano, J. C. *J. Am. Chem. Soc.* **1981,** *103,* 7739.
39. Burkhart, R. D. *J. Am. Chem. Soc.* **1968,** *90,* 273.
40. Rüchardt, C.; Weiner, S. *Tetrahedron Lett.* **1979,** 1311.
41. Lomas, J. S.; Dubois, J.-E. *J. Org. Chem.* **1982,** *47,* 4505.
42. Lomas, J. S.; Dubois, J.-E. *J. Org. Chem.* **1974,** *39,* 1776.
43. DeTar, D. F.; Tenpas, C. J. *J. Am. Chem. Soc.* **1976,** *98,* 4567.
44. Langley, D. R.; Doyle, T. W.; Beveridge, D. L. *J. Am. Chem. Soc.* **1991,** *113,* 4395.
45. Mendenhall, G. D.; Ingold, K. U. *J. Am. Chem. Soc.* **1973,** *95,* 3422.
46. Mendenhall, G. D.; Griller, D.; Lindsay, D.; Tidwell, T. T.; Ingold, K. U. *J. Am. Chem. Soc.* **1974,** *96,* 2441.

9

SELECTED ALKYL RADICALS

9.1 THE METHYL RADICAL

The methyl radical deserves special attention because it is the only member of the family of primary alkyl radicals with only hydrogen as a substituent. For typical primary alkyl radicals, reducing the chain length by one carbon is a minor perturbation, but the differences between methyl and ethyl are major ones.

9.1.1 Structure

The ESR [1] spectrum, the infrared spectrum [2], the strong absorption band at 216.0 nm [3], the photoelectron spectrum [4], and calculations (with configuration interaction [5,6]), all demand that $CH_3\cdot$ be essentially planar and that the umbrella vibration about the planar conformation have a single minimum. Of the charged species, CH_3^+ is also planar and CH_3^- is pyramidal [7].

The infrared-active "umbrella" vibration [6] that distorts the methyl radical into a symmetrical pyramid has a frequency of 617 cm^{-1}, which corresponds to a zero-point vibrational energy of about 0.88 kcal.

The vibrational energy needed to distort $CH_3\cdot$ all the way to a tetrahedral configuration is estimated to be about 10 kcal/mole [8], but the reactions of methyl with other radicals to give tetrahedral products are nevertheless fast and diffusion controlled. The energy needed to distort the radical is more than offset by the energy supplied by the simultaneous formation of a new bond.

The decision between single and double potential energy minima, the latter corresponding to a pair of shallow mirror-image pyramids, rests on a delicate balance of hybridization and nonbonded interaction energies. Even for radicals in which the minima are more distinct, the barrier between the two conformations is likely to be small, and chiral radicals long-lived enough to be freely diffusing always racemize before reaction occurs. If the barrier is less than the zero-point vibrational energy of the pyramidal inversion mode, the radical may be called quasi-planar, since it can never be assigned either of the pyramidal conformations within the limits of quantum mechanical uncertainty.

In contrast to $CH_3\cdot$, $CF_3\cdot$ has two shallow potential energy minima on both sides of the planar conformation. On the basis of the $^{13}CF_3\cdot$ ESR spectrum, the odd electron is in an orbital with about 23% carbon 2*s* character, the bond angle is about 111°, and the distortion from planarity is about 18° [9].

9.1.2 Ionization Potential and Electron Affinity

The ionization potential of the methyl radical is about 9.84 eV [4,10] or 232 kcal mole^{-1}, much higher than those of other alkyl radicals. For example, the ionization potential of the ethyl radical is about 8.39 eV or 193 kcal mole^{-1}. As a result, a radical pair $CH_3\cdot \cdot X$ is very unlikely to undergo electron transfer to form the ion pair $CH_3^+ \ ^-X$. This continues to be the case even when the electrostatic potential energy is reduced by the close proximity of the two ions, and still further reduced by a high dielectric constant solvent. The high ionization potential of $CH_3\cdot$ is probably why the thermolysis of acetyl peroxide, unlike that of other diacyl peroxides, appears to be a purely radical process [11] unaccompanied by polar reactions such as the carboxyinversion reaction [12,13].

The electron affinity of $CH_3\cdot$ is 1.8 ± 0.7 kcal mole^{-1} [14].

9.1.3 Reactions of $CH_3\cdot$

Methyl radicals give most of the reactions shown by other radicals except for disproportionation, which would give methane and methylene. Many other abstraction reactions do occur however, and the salient features are discussed in the next section.

Abstraction of H from C–H Bonds. Since the methyl radical is a relatively high-energy species, the transition state for the abstraction reactions tends to resemble methyl plus substrate rather than products. Despite this, the reactions show some selectivity for the formation of the more stable product radical, although not a great deal. As is the case for most radicals, the selectivity is governed mainly by the exothermicity of the reaction, abstraction of tertiary hydrogen being preferred over secondary, and secondary over tertiary. A plot of the energies of activation [15] for the abstraction of a series of alkane hydrogens against the strengths of the C–H bonds gives a fairly linear Evans–Polanyi or rate-equilibrium relationship [16,17]. The slope is about 0.49.

The hydrogen abstraction reactions of $CF_3\cdot$ also obey an Evans–Polanyi relationship. The slope of that relationship is about the same as for methyl, but the $CF_3\cdot$ reactions are uniformly faster and the points fall on a separate line.

The selectivity of methyl in abstracting hydrogen from a wide variety of molecular environments has been studied by comparing the yield of CH_4 from reactions with various substrates RH (Eq. 9.1) with that of CH_3D from the competing reaction with *t*-butyl mercaptan (Eq. 9.2), a convenient standard [18].

$$CH_3\cdot + RH \longrightarrow CH_4 + R\cdot \tag{9.1}$$

$$CH_3\cdot + t\text{-Bu—SD} \longrightarrow CH_3D + t\text{-Bu—S}\cdot \tag{9.2}$$

The selectivity relative to the abstraction of deuterium from *t*-butyl deuteromercaptan, k_{RH}/k_{SD}, is the slope of the plot of $[CH_4]/[CH_3D]$ against [RH]/[*t*-BuSD]. The relative reactivities, per hydrogen atom, toward primary, secondary, and tertiary alkyl hydrogens are 1:4.3:46. These selectivities of methyl for hydrogen abstraction are about the same as those of H·, phenyl radical, and *n*-hexyl. Several radicals, F·, Cl· (unsolvated), are much less selective than methyl, while Br· and $CCl_3\cdot$ are a great deal more selective.

Selectivity for abstraction of hydrogen, per hydrogen, from cycloalkanes depends on the ring size. For the methyl radical the relative rates are cyclohexane 1, cyclopentane 1.2, cycloheptane 1.5, and cyclooctane 2.2. Although these differences are small, the trend is the same for a number of other abstracting radicals as well.

Relative selectivities of methyl for primary benzylic, secondary benzylic and tertiary benzylic hydrogens are 1:4.2:13 [18].

Polar substituents in a toluene substrate have only a very slight effect on the rate of hydrogen abstraction by methyl radicals, giving a Hammett ρ of about −0.1 [18–20].

Abstraction of H from (n-Butyl)$_3$SnH. Methyl radicals, generated by a sequence of reactions that amounts to an S_H2 displacement by *t*-butoxy from trimethylarsine, abstract hydrogen from the Sn–H bond of tributyltin hydride [21].

$$CH_3\cdot + n\text{-}Bu_3SnH \longrightarrow CH_4 + n\text{-}Bu_3Sn\cdot \tag{9.3}$$

The rate constant of reaction 9.3, 1.06 10^7 M^{-1} sec^{-1} at 300 K, was obtained from the rise in the 390-nm absorption of the organotin radical. This rate constant is about four times that for the reaction of other alkyl radicals with tributyltin hydride.

Addition Reactions. Methyl radicals will add to either end of an unsymmetrical carbon–carbon double bond, but favor the end that gives the more stable radical, bonding, for example, to the beta carbon of acrylonitrile or vinyl methyl ketone [22]. Addition of methyl radical to allene is an exception to the usual preference for generating the more stable radical [23], since the product is the vinyl radical (Eq. 9.4) rather than the allylic radical (Eq. 9.5).

$$CH_3\cdot + CH_2{=}C{=}CH_2 \longrightarrow CH_2{=}C\cdot{-}CH_2CH_3 \qquad (9.4)$$

$$CH_3\cdot + CH_2{=}C{=}CH_2 \longrightarrow CH_2{=}C(CH_3){-}CH_2\cdot \qquad (9.5)$$

A probable explanation [23] for the formation of the less stable radical is that twisting of the molecule into the allylic conformation [24] is not far enough advanced at the transition state.

Although effects of polar substituents in the substrate are unimportant in the abstraction of hydrogen by methyl radicals, they are appreciable in the addition to alkenes. In addition reactions, methyl is nucleophilic. For example, the reaction, in the gas phase at 180°C, is about 11 times as fast with tetrafluoroethylene as with ethylene [25–27].

Another indication of nucleophilicity is the positive linear correlation of the logarithms of the addition rates of the methyl radical (free energies of activation) with the ionization potentials of the alkenes [28]. The alkenes with the higher ionization potentials are those in which the substituents are more electron-withdrawing, which makes the alkene more reactive toward nucleophiles.

The trifluoromethyl radical, in contrast to methyl, is distinctly electrophilic, and its addition rates show the opposite (negative) correlation with the ionization potentials of the substrate alkenes [29].

The relative addition rates of $CCl_3\cdot$ to ring-substituted 3-phenyl-1-propenes and 4-phenyl-1-butenes are correlated with the Hammett sigma values of the substituents. Although the addition of an electrophilic radical like $CCl_3\cdot$ is expected to be accelerated by electron-releasing substituents in the substrate, the observed rho values are too large for substituents so far away. A probable explanation is that the substituents are stabilizing an intermediate pi complex between the $CCl_3\cdot$ radical and the benzene ring; the addition to the double bond is an intramolecular reaction of that intermediate (Eq. 9.6) [30].

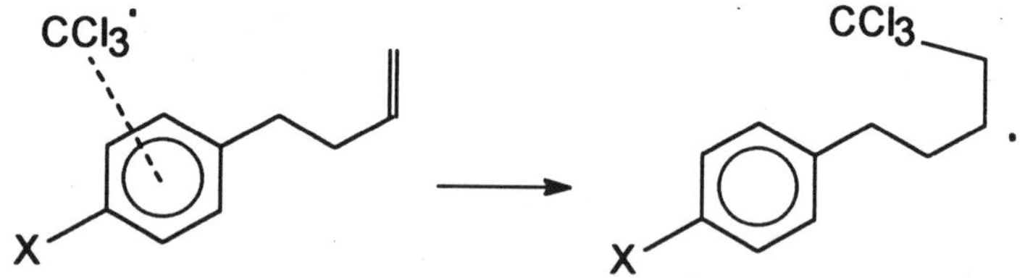

(9.6)

With some alkenes, addition and hydrogen abstraction reactions of the methyl radical occur at rates that are close enough for both reactions to be important at the same time. On a single reactivity scale [18,31] in which the rate of abstraction of primary aliphatic hydrogen is 1, the rate of abstraction of a primary allylic hydrogen (as in 2-butene) is 9, and the rate of addition to a symmetrically dialkyl substituted cis double bond is 59. Using the appropriate statistical corrections for six equivalent primary hydrogens and two equivalent double bond ends, the ratio of addition to abstraction is 2.2.

Reaction with Aromatic Compounds. Methyl radicals add to aromatic compounds to form substituted cyclohexadienyl radicals, and the subsequent removal of hydrogen, by another radical or an oxidant, gives the methyl-substituted arene. In the methylation of a monosubstituted benzene, substituents such as methoxy and chloro, that are o,p-directing in ionic electrophilic aromatic substitution, give a higher proportion of meta to para methyl derivative, again illustrating the nucleophilic character of the methyl radical [32–35].

A somewhat dubious quantity called the *methyl affinity* is derived from the decomposition of acetyl peroxide in isooctane solution in the presence of the aromatic substrate [36]. In isooctane, without the added aromatic substrate, most of the acetoxy radicals from the acetyl peroxide decomposition lose CO_2 and most of the methyl radicals react with the isooctane to give methane. Any difference between the CO_2 and methane yields is attributable to other reactions of the methyl radicals, for example, addition to the aromatic substrate. The methyl affinity is proportional to the decrease in the amount of methane formed in the presence of the aromatic substrate. One problem with the interpretation of the methyl affinities is the possibility that some of the methane is generated by the reaction of methyl with the intermediate cyclohexadienyl radical or with the methyl group of the product [37] or with a hydrogen-containing substituent such as methoxy. Despite these uncertainties in some of the methyl affinities, they are found to parallel the reactivities of the substrates toward phenyl radicals. The methyl affinities of a series of polynuclear aromatic hydrocarbons are correlated with the relative SCF–MO localization energies required to localize an electron at the reactive site [38].

9.2 THE ETHYL RADICAL

9.2.1 Structure

The infrared spectrum of the ethyl radical isolated in an argon matrix has bands at 3112 and 3033 (methylene stretching), 540 (umbrella bending), 2987, and 2920 (other methylene bending modes), 2853 and 2842 (methyl stretching), 1440, 1366, 1175, and 1138 cm^{-1} (methyl bending) [39]. Analysis of this spectrum and ab initio calculations [6,40] show that the bonds at the radical center are not quite planar, with an angle between the C–C bond and the plane defined by the other two bonds being about 11.9°.

The only symmetry element is a plane that bisects the CH_2 group and contains the C–C bond. A β C–H bond is eclipsed with the radical center, but the barrier between the equivalent eclipsed conformations is very small, only about 150 cal, in agreement with the ESR spectrum. The eclipsed β C–H bond is slightly longer than the other two, and the C–C bond is slightly shorter than normal. The lengths of the α C–H bonds are like those in ethylene.

The half-occupied molecular orbital resembles a *p* orbital on the CH_2 carbon, but has a large contribution from the β C–H bond that is eclipsed with the CH_2 *p* orbital, a consequence of hyperconjugation.

9.2.2 Formation of the Ethyl Radical

The ethyl radical is formed by the β cleavage of the propionyloxy radical, which can be made by the thermolysis or photolysis of propionyl peroxide [39,41,42] or *t*-butyl perpropionate [43,44] and by the photolysis of 1-naphthylmethyl propionate [45] (Eq. 9.7).

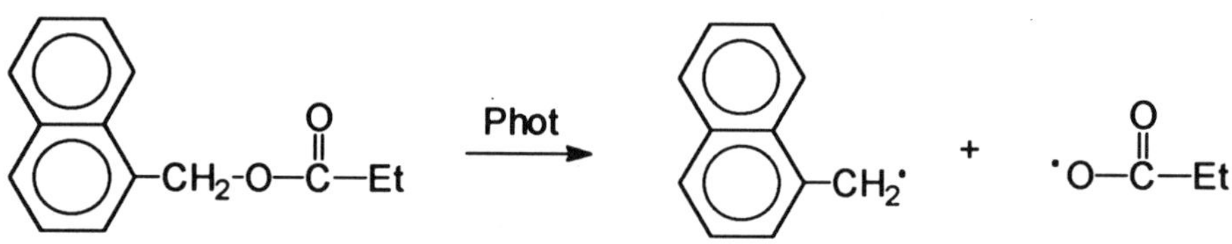

(9.7)

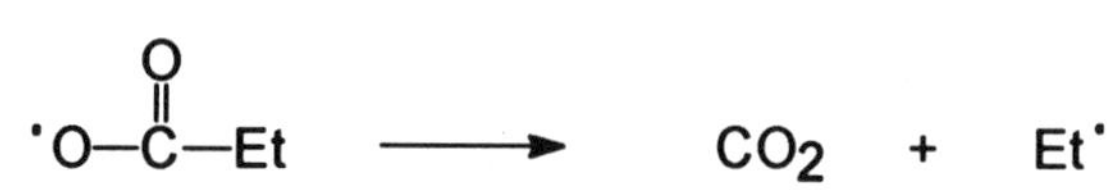

(9.8)

The decarboxylation of propionyloxy (Eq. 9.8) appears to be considerably faster than that of acetoxy [45].

Other β-cleavage reactions giving ethyl radical are the loss of nitrogen after photolysis of azoethane [46], the β cleavage of 1,1-dimethylpropyloxy [47] to ethyl radical and acetone, or propyloxy [4] to ethyl and formaldehyde, and the photolysis of propionaldehyde [48].

The photolysis of propionyl peroxide in an argon matrix gives ethyl radicals and CO_2, as expected. Rather surprisingly, further photolysis of the ethyl radicals is reported to give propionic acid [49]. The ethyl radical, in an unidentified excited electronic state, may be able to bond to CO_2, although the decarboxylation of propionyloxy radicals to give CO_2 and ground state ethyl radical is exothermic and quite fast.

9.2.3 Disproportionation and Combination of Ethyl Radicals

The total second-order rate constant for the disappearance of $CH_3CH_2\cdot$ in cyclohexane at 25°C is estimated to be about 6.4×10^9 M^{-1} sec^{-1}, which approximates the estimated rate of formation of encounter complexes by diffusion multiplied by the spin statistical factor of $\frac{1}{4}$ [50]. For the gas phase at room temperature the disproportionation rate constant is reported to be 1.37×10^9 and the recombination rate constant is $9.7 \times 10^9\ M^{-1}\ sec^{-1}$ [51].

Disproportionation is able to compete with other hydrogen abstraction reactions because the bond dissociation energy for the β C–H bond in the ethyl radical is only about 39 kcal/mole [52] as compared to 98 for the C–H bond in ethane.

$$2\ CH_3CH_2\cdot \longrightarrow CH_3CH_3 + CH_2{=}CH_2 \tag{9.9}$$

In solution the partitioning between disproportionation and combination reactions occurs after the encounter complex is formed. However, the ratio k_{dis}/k_{comb} depends on the medium nevertheless [46,53]. In the gas phase at room temperature k_{dis}/k_{comb} is about 0.14; in solvents it ranges from 0.16 for isooctane, the solvent with the lowest viscosity and the lowest internal pressure, to 0.24 for ethylene glycol, the solvent with the highest viscosity and highest internal pressure [46]. Applying hydrostatic pressure also increases k_d/k_c [54], but since external pressure also increases the viscosity, the attribution of the effect on k_d/k_c to a more compact transition state for the disproportionation is still uncertain.

9.2.4 Spin Trapping

Ethyl radicals generated in water by the reaction of the hydroxyl radical with diethyl sulfoxide are readily trapped by 2-methyl-2-nitrosopropane to give the adduct radical shown in Eq. 9.10 [55].

$$CH_3CH_2\cdot + t\text{-Bu}{-}N{=}O \longrightarrow t\text{-Bu}{-}\underset{\displaystyle CH_3CH_2}{\overset{}{N}}{-}O\cdot \tag{9.10}$$

The rate constant for the trapping of the ethyl radical, at 18°C, is $5.3 \times 10^7\ M^{-1}\ sec^{-1}$. The nitroso spin trap is electrophilic, which accounts for the much lower rate constant for methyl, $1.7 \times 10^7\ M^{-1}\ sec^{-1}$. The value for isopropyl, $4.6 \times 10^7\ M^{-1}\ sec^{-1}$, represents a balancing of electronic and steric effects.

9.3 OTHER PRIMARY ALKYL RADICALS

9.3.1 Unbranched Chains

n-Butyl and *n*-pentyl radicals have been made by photolyzing the corresponding diacyl peroxides in an argon matrix [56]. The infrared spectra can be separated into two parts: one is characteristic of the terminal $CH_2CH_2\cdot$; the other part differs negligibly from the spectrum of the corresponding alkane. This means that the structural effect of the terminal radical center is negligible beyond the β CH_2 group. The bands characteristic of *n*-alkyl radicals are at approximately 3105 and 3025 cm^{-1}, attributed to the α CH stretch modes, and at approximately 520 cm^{-1} attributed to the umbrella mode. The effect of the terminal $CH_2\cdot$ in lowering the frequency for the β CH stretch is consistent with a lower strength for those bonds and the ready occurrence of disproportionation.

n-Pentyl radicals are formed in the gas phase by the photolysis of *n*-pentylazomethane at 25°C [57]. The ratio of disproportionation to combination of *n*-pentyl radicals is about 0.14, not very different from that for the ethyl radical.

Some of the *n*-pentyl radicals undergo a unimolecular isomerization into 2-pentyl radicals, not by the forbidden 1,2 sigmatropic rearrangement, but by a 1,4 hydrogen atom transfer. The five-membered cyclic transition state leads to the breaking of a secondary C–H bond and the formation of a primary C–H bond. The extent (0.5–8%) to which this reaction occurs depends on how much excess vibrational energy the radicals have initially (i.e., on the wavelength of the light used in the photolysis) and how long they are able to retain that energy before it is dissipated by collisions. Another reaction of the pentyl radical in the gas phase is β cleavage to ethylene and a propyl radical (0.5–2.5%). Both of these reactions become less important as the pressure is increased.

n-Pentyl radicals have been made in solution by the thiol plus phosphite chain reaction (Eqs. 9.11 and 9.12, R = *n*-pentyl) [58].

$$R\cdot + RSH \longrightarrow RH + RS\cdot \qquad (9.11)$$

$$RS\cdot + P(OEt)_3 \longrightarrow R\cdot + SP(OEt)_3 \qquad (9.12)$$

The rate constant for the disappearance of the pentyl radical, combination plus disproportionation, is in the diffusion-controlled range, about 10^9 M^{-1} sec^{-1} in benzene at 25°C.

n-Hexyl radicals generated by the thermolysis of *n*-heptanoyl peroxide in CCl_4 are vibrationally unexcited because of collisions with the solvent. They undergo 1,5 internal hydrogen transfer either not at all or to an extent less than 3% [59]. The ratio of disproportionation to recombination is 0.15, similar to that for the gas phase ratios for ethyl and propyl radicals.

A small amount of rearrangement by abstraction of hydrogen from the 3 position has been observed for propyl radicals [60]. Thermolysis in CCl_4 of dibutyryl peroxide labeled with ^{13}C in the alpha position gave about 4% of the *n*-propyl chloride with the label in the 3 position as well as the normal product.

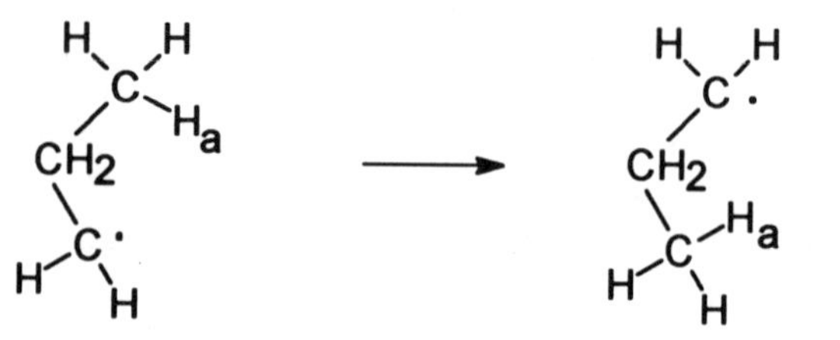

(9.13)

Primary alkyl radicals should be more nucleophilic than methyl because of the inductive effect of the alkyl substituent. Thus the abstraction of hydrogen from substituted toluenes by undecyl radicals, in solution at about 80°C, has a Hammett ρ of $+0.47$ [61] or 0.50 [62], in contrast to a value near -0.1 for the methyl radical [20]. The slightly negative value for methyl probably reflects the importance of the substituent effect on the bond dissociation energy; the positive values for radicals like undecyl and *t*-butyl are due to polar effects on the transition state.

9.3.2 The 5-hexenyl Radical

The 5-hexenyl radical, made by the photolysis of 6-heptenoyl peroxide, cyclizes to cyclopentylmethyl (Eq. 9.14) [1].

(9.14)

The rate constant for reaction 9.14 has been estimated as 2.56×10^{10} sec^{-1} at 298 K by competition with abstraction of H from tributyltin hydride, a reaction whose rate can be measured directly [21]. The temperature dependence is given by Eq. 9.15.

$$\log k(\text{sec}^{-1}) = 10.42 - 6.85(\text{kcal mole}^{-1})/RT \qquad (9.15)$$

The rate constant for the cyclization is an important quantity because the rates of competing reactions can be calculated from it using product ratios. For example, when 6-heptenoyl peroxide is decomposed in a solution of styrene in acetic acid, with Cu^{2+} present to prevent polymerization, there are two products, one incorporating the 5-hexenyl structure and the other incorporating the cyclopentylmethyl structure.

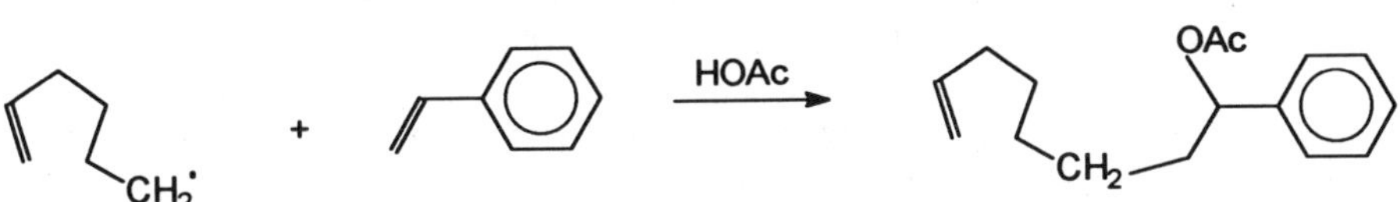

(9.16)

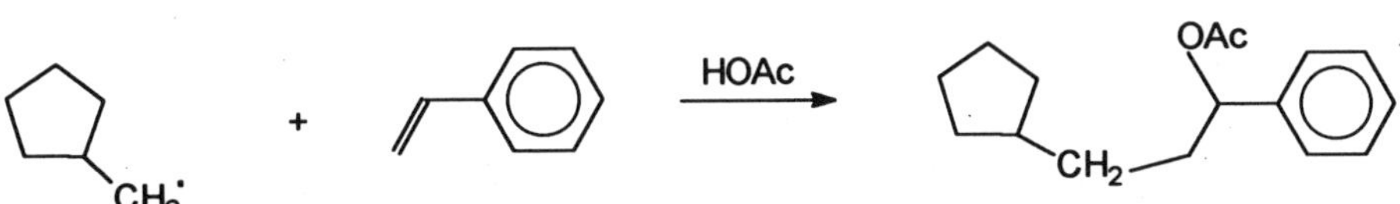

(9.17)

From the yields of the two products, the rate of reaction 9.14, and the rate of oxidation of the cyclopentylmethyl radical by Cu^{2+}, it was possible to estimate the rate of the addition of the 5-hexenyl radical to styrene, about $5.4 \times 10^4\ M^{-1}\ sec^{-1}$ at 25°C [63].

9.3.3 ω-Phenyl Substituents

n-Alkyl radicals with phenyl substituents at the ω position of a chain of 2, 3, or 4 methylenes have been made by thermolysis of diacyl peroxides [64]. In all cases the ^{13}C CIDNP spectra of the products of the radical part of the reaction are consistent only with the open chain structure and not with reaction by way of a radical that has spent any time in the spiro structure. The spiro[2.5]octadienyl radical, made by hydrogen abstraction from one of the corresponding dienes, opens (Eq. 9.18) to give the β-phenylethyl radical within nanoseconds [65]. The reverse process, which would have lead to scrambling of a terminal CD_2 label, does occur in the rearrangement of the neophyl radical (Eq. 9.19), where the creation of a tertiary radical gives the reaction additional driving force.

(9.18)

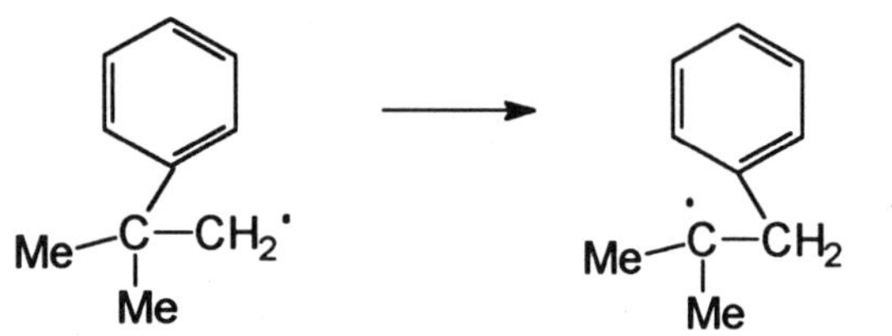

(9.19)

9.3.4 Cycloalkylmethyl Radicals

Cyclopropylmethyl. The cyclopropylmethyl radical can be generated by the attack of *t*-butoxy radicals on methylcyclopropane. As the temperature is allowed to rise above −100°C, the ESR spectrum of the cyclopropylcarbinyl radical is replaced by that of the allylcarbinyl radical because relief of strain in the three-membered ring favors the β-cleavage reaction (Eq. 9.20). The observation of two discrete ESR spectra removes the possibility that either radical has a nonclassical or rapidly equilibrating structure [66]. Line broadening in the low-temperature ESR spectrum indicates a barrier to rotation about the bond between the cyclopropane ring and the $CH_2\cdot$ group [67].

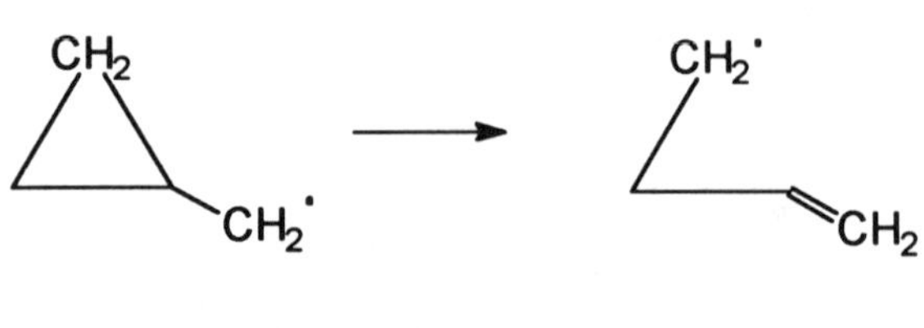

(9.20)

Cyclohexylmethyl Radicals. In the ESR spectra of cyclohexylmethyl radicals the hyperfine splitting factor of the adjacent ring hydrogen (H_β) is always about 30 G when the CH_2· is in the equatorial position, but is about 41 G when the CH_2· is axial [68]. At temperatures below 180 K unsubstituted cyclohexylmethyl shows only the equatorial conformer, **I**, with an H_β hyperfine splitting of 30.4 [69]. At 182 K a second minor conformer, **II**, with an axial CH_2· and a 41.2 G splitting can be seen. Very similar H_β splittings are seen when the CH_2· group is locked in a particular conformation by a cis or trans 4-*t*-butyl substituent or by being part of an adamantane structure.

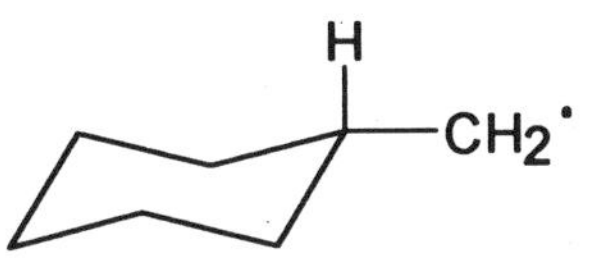

30.4 G

I

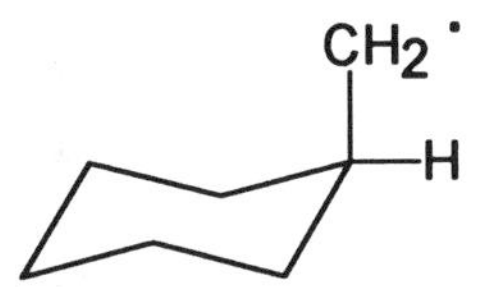

41.2 G

II

In both the equatorial and the axial conformation the preferred structure has the half-occupied orbital on C_α eclipsing the β H. The reason for the much larger hyperfine splitting constant in the axial radicals is that steric hindrance to rotation about the C_1–C_α bond holds the CH_2· more firmly in the eclipsed position.

9.4 ISOPROPYL RADICAL

9.4.1 Structure

The isopropyl radical, as indicated both by ab initio calculations and an analysis of its infrared spectrum, has a plane of symmetry containing the lone α-CH group with the two methyl groups on either side of it. Two β C–H bonds, one from each methyl group, are eclipsed with the *p* orbital on the central carbon and contribute substantially to the half-occupied molecular orbital. The barrier to rotation about the CH_3–C bonds is small, however, so that all six β hydrogens are equivalent in the ESR spectrum. That spectrum, in cyclopentane at −105°C, shows two splittings, a 22.06-G doublet due to coupling with the alpha proton and a 24.74-G septet from the beta protons [1]. The radical center is pyramidal, deviating 18.6° from planarity [6].

The infrared spectrum has an olefin-like band at 3058 cm^{-1} assigned to the α C–H stretch and a band at 369 cm^{-1} due to the out-of-plane deformation at the radical center [70].

The adiabatic ionization potential is 7.36 eV or 170 kcal $mole^{-1}$ [4]. The ultraviolet absorption spectrum has a lambda max at 220 nm [71].

9.4.2 Formation

The isopropyl radical can be produced in solution by the photolysis of azoisopropane [72], diacyl peroxides [73] or peresters [44]. In the gas phase it is more conveniently formed by photolysis of isobutyl nitrite [4] or 2,4-dimethyl-2-butyl hypochlorite [47] with a subsequent β cleavage (Eqs. 9.21–9.24, R = isopropyl).

$$RCH_2ONO \longrightarrow RCH_2O\cdot + \cdot NO \qquad (9.21)$$

$$RCH_2O\cdot \longrightarrow R\cdot + CH_2{=}O \qquad (9.22)$$

$$R(CH_3)_2COCl \longrightarrow R(CH_3)_2CO\cdot + \cdot Cl \qquad (9.23)$$

$$R(CH_3)_2CO\cdot \longrightarrow R\cdot + (CH_3)_2C{=}O \qquad (9.24)$$

The thermolysis of isobutyryl peroxide in hexachloroacetone at 80°C, although a considerable part of the peroxide is likely to be lost in a competing heterolytic reaction, does give some isopropyl radicals. This is shown by CIDNP signals from isopropyl chloride (*A*/*E* methyl doublet, *A*/*E* methine septet), as well as *E*/*A* signals from the disproportionation products propane and propene [42].

9.4.3 Reactions of Isopropyl

When azoisopropane is photolyzed in a mixture of a substituted toluene and a deuterated thiol, propane and 2-deuteropropane are formed by reactions 9.25–9.27 (R = isopropyl, R′ = phenyl or *t*-butyl) [72].

$$R\cdot + XC_6H_4CH_3 \longrightarrow RH + XC_6H_4CH_2\cdot \qquad (9.25)$$

$$R\cdot + R'SD \longrightarrow RD + R'S\cdot \qquad (9.26)$$

$$R\cdot + R'SH \longrightarrow RH + R'S\cdot \qquad (9.27)$$

From the dependence of the RH/RD yield ratio on the concentrations of the substituted toluene, the deuterated thiol, and the undeuterated thiol, it was possible to obtain rate constants k_x relative to the rate of abstraction of hydrogen from unsubstituted toluene. A plot of these against the Hammett sigma values for the substituents (X) gave a rho value of 0.8 ± 0.1 at 30°C. Thus the isopropyl radical is considerably more nucleophilic than methyl or primary alkyl radicals.

The rate of disappearance by combination and disproportionation, measured by means of the transient ultraviolet absorption, is equal within a few percent to $\frac{1}{4}$ of the theoretical encounter rate [71].

9.5 THE *t*-BUTYL RADICAL

9.5.1 Structure

The *t*-butyl radical has C_{3v} symmetry and the deviation from planarity of the bonds at the central carbon is 24.1° [6,74]. The half-occupied molecular orbital resembles those of the ethyl and isopropyl radicals except that three β C–H bonds are eclipsed with the radical center and contribute to the hyperconjugation.

The ESR spectrum in propane solution has a temperature-dependent hyperfine splitting factor for the central ^{13}C atom [75]. Rather than increasing steadily with temperature as would be expected if the radical were planar, it goes through a minimum at about 220 K as expected for a nonplanar radical with a barrier to inversion of about half a kilocalorie.

The adiabatic ionization potential of the *t*-butyl radical is 6.70 eV, or 154 kcal mole^{-1} [4]. The ultraviolet absorption spectrum has peaks at 220 and 307 nm [71].

9.5.2 Formation

The *t*-butyl radical has been made by the many of the standard reactions used for other alkyl radicals such as the photolysis of azoisobutane (Eq. 9.28) [76,77], the photolysis or thermolysis of *t*-butyl perpivalate (Eq. 9.29) [78–80], the photolysis of *t*-butyl ketones [81,82], and the photolysis of *t*-butylmercuric chloride [83].

$$t\text{-Bu—N=N—}t\text{-Bu} \longrightarrow t\text{-Bu}\cdot + N_2 + t\text{-Bu}\cdot \qquad (9.28)$$

$$t\text{-Bu—O—O—}\overset{\overset{\displaystyle O}{\|}}{C}\text{—}t\text{-Bu} \longrightarrow t\text{-BuO}\cdot + CO_2 + t\text{-Bu}\cdot \qquad (9.29)$$

Preparation by photolysis of dipivaloyl peroxide is possible but not practical, because this peroxide thermolyzes very rapidly even at 10°C, wasting most of the peroxide in a largely ionic reaction while it is being photolyzed [84].

Thermolysis of the peroxide of Eq. 9.30 at 130°C gives alkoxy radicals which then give *t*-butyl by β cleavage [42].

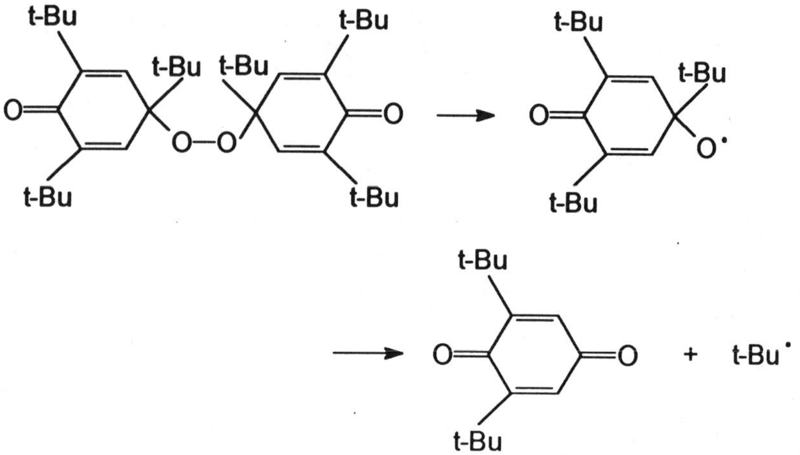

(9.30)

The products in thiophenol included isobutane, which had a CIDNP methyl doublet in emission.

Another source of *t*-butyl radicals is the pyrolysis of highly strained compounds such as tri-*t*-butylmethane [85] or tri-*t*-butylcarbinol [86,87], discussed in Chapter 8.

9.5.3 Reactions

Self-Termination The rate of disappearance of *t*-butyl radicals in various solvents and at various temperatures has been studied by ESR and found to be within a few percent of the encounter rate multiplied by the spin-statistical factor of $\frac{1}{4}$ [71]. To avoid complications due to side reactions the radicals were generated by the photolysis of di-*t*-butyl ketone, a particularly clean reaction. To avoid the inadequacies of current theoretical treatments of diffusion, the encounter rates were estimated using the diffusion of isobutane and butane as a model.

The self-termination reaction within the encounter complex consists of two competing processes, disproportionation and combination. In *n*-octane at 25°C k_d/k_c is about 5.3. In a series of *n*-alkanes from C_8 to C_{16} the k_d/k_c ratio increases with increasing chain length (increasing viscosity) of the solvent and decreases with increasing temperature (decreasing viscosity) [88]. It has been noted, however, that internal pressure varies the same way for these solvents [89]. In an attempt at resolving this ambiguity the k_d/k_c ratio was measured in *n*-octane at various pressures [81]. The k_d/k_c ratio and the viscosity of the *n*-octane both increase with the applied external pressure, but the correlation with viscosity is somewhat better than the correlation with pressure. If viscosity is the important parameter, then rotational motions of the *t*-butyl radicals that favor k_c are more interfered with by the viscous solvent than those that favor k_d. If pressure is the important parameter, then the transition state for disproportionation is more compact and occupies a smaller volume than that for combination.

9.6 HIGHLY HINDERED RADICALS

The di-*t*-butylmethyl radical can be made by photolyzing *t*-butyl peroxide in the presence of di-*t*-butylmethane (Eq. 9.31) and also by photolyzing hexamethylditin in the presence of di-*t*-butylmethyl chloride (Eq. 9.32) [90,91].

$$t\text{-BuO}\cdot + \text{Me}_3\text{C—CH}_2\text{—CMe}_3 \longrightarrow t\text{-BuOH} + \text{Me}_3\text{C—C}\cdot\text{H—CMe}_3 \qquad (9.31)$$

$$\text{Me}_3\text{Sn}\cdot + \text{Me}_3\text{C—CHCl—CMe}_3 \longrightarrow \text{Me}_3\text{SnCl} + \text{Me}_3\text{C—C}\cdot\text{H—CMe}_3 \qquad (9.32)$$

The *g* value of this radical is 2.0027, the hyperfine splitting factors are 21.89 G for the alpha hydrogen and 0.68 G for the methyl hydrogens. The radical decay follows a first-order rate law with a half-life of 58 sec at 25°C, and may involve β elimination of a methyl radical.

The 2,2,4,4,6,6-hexamethylcyclohexyl radical has been prepared by the photolysis of di-*t*-butyl peroxide in the presence of the corresponding hydrocarbon [91]. Like the di-*t*-butylmethyl radical, its bimolecular reactions are sterically hindered. Decay of the radical follows first-order kinetics with a 990-sec half-life at 25°C.

The tri-*t*-butylmethyl radical can be generated by the photolysis of di-*t*-butyl peroxide in the presence of tri-*t*-butylmethane [91]. It has a half-life of about 2 min at 25°C.

9.7 TRENDS WITH INCREASED BRANCHING: THE SERIES METHYL TO *t*-BUTYL

In this section we examine the trends in physical and chemical properties of a series of radicals consisting of the methyl radical and the prototypical primary, secondary, and tertiary radicals ethyl, isopropyl, and *t*-butyl.

9.7.1 Spin Densities

A steady decrease in the ESR beta coupling constants (Table 9.1) in the series ethyl, isopropyl, *t*-butyl [92] means that there is a corresponding decrease in the spin density on the central carbon.

TABLE 9.1 Effects of Branching in Alkyl Radicals

Radical	$a(\beta H)$ G	Ionization Potential	k_{Cl}/k_{Br}	k_{SnH}	k_d/k_c[a]	$k_d/_c$[b]
Me	—	9.84	—	10.6	—	—
Et	26.87	8.39	19.7	2.30	0.32	0.14
i-Pr	24.68	7.36	8.03	1.50	2.4	1.2
t-Bu	22.72	6.70	4.83	1.87	19	7.2

[a]For R· and $(CH_3)_3CO$·.
[b]For two R·.

9.7.2 Ionization Potentials

Adiabatic ionization potentials (Table 9.1) decrease abruptly from 9.84 eV for the methyl radical to 8.39 for ethyl, and then more gradually for isopropyl and *t*-butyl [4].

9.7.3 Cl/Br Abstraction Ratios

Alkyl radicals generated by the reaction sequence of Eqs. 9.33–9.34 in the presence of Cl_3CBr give two competing reactions: Eq. 9.33, which merely continues the photolysis chain reaction and gives RCl as a product, and Eq. 9.35, which gives RBr as a product [47]. The quantity of interest is the ratio of the rate constants for the two reactions, k_{Cl}/k_{Br}.

$$R\cdot + R(CH_3)_2COCl \longrightarrow RCl + R(CH_3)_2CO\cdot \qquad (9.33)$$

$$R(CH_3)_2CO\cdot \longrightarrow (CH_3)_2CO + R\cdot \qquad (9.34)$$

$$R\cdot + Cl_3CBr \longrightarrow RBr + Cl_3C\cdot \qquad (9.35)$$

For reaction at 0°C, k_{Cl}/k_{Br} decreases from 19.7 for ethyl, to 8.03 for isopropyl, to 4.83 for *t*-butyl (Table 9.1). The activation energy differences, which increase from 0.15 kcal mole^{-1} for ethyl to 0.92 for isopropyl and to 3.38 for *t*-butyl, actually favor bromine abstraction, but are more than offset by the less negative entropies of activation for chlorine abstraction.

9.7.4 Tin Hydride Rates

Another methyl through *t*-butyl sequence consists of the rates of the reactions with tri-*n*-butyltin hydride in isooctane (Eq. 9.36) [21].

$$R\cdot + (C_4H_9)_3SnH \longrightarrow RH + (C_4H_9)_3Sn\cdot \qquad (9.36)$$

The rate constants k_{SnH} (Table 9.1) are all for the reactions at 300 K and in units of M^{-1} sec^{-1} times 10^{-6}. The remarkable thing about this result is the fact that except for methyl, all of the radicals, including *n*-butyl (2.47) and cyclohexyl (2.1), are reacting at about the same rate at room temperature. The linear trend of the activation energies (except for methyl), as a function of the entropy of activation, is an example of the isokinetic relationship [93].

9.7.5 Disproportionation/Combination Ratios

The geminate alkyl radicals R· and alkoxy radicals $(CH_3)_3CO\cdot$ formed by photolysis of the peresters $RCO_2\text{-}OC(CH_3)_3$ in decalin at 30°C either combine to give the ethers $ROC(CH_3)_3$ or disproportionate to give the alcohol and alkene [44]. These k_d/k_c ratios and those for the alkyl with alkyl reactions [73] are given in Table 9.1. In each case the former ratios are about twice as large.

9.8 RADICAL THERMOCHEMISTRY

When a proposed reaction mechanism involves the homolysis of a bond, the strength of the bond being broken should be consistent with the observed reaction rate. Since the energy $\Delta H°(\text{A–B})$ for the dissociation of a bond A–B is rarely available, it is often derived [52] from tabulated heats of formation, ΔH_f°, as in Eq. 9.37.

$$\Delta H°(\text{A–B}) = \Delta H_f^\circ(\text{A}\cdot) + \Delta H_f^\circ(\text{B}\cdot) - \Delta H_f^\circ(\text{AB}) \qquad (9.37)$$

Although reliable values of $\Delta H_f^\circ(\text{AB})$ are easily available, either from heats of combustion or from group equivalent parameters [94], the same is not true of heats of formation of radicals.

9.8.1 Heats of Formation

Values of the heats of formation of alkyl radicals obtained in the 1970s or earlier arc now believed to be much too low [95–97].

Heats of formation of alkyl radicals have been derived from activation energies of gas phase radical chain reactions, usually iodination or bromination. In addition to the usual experimental difficulties, such as wall effects, there were problems with the interpretation of the results. For example, one of the reactions in the bromination or iodination of isobutane is the transfer of a hydrogen atom from the hydrogen halide to the *t*-butyl radical. These were assumed to be simple single-step processes (Eq. 9.38), with small positive activation energies near zero.

$$\text{HX} + (\text{CH}_3)_3\text{C}\cdot \longrightarrow (\text{CH}_3)_3\text{CH} + \text{X}\cdot \qquad (9.38)$$

However, more recent investigations [97–100] of these hydrogen transfer reactions have shown that they have substantial negative activation energies,

-1.5 ± 0.2 kcal mole^{-1} for the reaction with HI and -1.4 ± 0.2 kcal mole^{-1} for the reaction with HBr. The apparent negative activation energies for Eq. 9.38 are understandable if the atom transfer is actually a two-step process as in Eq. 9.39.

$$HX + (CH_3)_3C\cdot \rightleftharpoons (CH_3)_3C\cdot XH \longrightarrow (CH_3)_3CH + X\cdot \quad (9.39)$$

The use of the corrected activation energies gives the Seetula, Russell, and Gutman 1990 "best estimates" shown in Table 9.2 [97].

Heats of formation of ethyl, *i*-propyl, and *t*-butyl radicals are particularly important because these radicals can be used to make predictions about other primary, secondary, and tertiary radicals for which data are not available. For example, if the heat of formation of ethyl radical is +28.3 kcal mole^{-1} and the heat of formation of ethane is -20.2, the difference between the two is $+28 - (-20.2)$ or $+48.5$. Because of the cancellation of structural effects in an alkane RCH_2–H and the radical $RCH_2\cdot$, the difference should again be close to +48.5. The heat of formation of a primary radical $RCH_2\cdot$ can then be estimated by adding 48.5 to the easily obtained heat of formation of RCH_3. Similarly, the heats of formation of secondary and tertiary radicals can be estimated from the heats of formation of the corresponding alkane, propane, or isobutane and *i*-propyl or *t*-butyl radicals [96].

Reliable bond energies and heats of formation are very important in the investigation of reaction mechanisms. Combined with experimental enthalpies of activation, bond energies can be used to discriminate between plausible reaction mechanisms [95]. Inaccurate bond energies, on the other hand, can lead to the wrong and often bizarre conclusions. The use of the newer heats of formation has cleared up some of these errors [95,96,101].

A good example of the misleading effects of the older, low, heats of formation of radicals involves the thermal cis-trans rearrangement of cycloal-

TABLE 9.2[a] Heats, Entropies, and Bond Dissociation Energies

	ΔH_f° (kcal mole^{-1})			S° (cal mole^{-1} deg^{-1})	BDE (kcal mole^{-1})	
R	R·	RH	RR	R·	R–H	R–R
H	+52.09	[0]	[0]	27.391	104.2	104.2
Me	+34.9 ±0.2	−17.9	−20.2	51 ± 3	104.9	89.8
Et	+28.3 ±0.4	−20.2	−30.4	62 ± 3	100.6	87.0
i-Pr	+21.3 ±0.7	−24.8	42.6	70 ± 3	98.2	85.2
t-Bu	+11.6 ±0.4	−32.4	−53.8	75.5 ± 2	96.1	77.0

[a]For perfect gases at 1 atm and 25°C.

kanes [102]. In order to reconcile the calculated heats of formation of the diradical intermediates, it was necessary to postulate that the diradicals were metastable, that is, that the closure of the ring required an activation energy of several kilocalories per mole.

9.8.2 Entropies

The entropies of methyl, ethyl, *i*-propyl, and *t*-butyl radicals (Table 9.2) were determined experimentally, along with the heats of formation, from the temperature dependence of the forward and back reactions of the alkyl radicals with HBr (Eq. 9.40) [98].

$$R\cdot + HBr \rightleftharpoons RH + Br\cdot \quad (9.40)$$

For *t*-butyl radical and isobutane, the rate constants of Eq. 9.40 could be measured directly for both the forward and reverse directions [98]. For the other radicals the rate constants of the back reaction were determined relative to that for the formation of the *t*-butyl radical from isobutane [99].

REFERENCES

1. Kochi, J. K.; Krucic, P. J. *J. Am. Chem. Soc.* **1969,** *91,* 3940.
2. Pacansky, J.; Bargon, J. *J. Am. Chem. Soc.* **1975,** *97,* 6896.
3. Herzberg, G. *Proc. Royal Soc.* **1961,** *A 262,* 291.
4. Houle, F. A.; Beauchamp, J. L. *J. Am. Chem. Soc.* **1979,** *101,* 4067.
5. Marynick, D. S.; Dixon, D. A. *Proc. Natl. Acad. Sci. U.S.A.* **1977,** *74,* 410.
6. Pacansky, J.; Koch, W.; Miller, M. D. *J. Am. Chem. Soc.* **1991,** *113,* 317.
7. Koenig, T.; Balle, T.; Snell, W. *J. Am. Chem. Soc.* **1975,** *97,* 662.
8. Bass, C. D.; Pimentel, G. C. *J. Am. Chem. Soc.* **1961,** *83,* 3754.
9. Fessenden, R. W.; Schuler, R. H. *J. Chem. Phys.* **1965,** *43,* 2704.
10. Foner, S. N. *Adv. Mol. Phys.* **1966,** *2,* 385.
11. Kashiwagi, T.; Kozuka, S.; Oae, S. *Tetrahedron* **1970,** *26,* 3619.
12. Flowers, G. C.; Leffler, J. E. *J. Org. Chem.* **1989,** *54,* 3995.
13. Leffler, J. E.; Barbas, J. T.; Flowers, G. C. *J. Org. Chem.* **1982,** *47,* 2286.
14. Ellison, G. B.; Engelking, P. C.; Lineberger, W. C. *J. Am. Chem. Soc.* **1978,** *100,* 2556.
15. Jackson, W. M.; McNesby, J. R.; Darwent, B. D. *J. Chem. Phys.* **1962,** *37,* 1610.
16. Trotman-Dickenson, A. F. *Chem. Ind.* **1965,** 379.
17. Kerr, J. A. *Chem. Rev.* **1966,** *66,* 465.
18. Pryor, W. A.; Fuller, D. L.; Stanley, J. P. *J. Am. Chem. Soc.* **1972,** *94,* 1632.
19. Pryor, W. A.; Tonellato, U.; Fuller, D. L.; Jumonville, S. *J. Org. Chem.* **1969,** *34,* 2018.

20. Pryor, W. A.; Lin, T. H.; Stanley, J. P.; Henderson, R. W. *J. Am. Chem. Soc.* **1973,** *95,* 6993.
21. Chatgilialoglu, C.; Ingold, K. U.; Scaiano, J. C. *J. Am. Chem. Soc.* **1981,** *103,* 7739.
22. Giese, B. *Angew. Chem. Int. Ed.* **1983,** *22,* 753.
23. Dewar, M. J. S.; Olivella, S. *J. Am. Chem. Soc.* **1978,** *100,* 5290.
24. Shaik, S. S.; Hiberty, P. C.; Lefour, J.-M.; Ohanessian, G. *J. Am. Chem. Soc.* **1987,** *109,* 363.
25. Cvetanovic, R. J.; Irwin, R. S. *J. Chem. Phys.* **1967,** *46,* 1694.
26. Thynne, J. C. J. *Int. J. Chem. Kinet.* **1970,** *3,* 155.
27. Sangster, J. M.; Thynne, J. C. J. *Int. J. Chem. Kinet.* **1969,** *1,* 571.
28. Stefani, A. P.; Chuang, L.-Y. Y.; Todd, H. E. *J. Am. Chem. Soc.* **1970,** *92,* 4168.
29. Stefani, A. P.; Herk, L.; Szwarc, M. *J. Am. Chem. Soc.* **1961,** *83,* 4732.
30. Martin, M. M.; Gleicher, G. J. *J. Am. Chem. Soc.* **1964,** *86,* 233.
31. Pryor, W. A. *Free Radicals,* McGraw-Hill: New York, 1966.
32. Nonhebel, D. C.; Walton, J. C. *Free Radical Chemistry,* Cambridge University Press: Cambridge, 1974.
33. Corbett, G. E.; Williams, G. H. *J. Chem. Soc.* **1964,** 3437.
34. Corbett, G. E.; Williams, G. H. *J. Chem. Soc. (B)* **1966,** 877.
35. Cowley, B. R.; Norman, R. O. C.; Waters, W. A. *J. Chem. Soc.* **1959,** 1799.
36. Heilman, W. J.; Rembaum, A.; Szwarc, M. *J. Chem. Soc.* **1957,** 1127.
37. Eliel, E. L.; Rabindran, K.; Wilen, S. H. *J. Org. Chem.* **1957,** *22,* 859.
38. Dewar, M. J. S.; Thompson, C. C. J. *J. Am. Chem. Soc.* **1965,** *87,* 4414.
39. Pacansky, J.; Dupuis, M. *J. Am. Chem. Soc.* **1982,** *104,* 415.
40. Pacansky, J.; Dupuis, M. *J. Chem. Phys.* **1978,** *68,* 4276.
41. Smid, J.; Rembaum, A.; Szwarc, M. *J. Am. Chem. Soc.* **1956,** *78,* 3315.
42. Kaptein, R. *Chem. Phys. Lett.* **1968,** *2,* 261.
43. Koenig, T. In *Free Radicals,* J. K. Kochi, Ed. Wiley-Interscience: New York, 1973, Chapter 3.
44. Sheldon, R. A.; Kochi, J. K. *J. Am. Chem. Soc.* **1970,** *92,* 5175.
45. Hilborn, J. W.; Pincock, J. A. *J. Am. Chem. Soc.* **1991,** *113,* 2683.
46. Dixon, P. S.; Stefani, A. P.; Szwarc, M. *J. Am. Chem. Soc.* **1963,** *85,* 2551.
47. Zavitsas, A. A.; Ehrenson, S. *J. Am. Chem. Soc.* **1965,** *87,* 2841.
48. Chen, H. E.; Vaish, S. P.; Cocivera, M. *J. Am. Chem. Soc.* **1973,** *95,* 7586.
49. Pacansky, J.; Coufal, H. *J. Chem. Phys.* **1979,** *71,* 2811.
50. Ingold, K. U. In *Free Radicals,* J. K. Kochi, Ed. Wiley-Interscience: New York, 1973, Chapter 2.
51. Dobis, O.; Benson, S. W. *J. Am. Chem. Soc.* **1990,** *112,* 1023.
52. Benson, S. W. *J. Chem. Ed.* **1965,** *42,* 502.
53. Stefani, A. P. *J. Am. Chem. Soc.* **1968,** *90,* 1694.
54. Backman, C.-M.; Claesson, S.; Szwarc, M. *Trans. Faraday Soc.* **1970,** *66,* 3061.
55. Madden, K. P.; Taniguchi, H. *J. Am. Chem. Soc.* **1991,** *113,* 5541.

56. Pacansky, J.; Gutierrez, A. *J. Phys. Chem.* **1983,** *87,* 3074.
57. Watkins, K. W. *J. Am. Chem. Soc.* **1971,** *93,* 6355.
58. Burkhart, R. D. *J. Am. Chem. Soc.* **1968,** *90,* 273.
59. DeTar, D. F.; Wells, D. V. *J. Am. Chem. Soc.* **1960,** *82,* 5839.
60. Reutov, O. A. *Russ. Chem. Rev.* **1984,** *53,* 462.
61. Henderson, R. W.; Ward, R. D. J. *J. Am. Chem. Soc.* **1974,** *96,* 7556.
62. Pryor, W. A.; Davis, W. H. J. *J. Am. Chem. Soc.* **1974,** *96,* 7557.
63. Citterio, A.; Arnoldi, A.; Minisci, F. *J. Org. Chem.* **1979,** *44,* 2674.
64. Olah, G. A.; Krishnamurthy, V. V.; Singh, B. P.; Iyer, P. S. *J. Org. Chem.* **1983,** *48,* 955.
65. Effio, A.; Griller, D.; Ingold, K. U.; Scaiano, J. C.; Sheng, S. J. *J. Am. Chem. Soc.* **1980,** *102,* 6063.
66. Kochi, J. K.; Krusic, P. J.; Eaton, D. R. *J. Am. Chem. Soc.* **1969,** *91,* 1877.
67. Walton, J. C. *Magn. Res. Chem.* **1987,** *25,* 998.
68. Ingold, K. U.; Walton, J. C. *Acc. Chem. Res.* **1989,** *22,* 8.
69. Kemball, M. L.; Walton, J. C.; Ingold, K. U. *J. Chem. Soc. Perkin II* **1982,** 1017.
70. Pacansky, J.; Coufal, H. *J. Chem. Phys.* **1980,** *72,* 3298.
71. Huggenberger, C.; Fischer, H. *Helv. Chim. Acta* **1981,** *64,* 338.
72. Davis, W. H., Jr.; Pryor, W. A. *J. Am. Chem. Soc.* **1977,** *99,* 6365.
73. Sheldon, R. A.; Kochi, J. K. *J. Am. Chem. Soc.* **1970,** *92,* 4395.
74. Paddon-Row, M. N.; Houk, K. N. *J. Am. Chem. Soc.* **1981,** *103,* 5046.
75. Griller, D.; Ingold, K. U.; Krusic, P. J.; Fischer, H. *J. Am. Chem. Soc.* **1978,** *100,* 6750.
76. Tanner, D. D.; Rahimi, P. M. *J. Am. Chem. Soc.* **1982,** *104,* 225.
77. Pryor, W. A.; Tang, F. Y.; Tang, R. H.; Church, D. F. *J. Am. Chem. Soc.* **1982,** *104,* 2885.
78. Pryor, W. A.; Davis, W. H. J.; Stanley, J. P. *J. Am. Chem. Soc.* **1973,** *95,* 4754.
79. Ernst, J. A.; Thankachan, C.; Tidwell, T. T. *J. Org. Chem.* **1974,** *39,* 3614.
80. Lorand, J. P.; Chodroff, S. D.; Wallace, R. W. *J. Am. Chem. Soc.* **1968,** *90,* 5266.
81. Neuman, R. C., Jr.; Frink, M. E. *J. Org. Chem.* **1983,** *48,* 2430.
82. Maillard, B.; Ingold, K. U.; Scaiano, J. C. *J. Am. Chem. Soc.* **1983,** *105,* 5095.
83. Russell, G. A.; Khanna, R. K. *J. Am. Chem. Soc.* **1985,** *107,* 1450.
84. Leffler, J. E.; More, A. A. *J. Am. Chem. Soc.* **1972,** *94,* 2483.
85. Rüchardt, C.; Weiner, S. *Tetrahedron Lett.* **1979,** 1311.
86. Lomas, J. S.; Dubois, J.-E. *J. Org. Chem.* **1982,** *47,* 4505.
87. Lomas, J. S.; Dubois, J.-E. *J. Org. Chem.* **1974,** *39,* 1776.
88. Schuh, H.-H.; Fischer, H. *Helv. Chim. Acta* **1978,** *61,* 2463.
89. Neuman, R. C. J. *J. Org. Chem.* **1972,** *37,* 495.
90. Mendenhall, G. D.; Ingold, K. U. *J. Am. Chem. Soc.* **1973,** *95,* 3422.
91. Mendenhall, G. D.; Griller, D.; Lindsay, D.; Tidwell, T. T.; Ingold, K. U. *J. Am. Chem. Soc.* **1974,** *96,* 2441.

92. Fessenden, R. W.; Schuler, R. H. *J. Chem. Phys.* **1963,** *39,* 2147.
93. Leffler, J. E.; Grunwald, E. *Rates and Equilibria of Organic Reactions,* Wiley: New York, 1963.
94. Souders, M. J.; Matthews, C. S.; Hurd, C. O. *Ind. Eng. Chem.* **1949,** *41,* 1040.
95. Gutman, D. *Acc. Chem. Res.* **1990,** *23,* 375.
96. Doering, W. von E. *Proc. Natl. Acad. Sci. U.S.A.* **1981,** *78,* 5279.
97. Settula, J. A.; Russell, J. J.; Gutman, D. *J. Am. Chem. Soc.* **1990,** *112,* 1347.
98. Russell, J. J.; Seetula, J. A.; Timonen, R. S.; Gutman, D.; Nava, D. F. *J. Am. Chem. Soc.* **1988,** *110,* 3084.
99. Russell, J. J.; Seetula, J. A.; Gutman, D. *J. Am. Chem. Soc.* **1988,** *110,* 3092.
100. Richards, P. D.; Ryther, R. J.; Weitz, E. *J. Phys. Chem.* **1990,** *94,* 3663.
101. Tsang, W. *J. Am. Chem. Soc.* **1985,** *107,* 2872.
102. Bergman, R. G. In *Free Radicals,* J. K. Kochi, Ed., Wiley: New York, 1973, Chapter 5.

10

CONJUGATED CARBON RADICALS

Radicals like allyl, benzyl, and triphenylmethyl have received a good deal of attention because of their relative ease of formation, involvement as reactive intermediates, longer lifetimes, and ability to react at more than one position (ambident reactivity). They have also been important first as a proving ground for theories of valence, and later of resonance, and still later of molecular orbitals.

10.1 TRIPHENYLMETHYL AND HEXAPHENYLETHANE—A CAUTIONARY TALE [1]

The discovery of triphenylmethyl, the first moderately stable free radical, was announced by Moses Gomberg in 1900 in two papers, identical except for the language. One, received on October 4, appeared in the *Journal of the American Chemical Society* [2], the other, received on October 1, in the presumably more prestigious and more widely read *Berichte der deutschen chemischen Gesellschaft* [3].

Triphenylmethyl was made by treating trityl chloride with a metal, either "molecular" silver or zinc. Oxygen had to be excluded, or no hydrocarbon would be obtained, only a peroxide. Corks and even rubber stoppers were inadequate, ground glass joints had to be used.

Despite Gomberg's statement that he wished to reserve further research in this field for himself, customary in those days, in the next few years papers on triphenylmethyl or hexaphenylethane appeared from at least a dozen laboratories. To put it in a modern context, triphenylmethyl combined the novelty of something like bucky balls with the controversial nature of something like polywater or cold fusion.

Naturally, Gomberg's interpretation of his product as free triphenylmethyl was not generally accepted. Gomberg himself considered hexaphenylethane as an alternative, but rejected that hypothesis on the grounds that the ethane should be unreactive like other trityl-carbon compounds such as triphenylacetic acid.

An additional difficulty was that some laboratories did get a stable hydrocarbon with the same molecular weight and analysis as trityl dimer. This compound was the Ullmann hydrocarbon (**II**), readily formed by a hydrogen shift (Eq. 10.1) from the dimer (**I**), which we now know has the α-para structure as shown by Lankamp, Nauta, and MacLean's 1968 NMR spectrum [4].

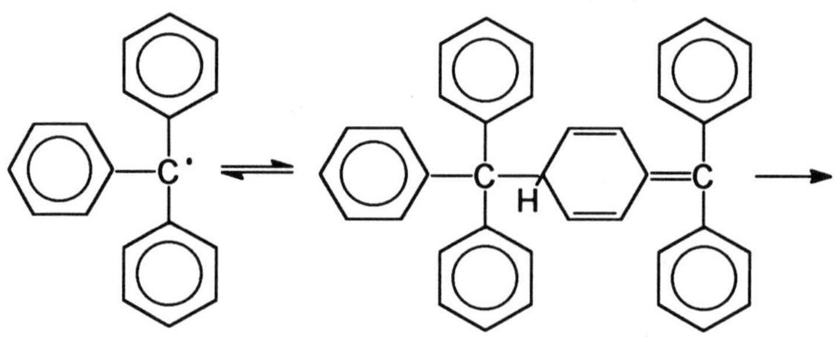

I

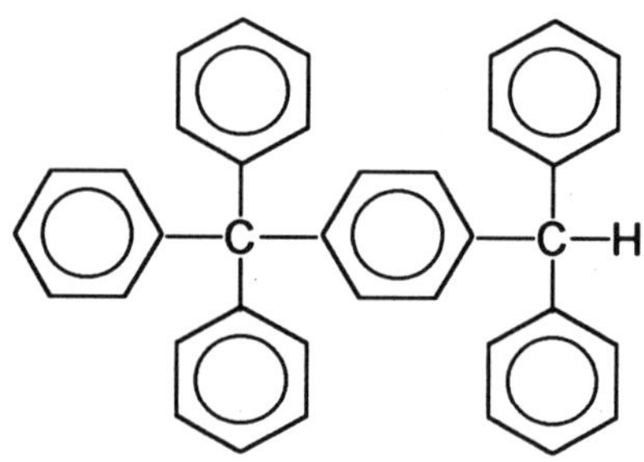

II

(10.1)

Ullman [5] made compound **II** by treating trityl chloride with zinc in acetic acid with added $SnCl_2$ and HCl. He was convinced that it was hexaphenylethane, and remarked: ". . .nicht die geringste Aehnlichkeit mit dem zersetzlichen Triphenylmethyl Gombergs zeigt."

The true structure of trityl dimer was actually suggested in 1905 by Jacobson [6], whose thoughts were probably led in that direction by his work on the semidine rearrangement. For a time Gomberg adopted Jacobson's structure (**I**) for the colorless solid dimer, but later, becoming persuaded by the quinonoid color theory that **I** would have to be colored, he favored the hexaphenylethane structure. This structure was accepted generally and without question from about 1924 to 1968.

All sorts of strong hints that should have caused the Jacobson structure to be reconsidered were ignored [1].

10.1.1 The Dissociation of Trityl Dimer

The equilibrium constant for the dissociation has been measured both by means of ESR and by spectrophotometry [7,8]. The value of K is 3.09×10^{-4} M in toluene at 20°C, and $\Delta H°$ is about 11.1 kcal mole^{-1}.

When a solution of trityl dimer is exposed to oxygen, the trityl radicals react with oxygen, causing it to be absorbed by the solution and forming peroxides, mostly ditrityl peroxide and *p*-trityl-triphenylmethylhydroperoxide [9,10]. The rate of production of trityl radicals by the dimer can be measured by the oxygen uptake, but the process is first order only in the presence of an inhibitor such as pyrogallol. The pyrogallol traps tritylperoxy radicals as tritylhydroperoxide, preventing them from reacting directly with the trityl dimer. The rate constant for the unimolecular dissociation of the trityl dimer is 2.5×10^{-3} sec^{-1} in toluene at 20°C. The energy of activation is about 19.5 kcal mole^{-1} [11].

The reaction with oxygen of trityl radicals trapped in a triphenylacetic acid crystalline matrix is reversible. The value of $\Delta H°$ for the addition to form the peroxy radical is -9 ± 1 kcal mole^{-1} [12].

Unlike the hexaphenylethane structure, the α-para structure predicts large steric rather than electronic effects on the dissociation of the dimer from either para or meta substituents. With the wrong structure for the dimer, it was thought that substituent effects on the free energy of the dimer itself could be neglected.

It was easy to see why para substituents could have a substantial effect on the equilibrium. Substituents in this position were adjacent to the odd electron in an important resonance structure. But how about the effects of substituents at the meta position? One para methyl substituent increased the dissociation constant by a factor of 5, but one meta methyl substituent increased it by a factor of 8 [13]!

The effects of para and meta phenyl substituents were of special interest, since resonance should be particularly important for this substituent. One para phenyl substituent increased the dissociation constant by a factor of about 17, but one meta phenyl substituent also had a substantial effect, a factor of about 13 [14].

To explain the effect of the meta substituent, dipolar structures like **III** were proposed. These structures were strictly ad hoc and not found useful in other contexts. In contrast to the strange substituent effects on the dissociation of "hexaphenylethane," substituent effects on the ionization of trityl chloride were quite as expected. One para phenyl substituent increased the ionization constant by a factor of 5.4 and one meta phenyl substituent decreased it by a factor of 0.73, an inductive effect uncompensated by resonance [14].

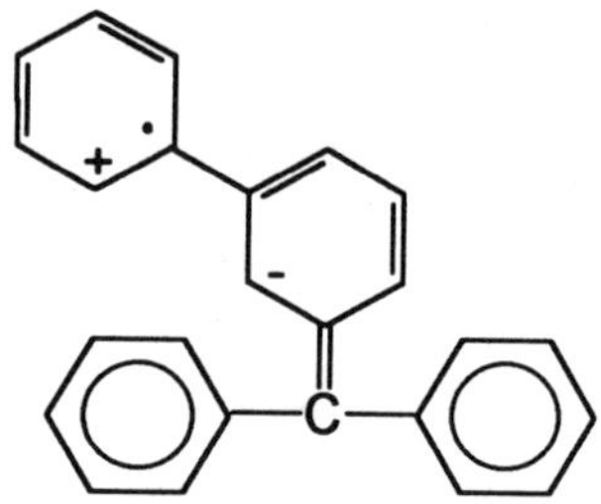

III

10.1.2 Reactivity at the Para Position

Besides the acid-catalyzed aromatization of the dimer to the Ullman hydrocarbon (**II**), there was the anomalous reactivity of halogen in the para position. One half, but only one half, of the bromine in *p*-bromophenyldiphenylmethyl radical is readily removed by molecular silver, a result easy to understand in terms of the α-para structure (Eq. 10.2) [15].

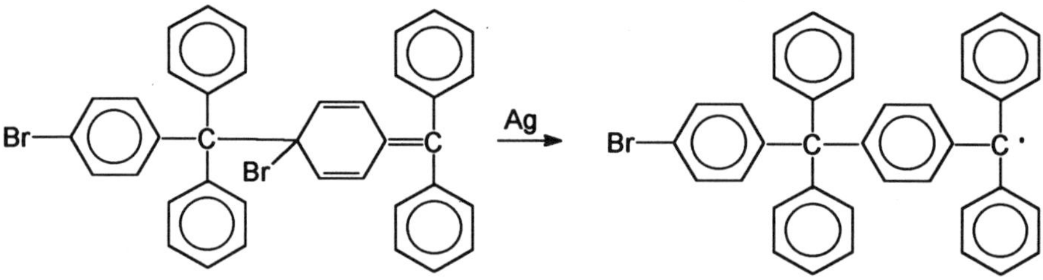

(10.2)

10.1.3 Hexaphenylethanes

An interesting question that remains is what hexaphenylethane would be like if it did exist. The α-para structure can be avoided in substituted hexaphenylethanes by two bulky flanking *t*-butyl groups in each ring [16]. Solutions of tris(3,5-di-*t*-butylphenyl)methyl in benzene give red crystals of the radical, but solutions in cyclohexane slowly deposit colorless crystals of the ethane on evaporation. An X-ray diffraction study of these crystals showed that the ethane has an elongated central bond of about 1.67 Å. A force field calculation gave 1.65 Å. Force field calculations predicted a similar value of 1.64 Å both for hexaphenylethane and for hexakis(3,5-di-*t*-butyl-4-phenyl-phenyl)ethane [17].

10.2 STRUCTURE OF THE TRITYL RADICAL

Because of repulsion of ortho hydrogens the planar conformation of triphenylmethyl radical is not an energy minimum. The benzene rings are rotated about 30–45°C about the phenyl to carbon bond axis, in a propeller conformation [18,19]. The absorption spectrum of triphenylmethyl has a peak in the ultraviolet [20] at 336 nm and a shoulder at 324 and a peak in the visible region [8,16] at 514 nm. If the radical is like the triphenylcarbenium ion, two isomeric propeller conformers, a symmetrical propeller and a propeller in which one of the blades is twisted the wrong way, may be contributing to the absorption.

Various attempts have been made to detect chirality in reaction products formed by way of chiral intermediate triarylmethyl radicals such as phenyl-*p*-biphenyl-α-naphthylmethyl, but the umbrella inversion has been too fast [21].

In the series of bond dissociation energies CH_3–H (105 kcal $mole^{-1}$), $PhCH_2$–H (88 kcal $mole^{-1}$), Ph_2CH–H (82 kcal $mole^{-1}$), and Ph_3C–H (81 kcal $mole^{-1}$), the phenyl effects decrease from 17 to 6 and then to 1 kcal because of a combination of saturation and steric effects [22].

10.3 SOME REACTIONS OF THE TRITYL RADICAL

10.3.1 Photodisproportionation

Although solutions of trityl radical are stable for a long time in dim light, in bright light the excited doublet undergoes a cyclization reaction (Eq. 10.3) [23,24]. Subsequent transfer of two hydrogens gives two molecules of triphenylmethane and a 9-phenylfluorenyl radical, which dimerizes [25,26]. In dioxane plus hydroxide ion the products are one molecule of triphenylmethane and one molecule of 9-phenylfluorene [27].

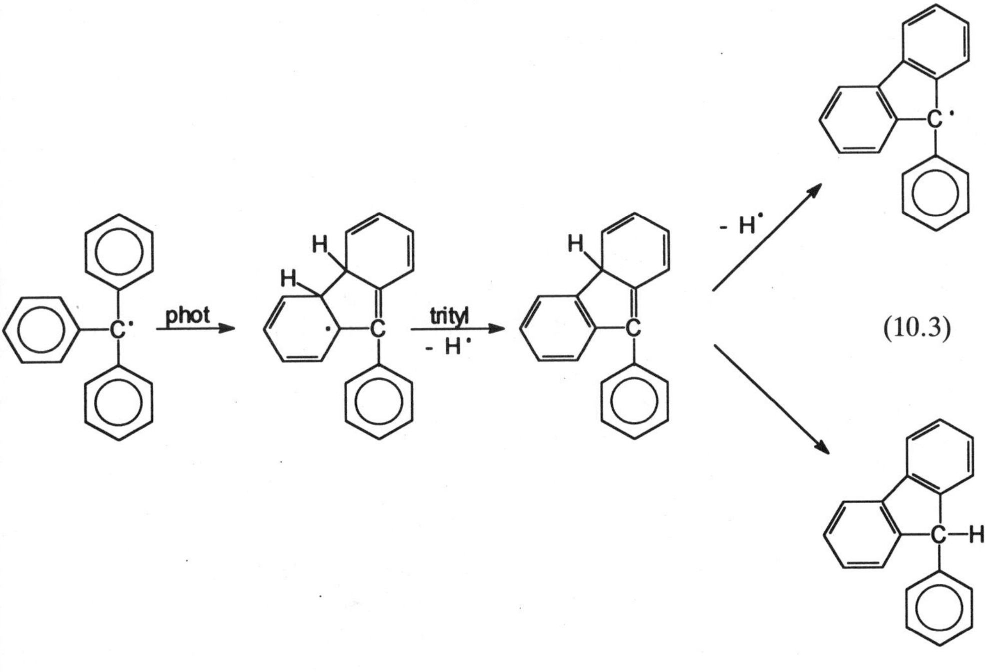

(10.3)

Laser irradiation of trityl radicals in aqueous solution gives a different reaction: photoionization to triphenylcarbenium ion and the hydrated electron [28].

10.3.2 Ambident Abstraction

The thermolysis of phenylazotriphenylmethane (PAT) in 2,2-dimethoxypropane in the presence of sodium methoxide gives, among other products, triphenylmethane formed by the abstraction of a hydrogen atom from the solvent, and diphenyl-*p*-biphenylylmethane formed by the combination of phenyl and trityl radicals. The base serves to catalyze the intramolecular aromatization of intermediate ortho or para semibenzenes.

When the trityl moiety in the PAT is completely deuterated, some of the triphenylmethane molecules are found to have hydrogen in a para or ortho position instead of the α position [29].

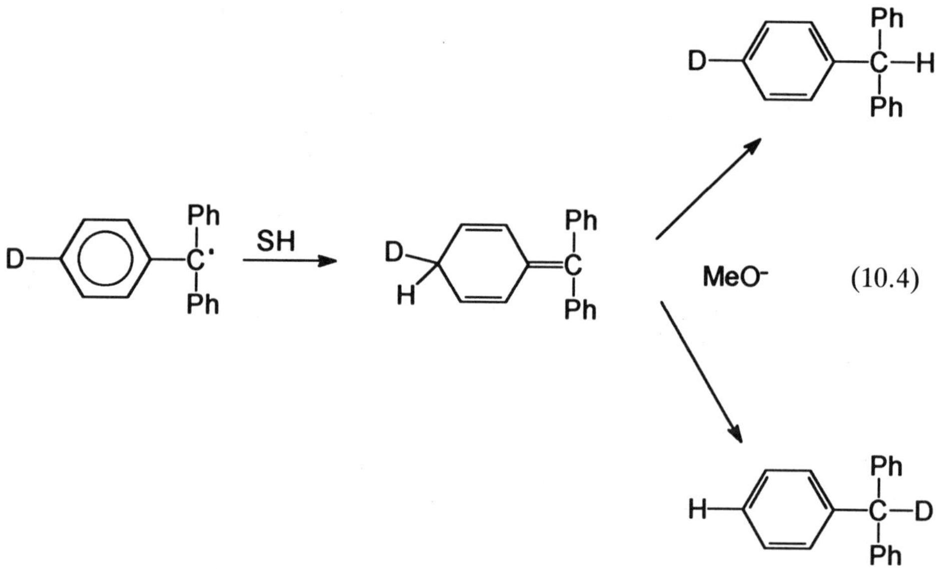

(10.4)

Some of the triphenylmethane is found to have deuterium in both the ring and α position. This happens when a trityl radical abstracts a deuterium atom from the para coupling product of a trityl and a phenyl radical (Eq. 10.5).

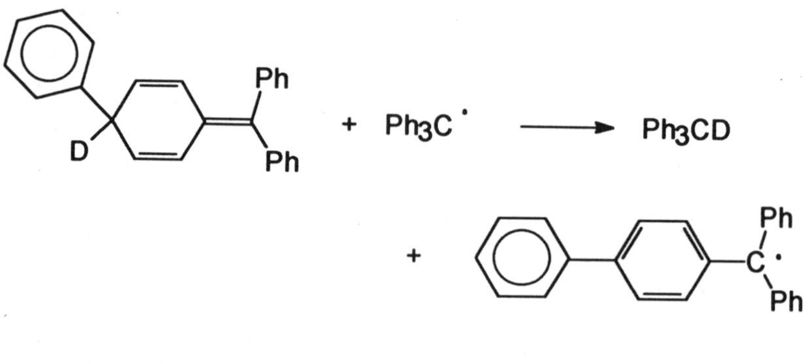

(10.5)

10.3.3 Ambident Coupling

The formation of the trityl dimer is an example of coupling restricted to the α-para choice of positions. The thermolysis of PAT gives the para-coupled cage product of the phenyl and trityl radicals, diphenyl-*p*-biphenylylmethane, accompanied by the α-coupled product, tetraphenylmethane.

The thermolysis of *t*-butyl triphenylperacetate in cumene at 25°C gives the para-coupled cage product, *p*-*t*-butoxyphenyl-diphenylmethane (Eq. 10.6) [30]. There is also spectroscopic evidence for the formation of *t*-butyl trityl ether, but this rearranged to the *p*-coupled product on chromatography. Attempts to make the α ether from trityl chloride and alkoxide also gave only the para isomer.

The thermolysis of the perester in styrene gave an α-coupled telomer (Eq. 10.7), but attempts to synthesize that compound by a nonradical route again gave only the para isomer.

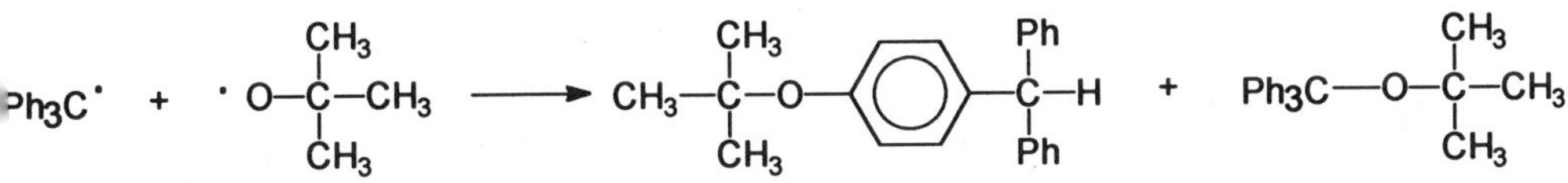

(10.6)

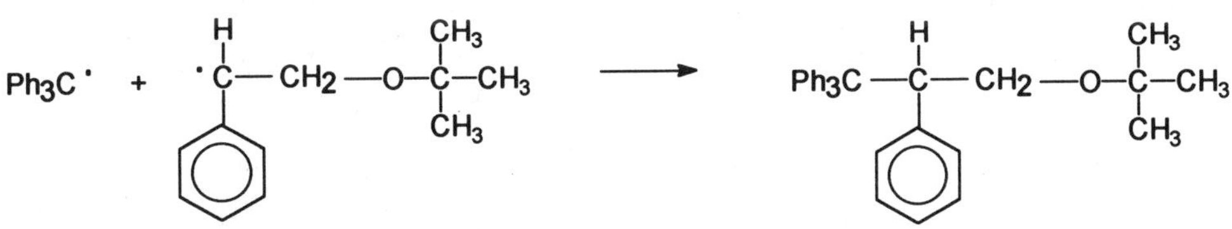

(10.7)

10.3.4 Reaction With Benzoyl Peroxide

Trityl radicals react with benzoyl peroxide in aromatic solvents ArH to give a mixture of trityl benzoate and triphenylarylmethane, Ph_3CAr [31,32]. The initial step (Eq. 10.8), giving trityl benzoate and a benzoyloxy radical, is accelerated by electron-withdrawing substituents in the peroxide and has a ρ value of +1.45 at 25°C [33].

$$Ph_3C\cdot + PhCOO{-}OOCPh \longrightarrow Ph_3C{-}OOCPh + PhCOO\cdot \quad (10.8)$$

The high positive ρ value, the large negative entropies of activation, and the fact that the reaction is faster in more polar solvents, mean that the transition state is quite polar. This is consistent with a nucleophilic displacement on oxygen and a transition state with considerable charge transfer character. A simple electron transfer mechanism (Eq. 10.8a) is ruled out by the results of a labeling experiment.

$$Ph_3C\cdot + PhCOO{-}OOCPh \longrightarrow Ph_3C^+ \ {}^-OOCPh + PhCOO\cdot \quad (10.8a)$$

With benzoyl peroxide labeled with ^{18}O in the carbonyl group, most of the label in the trityl benzoates is still in the carbonyl group [34]. The small amount of scrambling is expected if some of the trityl benzoate is formed by the combination of a free benzoyloxy radical with a trityl radical.

Most of the benzoyloxy radicals are trapped by addition to the aromatic solvent (Eq. 10.9) [31].

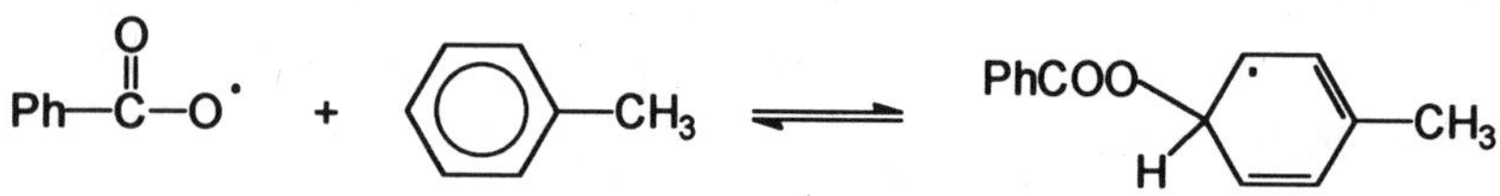

(10.9)

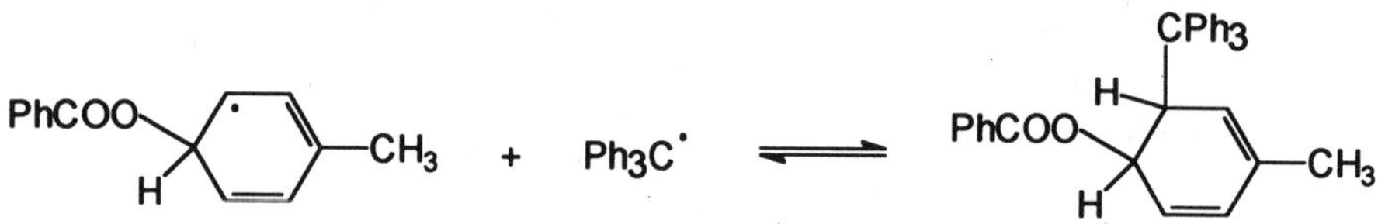

(10.10)

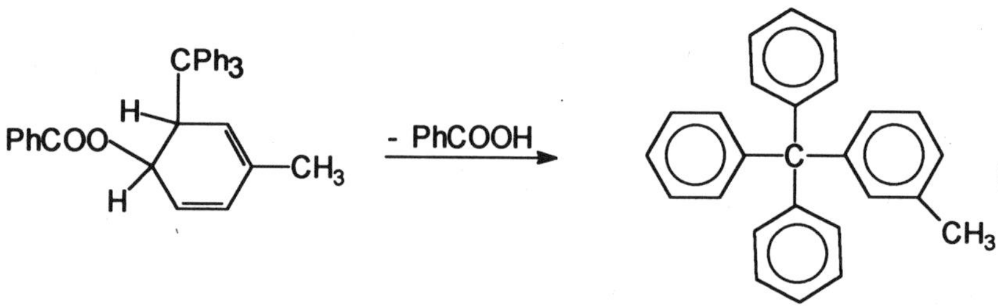

(10.11)

The substituted tetraphenylmethane, mostly the meta derivative, is formed by addition of trityl to the cyclohexadienyl radical formed by reaction 10.9, followed by loss of benzoic acid as in reaction 10.11 [31].

10.4 EFFECTS OF CONSTRAINED CONFORMATIONS

10.4.1 Triptycyl

The unpaired electron in the triptycyl radical (**IV**) has been shown by the ^{13}C splitting of its ESR line to be almost completely localized at position 1 [35]. The radical was made by irradiating a single crystal of the chloroform solvate of triptycene-1-formyl peroxide. The spin density is calculated to be 9.0% in the 2*s* atomic orbital and 77% in the 2*p* orbital, which is constrained to be orthogonal to the π orbitals of the benzene rings.

The lack of resonance stabilization of the triptycyl radical should make it much more reactive than trityl and harder to generate. Molecular silver, a

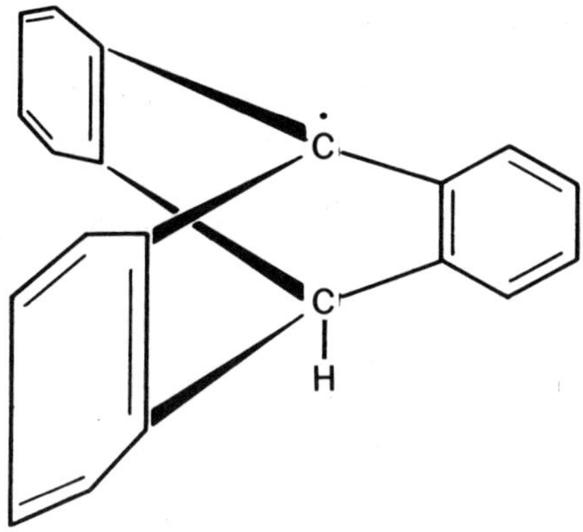

IV

reagent that quickly generates trityl radical from trityl halides at room temperature, failed to give any reaction at all with triptycyl bromide, even in mineral oil at 280°C [36]. Metallic sodium in mineral oil at 140°C reduced the bromide quantitatively to triptycene, but without formation of any organosodium derivative, or, on treatment with CO_2, any triptoic acid. If this reaction goes by a heterogeneous single electron transfer to form the triptycyl radical, the radical must abstract hydrogen very rapidly. Magnesium also reduced the bromide to triptycene.

Triptycene is not chlorinated by sulfuryl chloride in the presence of benzoyl peroxide [37]. Toluene, in contrast, gives benzyl chloride in high yields under the same conditions.

The rate constant for the decomposition of triptycene-1-formyl peroxide in benzene at 80°C is 14.2×10^{-5} sec^{-1}. This is about twice the rate for acetyl peroxide and four times that for benzoyl peroxide. But bicyclooctane-1-formyl peroxide, another bridgehead peroxide, decomposes at a much lower temperature, 30°C, with a rate constant of 17.1×10^{-5} sec^{-1} [38].

The products from the decomposition of triptycene-1-formyl peroxide in benzene at 80°C included up to 50 mole% CO_2 and 45 mole% triptycene as well as triptoic acid, triptycyl triptoate, and bitriptycyl [39]. When the peroxide is decomposed in the presence of iodine the triptycene is replaced by a nearly equivalent amount of triptycyl iodide. Unless the triptycycl iodide is formed by an S_Ni reaction of triptoyl iodide, some triptycyl radicals must be formed. If so, this is in striking contrast to the failure of the reaction between triptycyl bromide and molecular silver. Other reaction products were more easily explained in terms of a carboxyinversion intermediate. The carboxyinversion reaction is usually accompanied by radical products, but its occurrence in this instance, in which both the radical and the carbenium ion are of such high energy, is puzzling.

Bitriptycyl is a stable hydrocarbon melting at 577°C [40].

10.4.2 "Planar" Trityl

In the radicals **V** and **VI** the ortho hydrogens of trityl, whose steric repulsion causes trityl to adopt a nonplanar propeller conformation, have been replaced by structures tying the rings together, and making the radical more nearly planar. The sesquixanthhydryl radical **V** [41] exists as a colorless dimer in the

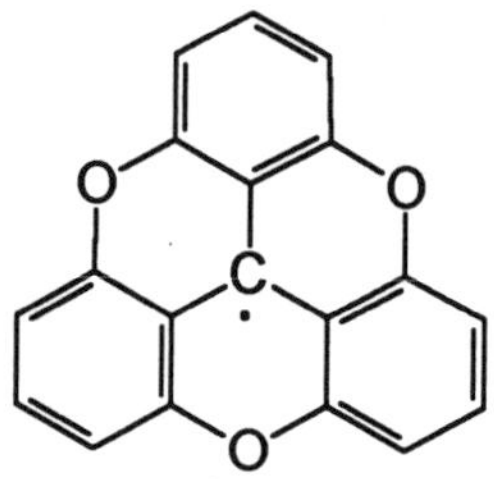

V

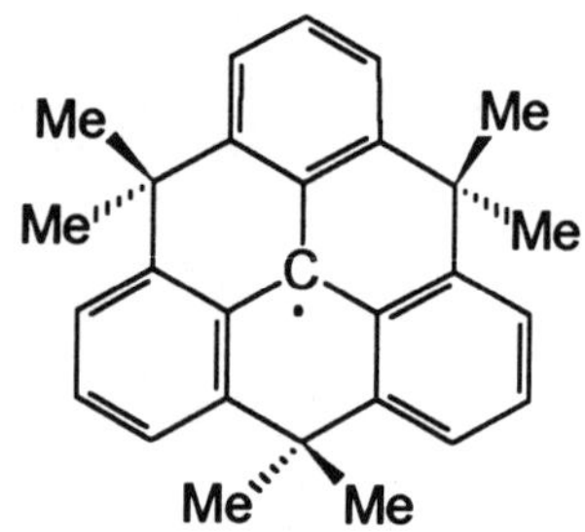

VI

crystals (mp > 300°C) and in solution at room temperature. The solutions become green on heating to about 150°C, and fade again on cooling. The solutions give an ESR signal consisting of a well-resolved quartet of septets, a signal whose intensity, like the color, also increases reversibly at higher temperature as the dimer becomes more dissociated. Hexa-*o*-methoxytrityl should be even less planar than trityl. In the series **V**, trityl, and hexa-*o*-methoxytrityl the hyperfine splitting constants of the para hydrogens are 3.17, 2.77, and 2.26 G, decreasing as the radical is made less planar.

Radical **VI** [42], unlike **V**, is monomeric in solution. Again the ESR spectrum is a quartet of septets. The hyperfine splitting factor of the para hydrogens is 3.15 G, about the same as that for **V**. The hyperfine splitting constant of the central ^{13}C (natural abundance) of **VI** is smaller, 18.23 G as compared to 23.5 for trityl. A bowl-shaped structure for **VI** should have given a larger ^{13}C splitting because of a lower *p* character of the central carbon orbital.

10.5 ISOMERIC DIMERS

In the 9-phenylfluorenyl radical two of the phenyl groups of triphenylmethyl have been tied back by a bond between the ortho positions. This modification reduces the steric strain in the α-α dimer to such an extent that it is the stable form, as shown by the ^{13}C NMR spectrum [43]. In bisfluoradenyl (**VII**) two of the rings are tied back. The dimer not only has the α–α structure, it also shows no dissociation by ESR [44].

A force field calculation shows that hexaphenyldisilane, because of the longer bonds to silicon, is almost strain free [45].

In tertiary benzylic radicals with only one or two phenyl groups the dimer structure depends on the size of the alkyl groups. The dimers of $Ph_2CR\cdot$

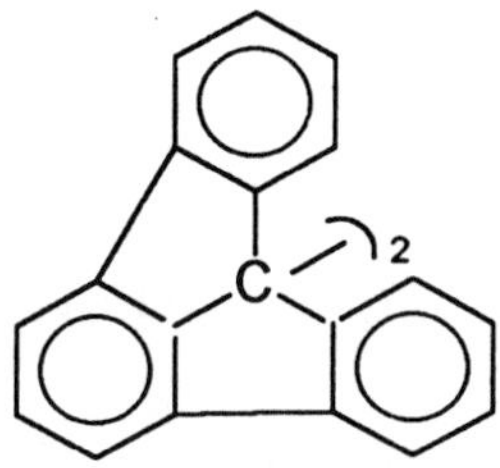

VII

radicals have the α–α structure when R is methyl, ethyl or *n*-propyl, but mainly the α-para structure when R is cyclopentyl, cyclohexyl, isopropyl, or *t*-butyl [46].

The dimers of $PhCR_2\cdot$ radicals are mostly α–α but with small amounts of the other isomers, as shown in Table 10.1 [10].

9,9,10,10-Tetraphenyl-9,10-dihydrophenanthrene, which can be regarded as hexaphenylethane in which two phenyl groups in *different moieties* are linked at the ortho positions, melts at 340°C and is unreactive toward O_2 and Br_2 [47]. It is made by treating the dipotassium salt of the corresponding ortho-bonded triphenylmethane with tetramethylethylene dibromide. The potassium salt is regenerated when the 9,9,10,10-tetraphenyl-9,10-dihydrophenanthrene in dioxane is treated with a Na/K alloy.

Tris(biphenylene)ethane (**VIII**) [48], with all three phenyl rings linked at

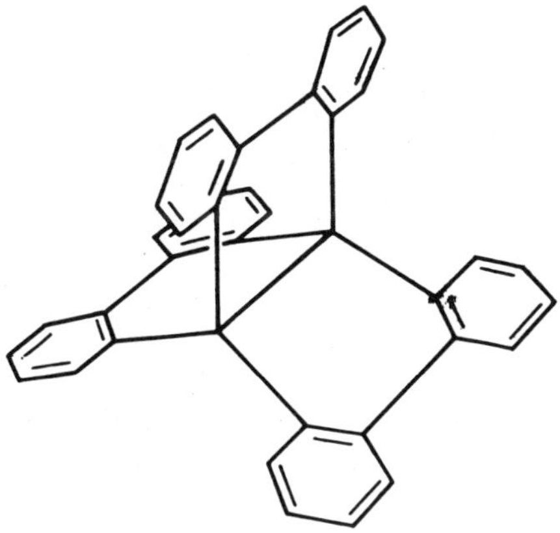

VIII

TABLE 10.1 Relative Amounts of Isomeric Dimers in PhCRR′·

R	R′	α-α	α-ortho	α-para
Methyl	Methyl	1	0.1	0.2
Ethyl	Ethyl	1	0.04	0.6
Methyl	Isopropyl	1	0.02	0.6

the ortho positions to phenyls in the other "trityl" moiety, is a very stable hydrocarbon melting at 473°C.

10.6 BENZYL AND RELATED RADICALS

10.6.1 Formation by Thermal Reactions

Phenylacetyl peroxide thermolyzes at low temperatures and gives benzyl radicals in concurrence with an ionic reaction leading to benzyl phenylacetate [49]. In CCl_4/CCl_3Br, benzyl bromide and 1,1,1-trichloro-2-phenylethane are formed, both showing CIDNP of the α hydrogens [50,51].

The decomposition of *o*-toluyl peroxide in hexachloroacetone gives *o*-toluyloxy/*o*-tolyl radical pairs. Intramolecular hydrogen transfer converts the *o*-toluyloxy radical into the *o*-carboxybenzyl radical and the *o*-tolyl radical (Eq. 10.12) into the benzyl radical. The final products include *o*-chloromethylbenzoic acid, which shows an enhanced emission CIDNP methylene signal, and benzyl chloride, showing enhanced absorption [52].

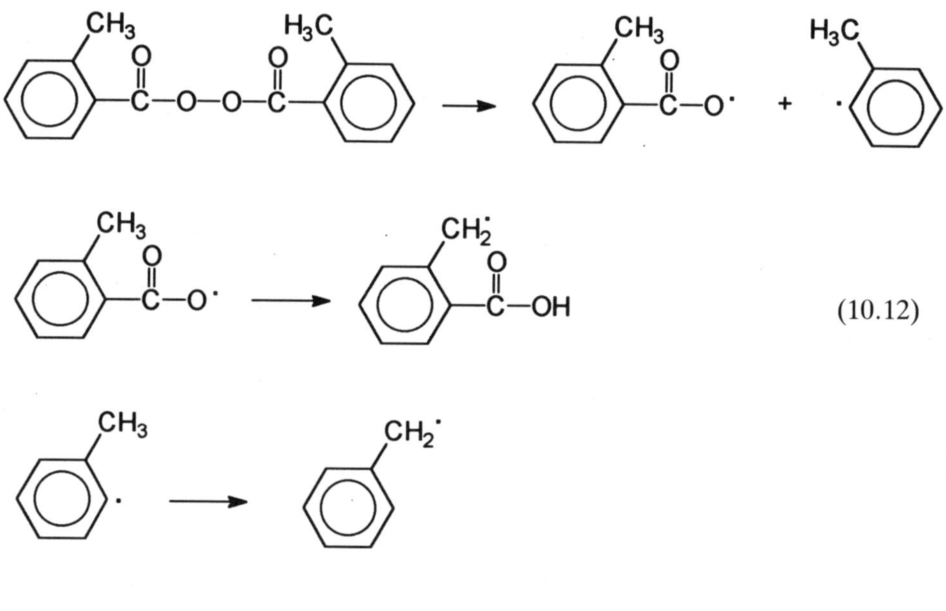

(10.12)

Benzyl radicals generated by the thermolysis of dibenzylmercury in an autoclave at 170°C abstract H atoms α to aromatic rings. The rates were obtained relative to the rate of deuterium abstraction from triphenylsilane-d using the toluene/toluene-α-d product ratios [53]. In a limited series of donors, consisting of mesitylene, α-methylnaphthalene, diphenylmethane, and triphenylmethane, the rate constants were linearly related to the resonance stabilization energy [54] of the resulting donor radical. However, when the benzylic carbon of the donor is part of a five- or six-membered ring, the rates are higher than predicted from the resonance energy alone, possibly because the developing *p* orbital is locked into a favorable conformation for overlap with the aromatic ring.

10.6.2 Formation by Cleavage of Radicals

Potassium graphite, KC_8, a potassium–graphite intercalation compound, can serve as a one-electron source for the reduction of aryl dialkyl sulfonium salts in THF [55].

$$C_6H_5R_2S^+ + e^-\cdot \longrightarrow [C_6H_5R_2S\cdot] \longrightarrow C_6H_5\text{—}S\text{—}R + R\cdot \quad (10.13)$$

This is a general reaction in which the cleaved radical R· gives typical radical products. With two different alkyl groups R and R′ in the same sulfonium salt, both radicals are formed, but not in equal amounts. The relative amounts of cleavage for radicals of various kinds are benzyl 28, secondary alkyl 6, primary alkyl [1], methyl 0.5, and phenyl less than 0.05.

Alkoxy radicals from the thermolysis of dibenzylmethylcarbinyl hypochlorites cleave to a corresponding acetophenone and benzyl radical or substituted benzyl radical [56].

$$(XC_6H_4CH_2)(C_6H_5CH_2)(CH_3)C\text{—}O\cdot \longrightarrow ArCH_2\text{—}CO\text{—}CH_3 + ArCH_2\cdot \quad (10.14)$$

The Hammett plot of the relative cleavage rates, log k_X/k_H, against σ^+ has a slope of -1.04.

10.6.3 Photolysis of Carbonyl Compounds

Carbonyl compounds of several types having a benzylic group bonded to a carbonyl give benzyl radicals by photolysis. Examples include ketones, esters, anhydrides, and salts.

Ketones. Type I photolysis of dibenzyl ketones gives pairs of benzyl and phenylacetyl radicals. Decarbonylation of the phenylacetyl radical gives a pair of benzyl radicals. If the dibenzyl ketone is unsymmetrical, with a substituent in just one of the phenyl rings, geminate and nongeminate combination gives a mixture of all three dibenzyls [57].

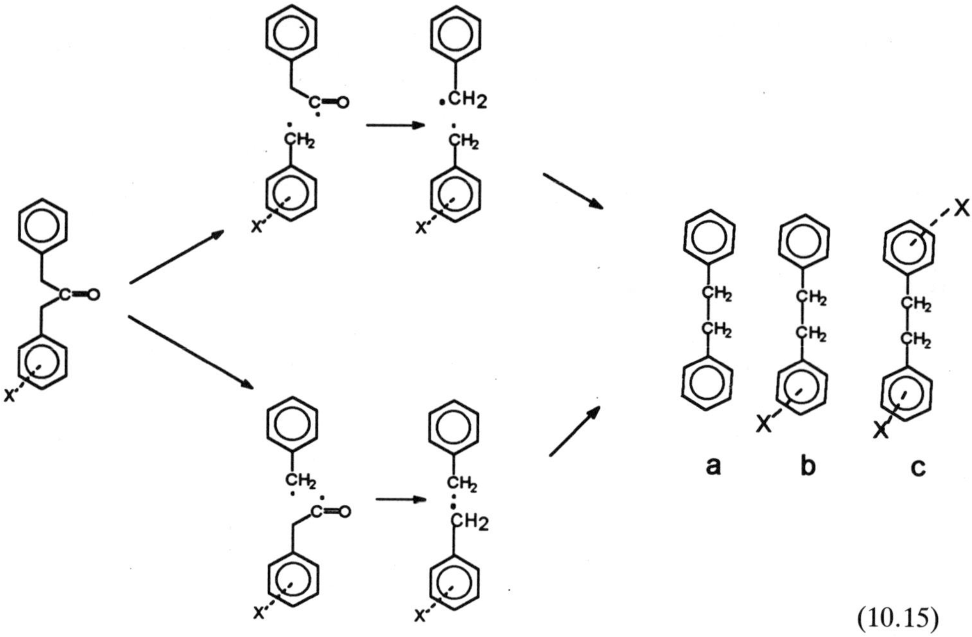

(10.15)

The relative rates of the two competing initial cleavage modes can be obtained from CIDNP. Taking *p*-methoxybenzyl benzyl ketone as an example, the ^{1}H CIDNP spectrum of the starting ketone during irradiation showed enhanced emission from both of the methylene groups. However, the intensity of the emission from the *p*-methoxybenzyl methylene was greater than that of the other by a factor of 3.2, indicating faster cleavage on that side of the carbonyl group in the dibenzyl ketone triplet state. A plot of log k_X/k_H against sigma had a ρ^+ value of -0.65. This value, and the fact that the fit is much better using σ^+ than using σ, implies a polar transition state with a considerable positive charge on the methylene group of the developing benzyl radical.

Photolysis, on a silica surface, of dibenzyl ketone partly labeled with ^{17}O at the carbonyl group, gives recovered ketone further enriched with the magnetic ^{17}O isotope [58]. An even greater enrichment of the magnetic isotope occurs for ^{13}C at the carbonyl, since the coupling of the nuclear and electronic spins is greater. The proportion of geminate combination of benzyl radicals from the photolysis of dibenzyl ketones on silica [59] and in micelles [60] is at a maximum in the weak magnetic field of the earth and decreases as the strength of an applied external field is increased.

Esters and Acids. Photolysis of α-naphthylmethyl esters of phenylacetic acid in methanol gives α-naphthylmethyl and acyloxy radical pairs [61,62]. With certain substituents in the naphthyl group, ion pairs may also be produced, probably by electron transfer in the radical pair. In methanol the naphthylacetoxy radical loses CO_2 with an estimated rate constant of 5×10^9 sec^{-1}.

Flash photolysis of either methyl phenylacetate or phenylacetic acid in methanol can cleave either the benzyl to carbonyl bond or the benzoyl to oxygen bond [63]. Either process can give the benzyl radical, a transient absorption of which has maxima at 314 and 302 nm, and eventually dibenzyl and 2-phenylethanol and other radical products. There is more cleavage of the carbonyl to oxygen bond in the ester than in the acid.

The photolysis of benzyl phenylacetate in methanol probably cleaves the benzyl to phenylacetoxy bond in the first step [64]. The products are 91% bibenzyl, 85% CO_2, and 3% CO. A *p*-methyl substituent in the alcohol moiety of the ester, where it can interact directly with the developing benzyl radical, substantially increases the quantum yield, while in the other moiety it does not.

Salts. Flash photolysis of sodium phenylacetate in methanol predominantly cleaves the benzyl to carbonyl bond, giving benzyl anion and benzyl radical, and eventually toluene (97%) from the anion, and dibenzyl (2%) from the radical [65]. Although the spectrum of the benzyl anion was not detected in these experiments, the nitrobenzyl anion has been seen in the photolysis of nitrophenylacetate ion in water [66]. The photolysis of sodium phenylacetate in water, rather than methanol, gives an insoluble polymeric acid [65].

Anhydrides. Photolysis of *p*-methoxyphenylacetic anhydride in acetonitrile gives a mixture of the corresponding ester *p*-methoxybenzyl *p*-methoxyphenylacetate, *p,p'*-dimethoxydibenzyl ketone, and *p,p'*-dimethoxybibenzyl [67]. When an asymmetrical anhydride, with only one *p*-methoxy substituent is used, the asymmetrical ester *p*-methoxybenzyl phenylacetate is the predominant ester, indicating an ion pair source for this product. All three possible bibenzyls are formed, in a nearly statistical ratio, probably from the benzyl radicals. Photolysis of *p*-methoxybenzyl phenylacetate also gives the three bibenzyls in a nearly statistical ratio.

10.6.4 Photolysis of Halomethylarenes

Excitation of 1-(halomethyl)naphthalenes with a pulsed 266-nm laser in hexane at room temperature gives a vibrationally excited level of the $S_2(\pi,\pi^*)$ state; this populates an upper triplet state and results in the formation of the 1-naphthylmethyl radical [68]. If the 266-nm pulse is followed by a later 355-nm pulse, the emission spectrum of the excited 1-naphthylmethyl radical is seen. The emission spectra from the experiments with the chloromethyl and bromomethylnaphthalenes are identical. They have two prominent bands near 600 nm and a characteristic vibrational structure.

Similar experiments with the 2-(halomethyl)naphthalenes gave the emission spectrum of the 2-naphthylmethyl radical, also consisting of bands near 600 nm but easily distinguishable from those of the 1-naphthylmethyl radical.

The transient absorption spectra of the 1- and 2-napthylmethyl radicals have broad bands with a maximum near 365 and 380 nm, respectively, with vibrational structure too unresolved for comparison with the emission spectra.

Photolysis of 1-(chloromethyl)naphthalene in cyclohexane gives only the radical products methylnaphthalene and 1-(cyclohexylmethyl)naphthalene. Photolysis in methanol gave both radicals and ions [69]. The products were dinaphthylethane and 2-(1-naphthyl)ethyl alcohol from the radical branch of the reaction, and 1-(methoxymethyl)naphthalene from the ionic branch. In methanol saturated with O_2, 1-(methoxymethyl)naphthalene was still produced in about the same yield but the radicals were diverted to give naphthaldehyde. The ion pairs could be formed either directly or by electron transfer between the geminate pair of radicals, since oxygen would be unlikely to intervene very efficiently.

Photolysis of benzyl chloride in *t*-butanol gave a mixture of homolysis products sich as bibenzyl, and photosolvolysis products such as benzyl *t*-butyl ether [70]. Both types of product are formed both in ketone-sensitized and in direct photolysis reactions, but the direct photolysis gives a higher proportion of homolysis products.

10.6.5 Energetics

Like the heat of formation of the *t*-butyl radical, the heat of formation of the benzyl radical used in the 1970s was probably too low.

Benzyl radicals were formed by reactions 10.16 and 10.17 in the gas phase and the ionization potential measured [71]. From the ionization potential, 7.20 eV, and other thermochemical quantities, the heat of formation of the benzyl radical is estimated to be 48.1 kcal mole^{-1} and the benzyl–hydrogen bond dissociation energy to be 88.2 kcal mole^{-1} [72].

$$C_6H_5CH_2CH_2ONO \longrightarrow C_6H_5CH_2CH_2O\cdot + NO\cdot \qquad (10.16)$$

$$C_6H_5CH_2CH_2O\cdot \longrightarrow C_6H_5CH_2\cdot + CH_2O \qquad (10.17)$$

The oxidation potential is 0.40 ± 0.03 V and the reduction potential is -1.78 ± 0.02 V [73].

10.6.6 Reactions

From the sinusoidally modulated UV/vis absorption of the benzyl radical, produced by modulated photolysis of dibenzyl ketone, it is possible to extract $(1.8 \pm 0.1) \times 10^9$ M^{-1} sec^{-1} as the rate constant for the dimerization of benzyl radicals in cyclohexane at 24°C [74]. This is equal within experimental error to the diffusion-controlled value predicted by the Smoluchowsky equation using the spin-statistical factor of $\frac{1}{4}$.

The rate constant for the reaction of benzyl radicals with O_2, measured by laser flash photolysis of the same ketone in cyclohexane at 27°C, is $(2.36 \pm 0.22) \times 10^9$ M^{-1} sec^{-1} [75]. This is slightly slower than the diffusion-controlled rate calculated using the $\frac{1}{3}$ spin-statistical factor for a doublet reacting with a triplet to give a doublet product.

The fluorescent excited state of the benzyl radicals is very short lived owing to a temperature-dependent nonradiative relaxation. However, the excited states of para-substituted benzyl radicals are longer lived by as much as two orders of magnitude, enough to undergo bimolecular reactions such as abstraction of a hydrogen atom from 1,4-cyclohexadiene [76]. The substituted benzyl radicals were produced by 248-nm laser pulses from the corresponding benzyl halides and then excited by 308-nm pulses. The rate constants for the abstraction reactions ranged from 3.2×10^9 M^{-1} sec^{-1} for *p*-cyano to 9.4×10^5 for *p*-methoxy.

10.7 ALLYL

The allyl radical is calculated to have the C_{2V} symmetrical structure [77] in agreement with ESR [78], photoelectron spectroscopy [71], and electron diffraction [79] data. The heat of formation is about 38 kcal $mole^{-1}$ [72].

It can be made by the photolysis of di-*t*-butyl peroxide in propylene. A similar procedure gives a cis methallyl from *cis*-butene-2 and a trans methallyl from *trans*-butene-2 [78].

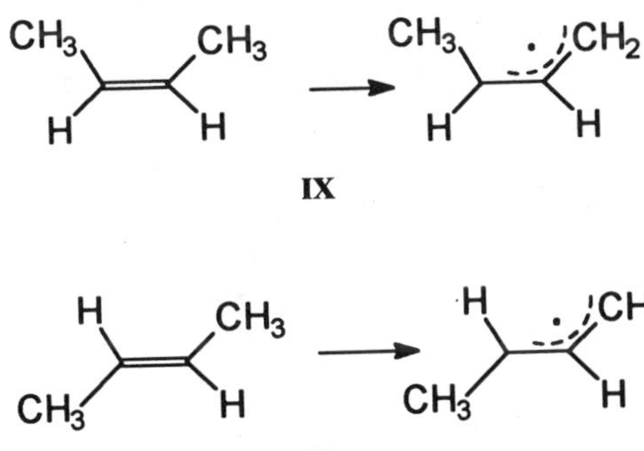

IX

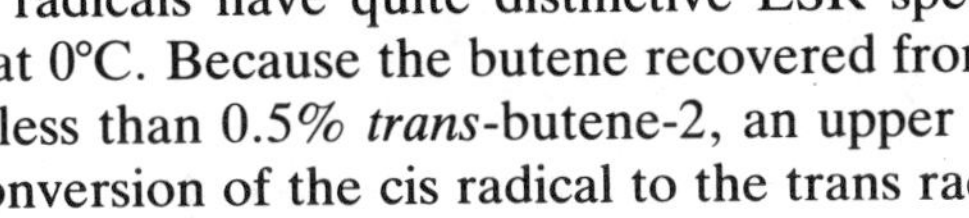

X

The two allylic radicals have quite distinctive ESR spectra, not only at −130°C, but also at 0°C. Because the butene recovered from the *cis*-butene-2 experiment had less than 0.5% *trans*-butene-2, an upper limit for the rate constant for the conversion of the cis radical to the trans radical is estimated to be 10^2 sec^{-1}.

10.8 PERCHLORO RADICALS

10.8.1 Perchlorotrityl

The perchlorotrityl radical $C_{19}Cl_{15}\cdot$ (mp 305°C) is made by the oxidation of the corresponding carbanion with iodine [80]. It is deep red and completely dissociated both in the solid state and in solution. Solutions exposed to air for months show no change in the optical spectrum. If there is any reaction, the half-life must be several decades at least. The radical is also inert to Br_2, Cl_2, boiling toluene, nitric oxide, and concentrated nitric acid.

The ESR spectrum consists of a single line at $g = 2.0026$, about the same position as the center of the triphenylmethyl ESR spectrum. This line is flanked by lines due to splittings in radicals containing ^{13}C, 29.5 G for the central carbon, 12.5 G for the adjacent carbon, and 10.7 G for the ortho carbon. The ^{13}C splitting of the central carbon of the trityl radical, is only 23.5 G, indicating that the odd electron in trityl is more delocalized than the odd electron in perchlorotrityl. This conclusion is supported by the ESR spectrum of a modified perchlorotrityl radical [81], with hydrogen in place of the one of the para chlorines. That radical gives an a_H of 1.8 G compared to 2.75 G for the p-H of the trityl radical.

Reactions of Perchlorotrityl. The space-filling propeller-shaped model of the perchlorotrityl radical shows the central carbon atom overhung by the ortho chlorines to such an extent that reagents cannot approach to within covalent bonding distance. But single-electron transfers do not require such a close approach. These include the reduction of the radical to the carbanion in ether by potassium metal [80] or by ascorbic acid [82], and the oxidation to the carbenium ion by $SbCl_5$ [83].

Covalent bond formation at the para positions is discouraged by the low extent of delocalization of the odd electron into the nearly perpendicular rings.

Irradiation [84] of the perchlorotrityl doublet at 532 nm in CCl_4 gives an excited doublet. The 605-nm fluorescence is quenched by single electron donors such as triphenylamine. The major photoproduct from perchlorotrityl is the known perchloro-9-phenylfluorenyl [85], which is a green solid melting at about 300°C and unreactive toward O_2.

The mechanism of the conversion of the excited doublet involves a sequence of two intermediates. The first of these may be a charge transfer isomer $Ar_2C^+{-}Ar^{-}\cdot$; the second is probably analogous to the first intermediate in the photodisproportionation of trityl (Eq. 10.3).

10.8.2 Perchlorobenzhydryl and Perchlorobenzyl

Although perchlorobenzhydryl and perchlorobenzyl radicals are remarkably unreactive compared to benzhydryl and benzyl, they are more reactive than perchlorotrityl.

Perchlorobenzhydryl [80] can be made by oxidation of the corresponding perchlorodiphenylmethide carbanion with I_2 or reduction of perchlorodiphenylmethane with $FeCl_2$ in ether. This radical has a half-life of 2–3 days in solution under air. It reacts slowly with Br_2 or Cl_2. Like perchlorotrityl it is reduced to the corresponding carbanion by potassium metal or by potassium hydroxide in dimethylsulfoxide.

Perchlorobenzyl, in equilibrium with its dimer, can be made by the irradiation of perchlorotoluene in CCl_4 (Eq. 10.18) [86].

$$2\ C_6Cl_5CCl_3 \underset{}{\overset{h\nu}{\rightleftharpoons}} 2\ C_6Cl_5CCl_2\cdot + Cl_2 \qquad (10.18)$$

The reaction can be driven to the right by adding copper filings to remove the Cl_2. This solution shows an ESR signal with an intensity which increases reversibly on warming. It was possible to isolate a derivative of perchlorodibenzyl in which each ring had one ortho hydrogen. Perchlorodibenzyl itself is presumed to have the α–α structure since it can be reduced to perchlorostilbene.

10.9 HETEROLYSIS AND HOMOLYSIS OF C–C BONDS

Solvents which, like sulfolane (tetramethylene sulfone) [87,88], combine a high dielectric constant with a reasonably low reactivity, make it possible for radical pairs and ion pairs to coexist (Eq. 10.19) [87–89].

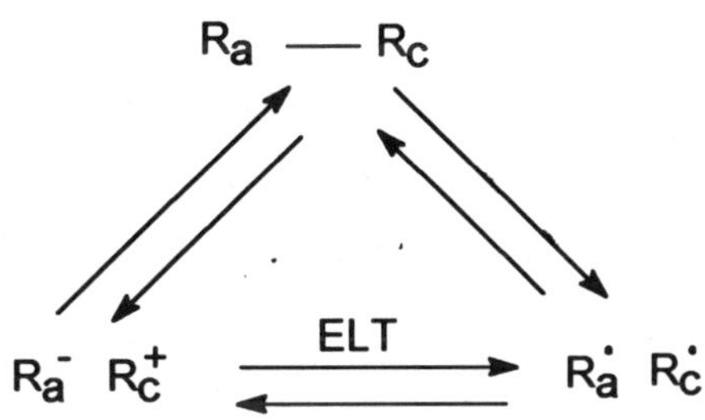

(10.19)

The heat of heterolysis can be measured directly by mixing solutions of the resonance-stabilized carbenium and carbanion solutions in a calorimeter [90]. For example, ΔH_{het} for the dissociation of covalently bonded trityl and 9-phenylfluorenyl at 25°C in 95% sulfolane–5% 3-methylsulfolane is 35.51 ± 0.56 kcal mole^{-1}.

The heat of homolysis cannot be obtained in the same way because solutions of the corresponding unassociated radicals are not available, but it could be calculated from ΔH_{het} and the enthalpy of electron transfer. Unfortunately the *enthalpy* of electron transfer in *solution* is not available. What is available is the *free energy* of electron transfer,

$$23.06[E_{red}(R_c^+) - E_{ox}(Ra_a^-)]$$

obtained from electrode potentials. This quantity differs from the needed enthalpy of electron transfer by $T\Delta S°$. The $T\Delta S°$ term for the reaction in solution should be substantial, but might be some constant proportion of ΔG_{ET} when systems in a related series are compared. With this caveat, the heat of homolysis is approximated by Eq. 10.20 [90].

$$\Delta H_{hom} = \Delta H_{het} - 23.06[E_{red}(R_c^+) - E_{ox}(R_a^-)] \qquad (10.20)$$

For trityl as R_c, and 9-phenylfluorenyl as R_a, ΔG_{ET} is -22.32 and ΔH_{hom} is 13.18. When R_c is 9-phenylxanthyl and R_a is 2,7-dibromo-9-carbomethoxyfluorenyl, ΔG_{ET} is only 1.52, and homolysis and heterolysis are nearly balanced, ΔH_{hom} and ΔH_{het} being 17.44 ± 0.43 and 18.96, respectively.

An even closer balance is achieved when R_c is 1,3,5-tricyclopropyltropylium (**XI**) and R_a is Kuhn's carbanion (**XII**) [89].

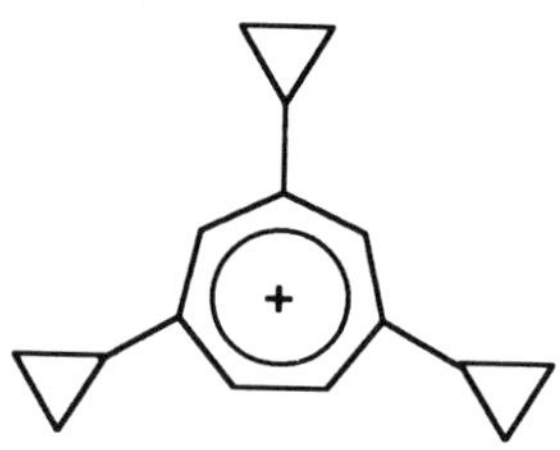

XI

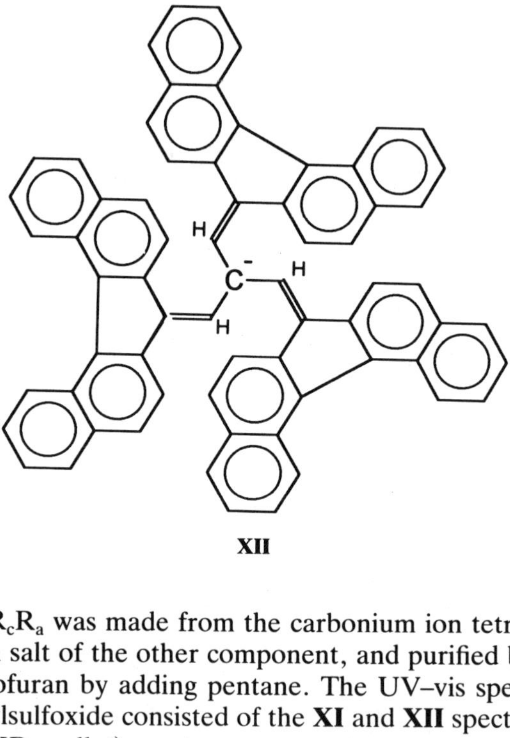

XII

The dark green ionic solid R_cR_a was made from the carbonium ion tetrafluoroborate and the potassium salt of the other component, and purified by precipitating it from tetrahydrofuran by adding pentane. The UV–vis spectrum of this product in dimethylsulfoxide consisted of the **XI** and **XII** spectra superimposed, as did the IR (KBr pellet) spectrum.

The behavior of R_cR_a in solvents depends on the dielectric constant. In DMSO (ϵ 46.5) it is dark green and consists of the two ions. In $CHCl_3$ (ϵ 4.81) it is orange and the absorption spectrum is that of the covalent compound. On standing, the chloroform solution changes color and shows absorption due to the R_a radical. In THF (ϵ 7.58) the visible spectrum of R_cR_a changed from that of the ions and rapidly approached that of an equilibrium mixture of the ions, radicals, and the covalent compound R_a–R_c.

A slow single-electron transfer converts a mixture of perchlorotriphenylmethyl cation and anion entirely into the perchlorotriphenylmethyl radical [91]. Equilibrium constants have also been obtained for the single electron transfers between various anions $XC_6Cl_4(C_6Cl_5)_2C^-$ and radicals $YC_6Cl_4(C_6Cl_5)_2C\cdot$ in THF [91]. The position of the equilibrium (Eq. 10.21) is usually such that both radicals are present in concentrations measurable by means of ESR.

$$XC_6Cl_4(C_6Cl_5)_2C^- + YC_6Cl_4(C_6Cl_5)_2C\cdot \overset{K}{\rightleftharpoons} XC_6Cl_4(C_6Cl_5)_2C\cdot + YC_6Cl_4(C_6Cl_5)_2C^- \quad (10.21)$$

The Hammett rho value for the equilibrium constants with X = H and Y H, *p*-methyl, *p*-chloro, and *p*-bromo, is about 4.

The ESR spectrum of the cation radical, **XIII**, showed intramolecular electron transfer between a radical site and a cationic site [92]. The spectrum indicated half a unit spin and half a unit charge on each end of the molecule. Since the ^{13}C coupling constant was $\frac{1}{2}$ that in the analogous simple radical $[C_6Cl_5)_3C\cdot]$, spin/charge exchange was fast.

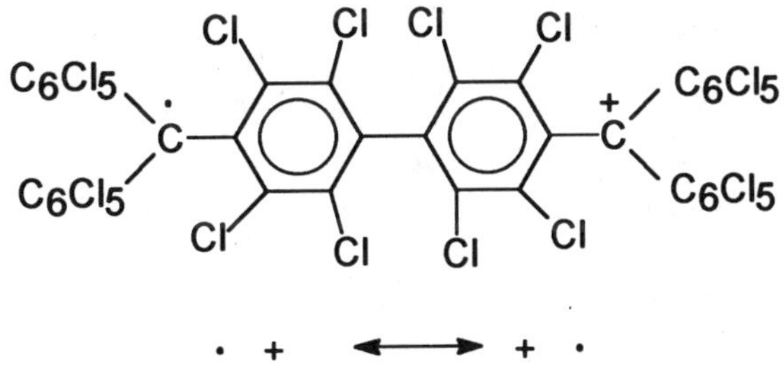

XIII

10.10 THE KOELSCH RADICAL

The Koelsch radical (**XIV**), the famous "delayed radical," is of historical interest because of what happened to the original manuscript [93]. The manuscript was first submitted to the *Journal of the American Chemical Society* in June of 1932 and rejected.

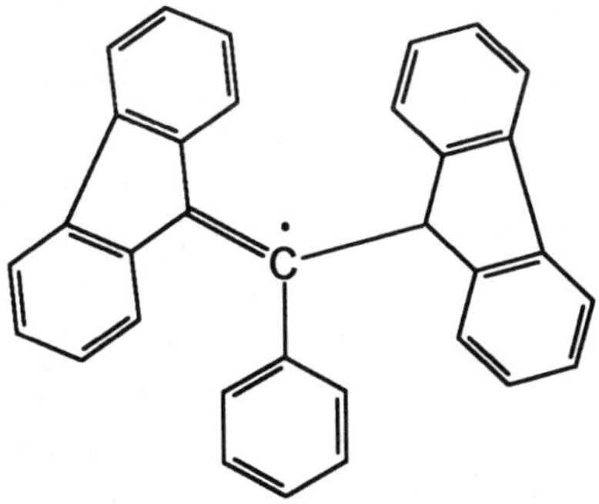

XIV

The paper was finally published, unchanged except for the addition of footnotes about quantum mechanical calculations and ESR, in August of 1957, twenty-five years later. Unlike trityl radical, **XIV** is very resistant to oxygen, part of it being recovered unchanged after passing oxygen through a boiling benzene solution for six hours. The ESR spectrum was not available in 1932, but in 1957 a sample that had been kept in air for 25 years was found still to have a high radical content.

REFERENCES

1. McBride, J. M. *Tetrahedron* **1974,** *30,* 2009.
2. Gomberg, M. *J. Am. Chem. Soc.* **1900,** *22,* 757.
3. Gomberg, M. *Ber. deutsch. chem. Ges.* **1900,** *33,* 3150.
4. Lankamp, H.; Nauta, W. T.; MacLean, C. *Tetrahedron Lett.* **1968,** 249.
5. Ullmann, F.; Borsum, W. *Ber. deutsch. chem. Ges.* **1902,** *35,* 2877.
6. Jacobson, P. *Ber. deutsch. chem. Ges.* **1905,** *38,* 196.
7. Colle, T. H.; Glaspie, P. S.; Lewis, E. S. *J. Org. Chem.* **1978,** *43,* 2722.
8. Ziegler, K.; Ewald, L. *Liebigs Ann. Chem.* **1929,** *473,* 163.
9. Lichtin, N. N.; Thomas, G. R. *J. Am. Chem. Soc.* **1954,** *76,* 3020.
10. Skinner, K. J.; Hochster, H. S.; McBride, J. M. *J. Am. Chem. Soc.* **1974,** *96,* 4301.
11. Ziegler, K.; Seib, A.; Knoevenagel, K.; Herte, P.; Andreas, F. *Liebigs Ann. Chem.* **1942,** *551,* 150.
12. Janzen, E. G.; Johnston, F. J.; Ayers, C. L. *J. Am. Chem. Soc.* **1967,** *89,* 1176.
13. Marvel, C. S.; Kaplan, J. F.; Himel, C. M. *J. Am. Chem. Soc.* **1941,** *63,* 1892.
14. Lichtin, N. N.; Glazier, H. *J. Am. Chem. Soc.* **1951,** *73,* 5537.
15. Bowden, S. T.; Watkins, T. F. *J. Chem. Soc.* **1940,** 1249.
16. Neumann, W. P.; Uzick, W.; Zarkadis, A. K. *J. Am. Chem. Soc.* **1986,** *108,* 3762.
17. Osawa, E.; Onuki, Y.; Mislow, K. *J. Am. Chem. Soc.* **1981,** *103,* 7475.
18. Adrian, F. J. *J. Chem. Phys.* **1958,** *28,* 608.
19. Andersen, P. *Acta Chem. Scand.* **1965,** *19,* 622.
20. Taub, I. A.; Harter, D. A.; Sauer, M. C. J.; Dorfman, L. M. *J. Chem. Phys.* **1964,** *41,* 979.
21. Wallis, E. S.; Adams, F. H. *J. Am. Chem. Soc.* **1933,** *55,* 3838.
22. Bordwell, F. G.; Cheng, J.-P.; Ji, G.-Z.; Satish, A. V.; Zhang, X. *J. Am.Chem. Soc.* **1991,** *113,* 9790.
23. Bromberg, A.; Schmidt, K. H.; Meisel, D. *J. Am. Chem. Soc.* **1985,** *107,* 83.
24. Schmidt, J. A.; Hilinski, E. F. *J. Am. Chem. Soc.* **1988,** *110,* 4036.
25. Gomberg, M.; Cone, L. H. *Ber. deutsch. chem. Ges.* **1904,** *37,* 3538.
26. Bowden, S. T.; Jones, W. J. *J. Chem. Soc.* **1928,** *1149.*
27. Letsinger, R. L.; Collat, R.; Magnuson, M. *J. Am. Chem. Soc.* **1954,** *76,* 4185.
28. Faria, J. L.; Steenken, S. *J. Am. Chem. Soc.* **1990,** *112,* 1277.

29. Huszthy, P.; Lempert, K.; Simig, G.; Tamás, J.; Hegedüs-Vajda, J.; Tóth, G. *J. Chem. Soc. Perkin II* **1985,** 491.
30. Lorand, J. P.; Bartlett, P. D. *J. Am. Chem. Soc.* **1966,** *88,* 3294.
31. Suehiro, T.; Igeta, S.; Kuwabara, O.; Hirai, M. *Tetrahedron* **1970,** *26,* 963.
32. Hammond, G. S.; Rudesill, J. T.; Modic, F. J. *J. Am. Chem. Soc.* **1951,** *72,*3929.
33. Suehiro, T.; Kanoya, A.; Hara, H.; Nakahama, T.; Omori, M.; Komori, T. *Bull. Chem. Soc. Jpn.* **1967,** *40,* 668.
34. Doering, W. von E.; Okamoto, K.; Krauch, H. *J. Am. Chem. Soc.* **1960,** *82,* 3579.
35. Reichel, C. L.; McBride, J. M. *J. Am. Chem. Soc.* **1977,** *99,* 6758.
36. Bartlett, P. D.; Lewis, E. S. *J. Am. Chem. Soc.* **1950,** *72,* 1005.
37. Bartlett, P. D.; Ryan, M. J.; Cohen, S. G. *J. Am. Chem. Soc.* **1942,** *64,* 2649.
38. Leffler, J. E.; More, A. A. *J.Am. Chem. Soc.* **1972,** *94,* 2483.
39. Bartlett, P. D.; Greene, F. D. *J. Am. Chem. Soc.* **1954,** *76,* 1088.
40. Koukotas, C.; Mehlman, S. P.; Schwartz, L. H. *J. Org. Chem.* **1966,** *31,* 1970.
41. Müller, E.; Moosmayer, A.; Rieker, A.; Scheffler, K. *Tetrahedron Lett.* **1967,** 3877.
42. Neugebauer, F. A.; Hellwinkel, D.; Aulmich, G. *Tetrahedron Lett.* **1978,** 4871.
43. Staab, H. A.; Rao, K. S.; Brunner, H. *Chem. Ber.* **1971,** *104,* 2634.
44. Rapoport, H.; Smolinsky, G. *J. Am. Chem. Soc.* **1960,** *82,* 934.
45. Hounshell, W. D.; Dougherty, D. A.; Hummel, J. P.; Mislow, K. *J. Am. Chem. Soc.* **1977,** *99,* 1916.
46. Beckhaus, H.-D.; Schaetzer, J.; Rüchardt, C. *Tetrahedron Lett.* **1983,** *24,* 3307.
47. Wittig, G.; Petri, H. *Liebigs Ann. Chem.* **1933,** *505,* 17.
48. Wittig, G.; Schoch, W. *Liebigs Ann. Chem.* **1971,** *749,* 38.
49. Bartlett, P. D.; Leffler, J. E. *J. Am. Chem. Soc.* **1950,** *72,* 3030.
50. Walling, C.; Lepley, A.R. *J. Am. Chem. Soc.* **1971,** *93,* 546.
51. Walling, C.; Lepley, A. R. *J. Am. Chem. Soc.* **1972,** *94,* 2007.
52. Childress, B. C.; Rice, A. C.; Shevlin, P. B. *J. Org. Chem.* **1974,** *39,* 3056.
53. Bockrath, B.; Bittner, E.; McGrew, J. *J. Am. Chem. Soc.* **1984,** *106,* 135.
54. Stein, S. E.; Golden, D. M. *J. Org. Chem.* **1977,** *42,* 839.
55. Beak, P.; Sullivan, T. A. *J. Am. Chem. Soc.* **1982,** *104,* 4450.
56. Walling, C.; Clark, R. T. *J. Am. Chem. Soc.* **1974,** *96,* 4530.
57. Johnston, L. J.; de Mayo, P.; Wong, S. K. *J. Am. Chem. Soc.* **1982,** *104,* 307.
58. Turro, N. J.; Paczkowski, M. A.; Wan, P. *J. Org. Chem.* **1985,** *50,* 1399.
59. Turro, N. J.; Cheng, C.-C.; Mahler, W. *J. Am. Chem. Soc.* **1984,** *106,* 5022.
60. Turro, N. J.; Weed, G. C. *J. Am. Chem. Soc.* **1983,** *105,* 1861.
61. Hilborn, J. W.; Pincock, J. A. *J. Am.Chem. Soc.* **1991,** *113,* 2683.
62. DeCosta, D. P.; Pincock, J. A. *J. Am. Chem. Soc.* **1989,** *111,* 8948.
63. Meiggs, T. O.; Miller, S. I. *J. Am. Chem. Soc.* **1972,** *94,* 1989.
64. Meiggs, T. O.; Grossweiner, L. I.; Miller, S. I. *J. Am. Chem. Soc.* **1972,** *94,* 7986.

65. Meiggs, T. O.; Grossweiner, L. I.; Miller, S. I. *J. Am. Chem. Soc.* **1972,** *94,* 7981.
66. Margerum, J. D.; Pertusis, C. D. *J. Am. Chem. Soc.* **1969,** *91,* 2467.
67. Roof, A. A. M.; Woerden, H. F. van; Cerfontain, H. *J. Chem. Soc. Perkin II* **1980,** 838.
68. Hilinski, E. F.; Huppert, D.; Kelley, D. F.; Milton, S. V.; Rentzepis, P. M. *J. Am. Chem. Soc.* **1984,** *106,* 1951.
69. Slocum, G. H.; Schuster, G. B. *J. Org. Chem.* **1984,** *49,* 2177.
70. Cristol, S. J.; Bindel, T. H. *J. Org. Chem.* **1980,** *45,* 951.
71. Houle, F. A.; Beauchamp, J. L. *J. Am. Chem. Soc.* **1978,** *100,* 3290.
72. Hayashibara, K.; Kruppa, G. H.; Beauchamp, J. L. *J. Am. Chem. Soc.* **1986,** *108,* 5441.
73. Wayner, D. D. M.; Griller, D. *J. Am. Chem. Soc.* **1985,** *107,* 7764.
74. Huggenberger, C.; Fischer, H. *Helv. Chim. Acta.* **1981,** *64,* 338.
75. Maillard, B.; Ingold, K. U.; Scaiano, J. C. *J. Am. Chem. Soc.* **1983,** *105,* 5095.
76. Tokumura, K.; Ozaki, T.; Itoh, M. *J. Am. Chem. Soc.* **1989,** *111,* 5999.
77. Takada, T.; Dupuis, M. *J. Am. Chem. Soc.* **1983,** *105,* 1713.
78. Kochi, J. K.; Krusic, P. J. *J. Am. Chem. Soc.* **1968,** *90,* 7157.
79. Vajda, E.; Tremmel, J.; Rozsondai, B.; Hargittai, I.; Maltsev, A. K.; Kagramonov, N. D.; Nefedov, O. M. *J. Am. Chem. Soc.* **1986,** *108,* 4352.
80. Ballester, M.; Riera, J.; Castañer, J.; Badía, C.; Monsó, J. M. *J. Am.Chem. Soc.* **1971,** *93,* 2215.
81. Ballester, M.; Castañer, J.; Riera, J.; Ibáñez, A.; Pujadas, J. *J. Org. Chem.* **1982,** *47,* 259.
82. Ballester, M.; Riera, J.; Castañer, J.; Casulleras, M. *Tetrahedron Lett.* **1978,** 643.
83. Ballester, M.; Riera, J.; Castañer, J.; Rodriguez, A.; Rovira, C.; Veciana, J. *J. Org. Chem.* **1982,** *47,* 4498.
84. Fox, M. A.; Gaillard, E.; Chen, C.-C. *J. Am. Chem. Soc.* **1987,** *109,* 7088.
85. Ballester, M.; Castañer, J.; Pujadas, J. *Tetrahedron Lett.* **1971,** 1699.
86. Olivella, S.; Ballester, M.; Castañer, J. *Tetrahedron Lett.* **1974,** 587.
87. Arnett, E. M.; Amarnath, K.; Harvey, N. G.; Cheng, J. P. *Science* **1990,** *247,* 423.
88. Arnett, E. M.; Molter, K. E.; Marchot, E. C.; Donovan, W. H.; Smith, P. *J. Am. Chem. Soc.* **1987,** *109,* 3788.
89. Komatsu, K.; Aonuma, S.; Takeuchi, K.; Okamoto, K. *J. Org. Chem.* **1989,** *54,* 2038.
90. Arnett, E. M.; Amarnath, K.; Harvey, N. G.; Cheng, J.-P. *J. Am. Chem. Soc.* **1990,** *112,* 344.
91. Ballester, M.; Pascual, I.; Torres, J. *J. Org. Chem.* **1990,** *55,* 3035.
92. Ballester, M.; Castañer, J.; Riera, J.; Pascual, I. *J. Am. Chem. Soc.* **1984,** *106,* 3365.
93. Koelsch, C. F. *J. Am. Chem. Soc.* **1957,** *79,* 4439.

11

SYNTHETIC APPLICATIONS

Free radical reactions often are tolerant of functional groups and give products not accessible by means of the conventional polar routes. They can also be highly selective, and often proceed under milder conditions. Because of these considerations, synthetic organic chemists have been responsible for some of the most important advances in the field of free radical chemistry. In fact, the literature of free radical reactions applied to organic synthesis is so large [1,2] that only a relatively few examples, with some emphasis on more recent developments, are given here.

11.1 SELECTIVE HALOGENATION

For radical halogenation to be useful in synthesis the halogen has to be directed to the proper position. If the desired position is already somewhat favored, it may be sufficient to use a more selective halogenating agent, for example Br· instead of Cl·, or Cl· complexed with pyridine [3,4] instead of Cl· alone. While the tert/primary selectivity for the chlorination of 2,3-dimethylbutane is only 50/1 in CCl_4, it climbs to 200/1 in the presence of 4 M pyridine [3].

11.1.1 Tethered Reagents

Still greater selectivity can be achieved by tethering the pyridine or other complexing agent in a position where the chlorinating reagent can reach only certain positions [5,6].

An example is the use of a nicotinate ester (Eq. 11.1). The yield of the 9-chloro product is 98% despite the fact that chlorination at that position has to compete with chlorination at activated positions near the enone system.

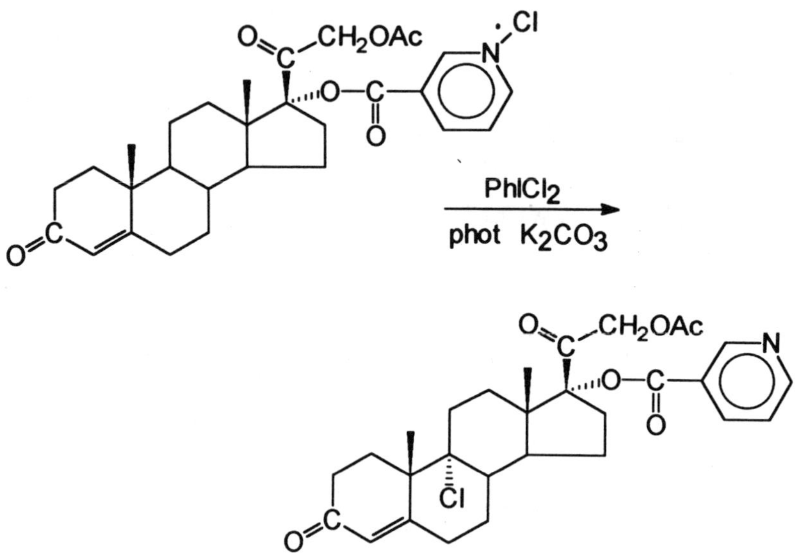

(11.1)

In the example shown in Eq. 11.1 the chlorine is transferred to the pyridine nitrogen by a "radical relay" chain (Eqs. 11.2–11.4) from PhICl·. After the hydrogen abstraction, PhICl· is regenerated when the carbon radical $R_C^{\cdot}$ abstracts Cl from $PhICl_2$ (Eq. 11.4).

$$PhICl_2 \xrightarrow{h\nu} PhICl\cdot + Cl\cdot \tag{11.2}$$

$$PhICl\cdot + \text{R-pyridine} \longrightarrow PhI + \text{R-pyridine-Cl}\cdot \tag{11.3}$$

$$\text{R-pyridine-Cl}\cdot \longrightarrow R_C^{\cdot} + HCl \tag{11.3a}$$

$$R_C^{\cdot} + PhICl_2 \longrightarrow R_C\text{—}Cl + PhICl\cdot \tag{11.4}$$

Other tethered carriers of Cl· that have been used include *m*-iodobenzoate esters [5], thiophenes [7], and diaryl sulfides [8].

Completely selective chlorination of 3α-cholestanol at the C-9 position is achieved with a tethered *m*-iodobenzoate bearing a Cl on the iodine (Eq. 11.5) [6]:

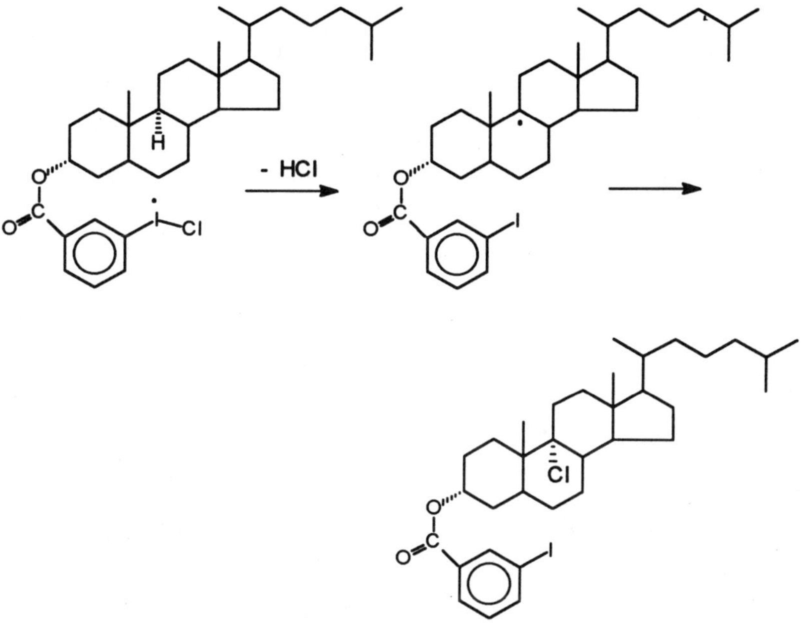

(11.5)

With *p*-iodophenylacetate, a more extended Cl carrier, selective chlorination occurs at C-14, a more remote position, as shown in Eq. 11.5a.

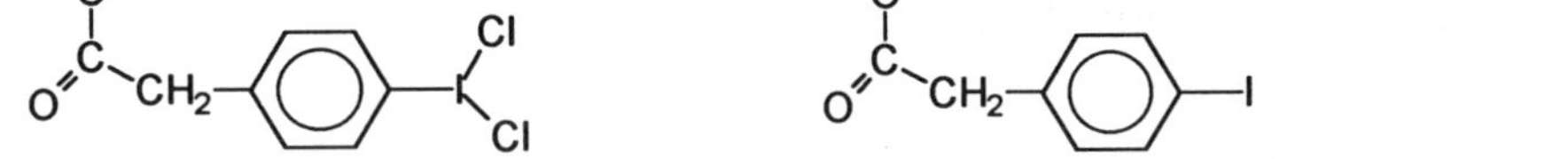

(11.5a)

11.2 REMOVAL OF FUNCTIONAL GROUPS

11.2.1 Halogen

Alkyl halides are conveniently reduced to the hydrocarbon by tri-*n*-butyltin hydride in a two-step chain reaction [9,10].

$$R\cdot + n\text{-}Bu_3SnH \longrightarrow RH + n\text{-}Bu_3Sn\cdot \tag{11.6}$$

$$n\text{-}Bu_3Sn\cdot + RX \longrightarrow n\text{-}Bu_3SnX + R\cdot \tag{11.7}$$

11.2.2 Deoxygenation

Alcohols can be deoxygenated by a chain reaction between Bu_3SnH and a thionoester derivative as in Scheme I [11]. The chain reaction is diagrammed using the traffic circle notation introduced by biochemists to represent enzyme-catalyzed reactions. At the intersections of the curved arrows, reagents join the sequence of reactions, and products leave. For example, R· and a hydrogen from the Bu_3SnH leave as R–H while $Bu_3Sn\cdot$ joins the circle as a new reagent. Various thionoesters can be used corresponding to X equal to phenyl, imidazole, and S–CH_3.

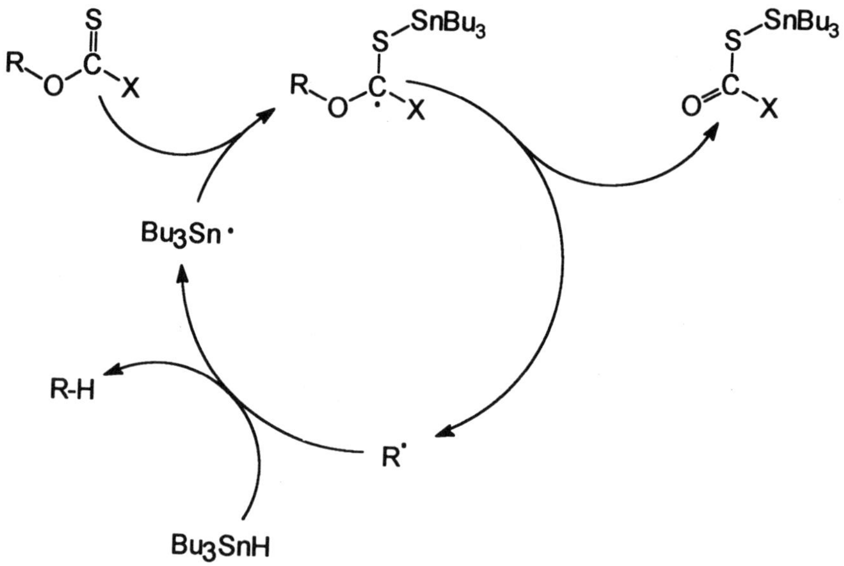

Scheme I

The reaction is quite tolerant of the presence of other functional groups. The thionoester group X = O—(CS)—OCH_3 in **I** (page 209) was replaced by H in 82% yield [11,12].

The hydroxyl group may also be replaced by an amino group, as indicated in Scheme Ia. Photolysis of the thiopyridone derivative gives methyl radicals. These react with the telluride to give the radical R·, which then adds to the diazirine [13].

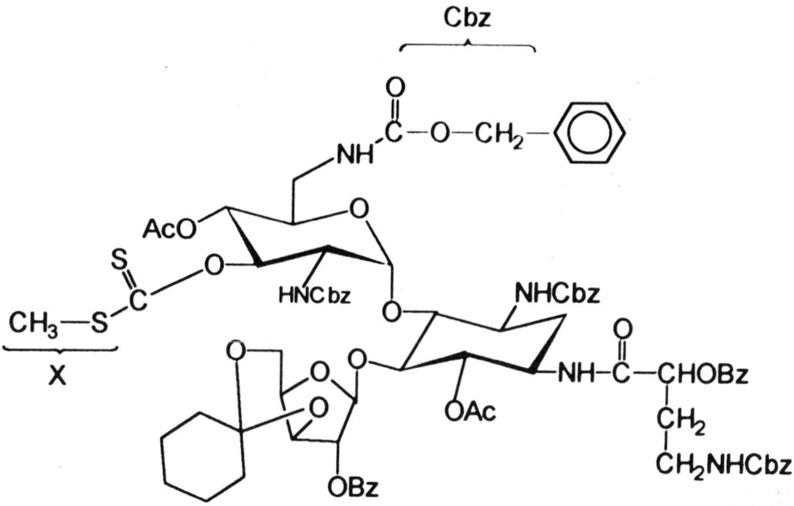

I

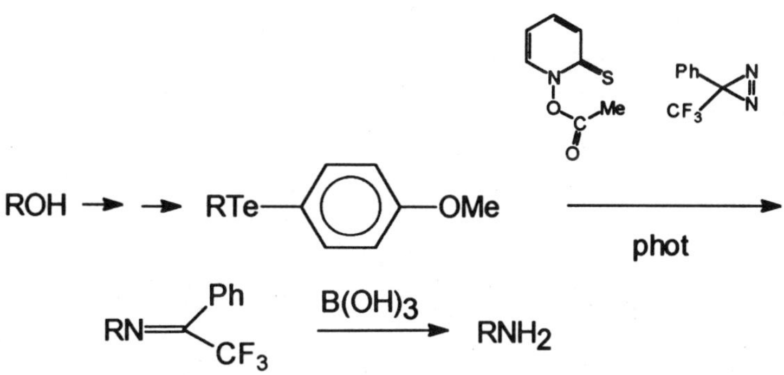

Scheme Ia

11.2.3 Deamination

Deamination is achieved by first converting the amino group into an isonitrile, an isothiocyanate, or an isoselenocyanate, then replacing it by H in a chain reaction like that used for deoxygenation [11]. With **II**, for example, the conversion yield was 90%.

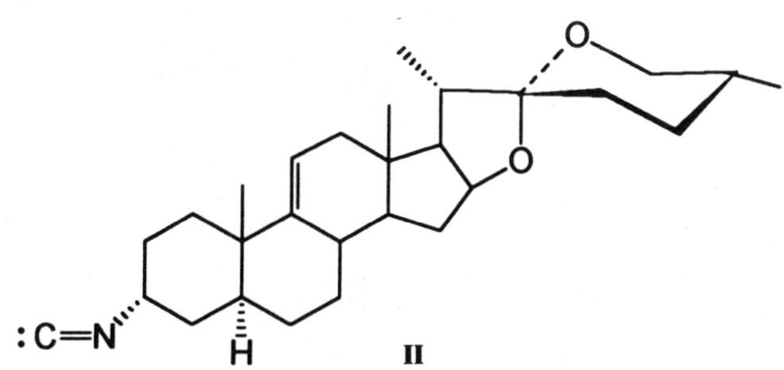

II

11.2.4 Decarboxylation

A reagent useful for decarboxylation is *N*-hydroxythiopyridone which is readily converted into the ester as in Eq. 11.8.

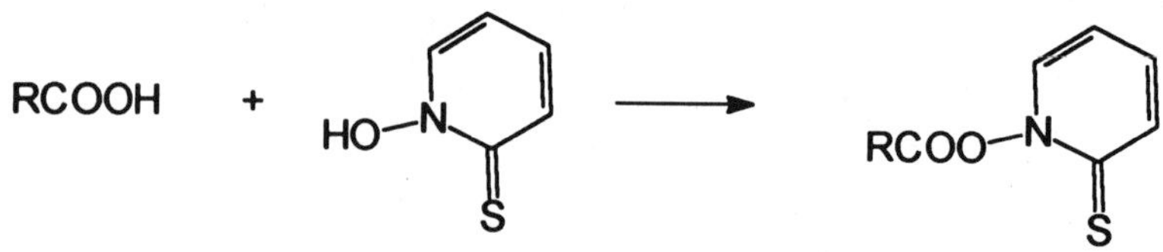
(11.8)

The decarboxylation of the ester is achieved by a chain reaction with *t*-butyl mercaptan and the *t*-butylthiyl radical (Scheme II).

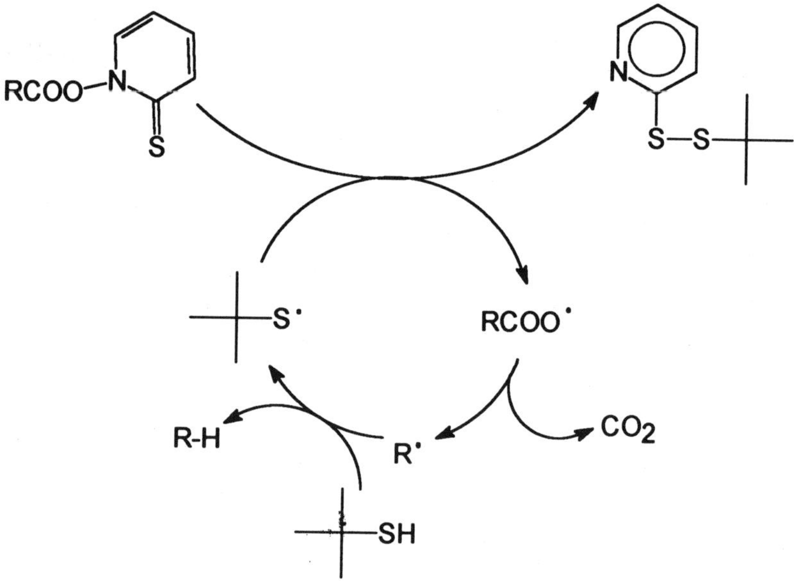

Scheme II

Replacement of carboxyl by amino is achieved by Scheme IIa [13].

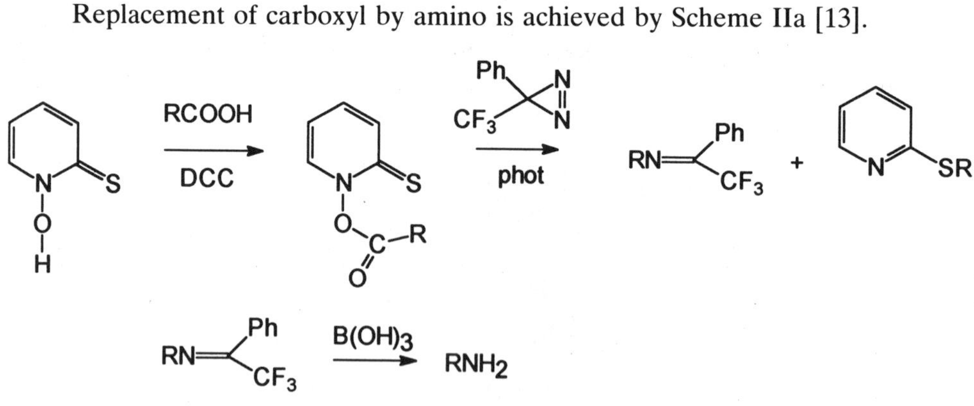

Scheme IIa

11.3 SELECTIVE HYDROGEN TRANSFER

The triplet state of a tethered benzophenone, represented by the diradical in Eq. 11.9, oxidizes the steroid to the product with a double bond specifically in the 14,15 position in 65% yield [6].

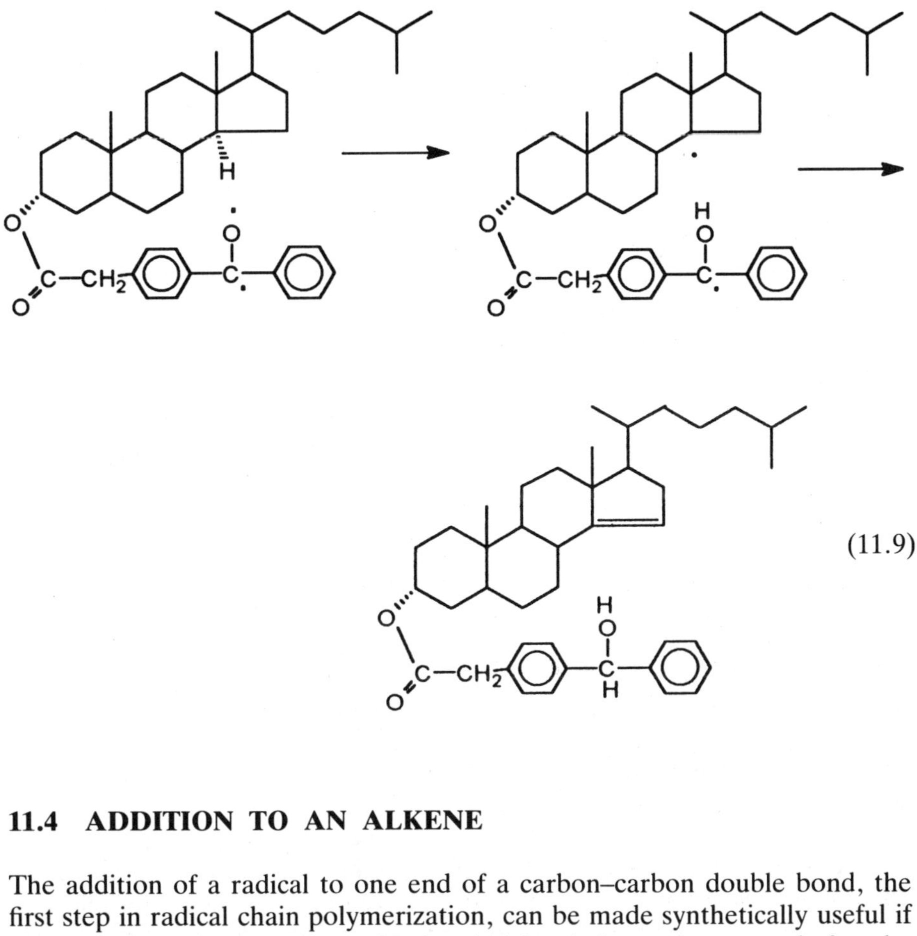

(11.9)

11.4 ADDITION TO AN ALKENE

The addition of a radical to one end of a carbon–carbon double bond, the first step in radical chain polymerization, can be made synthetically useful if competing reactions can be avoided and if the chain can be stopped after the first addition.

$$\text{R}\cdot + \text{C}{=}\text{C} \longrightarrow \text{R—C—C}\cdot \tag{11.10}$$

$$\text{R—C—C}\cdot + \text{X—Y} \longrightarrow \text{R—C—C—X} + \text{Y}\cdot \tag{11.11}$$

11.4.1 Intermolecular Reactions

An example of a process that amounts to the addition of R–H across a double bond is the chain reaction of cyclohexyl iodide, acrylonitrile, and tri-*n*-butyltin hydride (Scheme III) which gives β-cyclohexylpropionitrile in 95% yield [14]. The reaction is initiated by the photolysis of the cyclohexyl iodide.

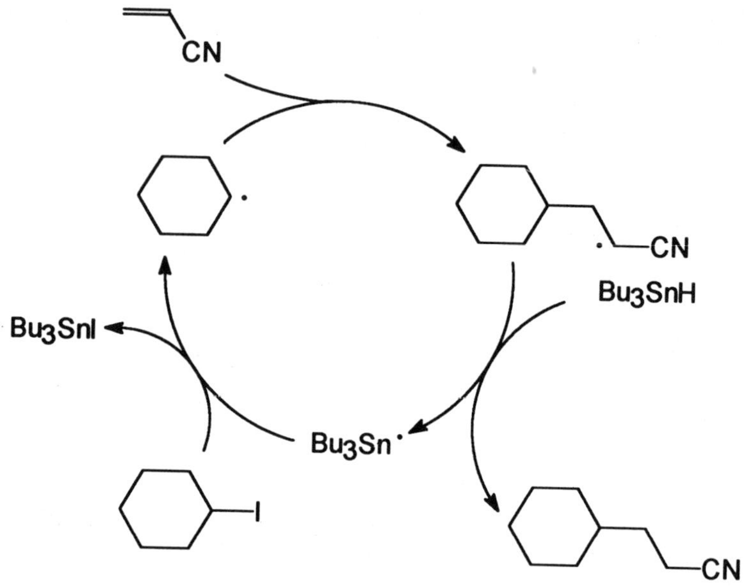

Scheme III

For a process like that in Scheme III to work, the rates of the desirable reactions must be much greater, by orders of magnitude if possible, than those of other reactions [1,9].

The cyclohexyl radical needed to react faster with acrylonitrile than with tributyltin hydride, which would otherwise divert it to cyclohexane.

The 1-cyano-2-cyclohexylethyl radical needed to react faster with the tributyltin hydride than with another molecule of acrylonitrile, which would have diverted it to a telomer.

These selectivities, inequalities in rates, are achieved in various ways:

First, to the extent that it is compatible with the objectives of the synthesis, the adding radical and the alkene should have opposite nucleo- or electrophilicities. In this example the use of the nucleophilic cyclohexyl radical with an alkene bearing an electron-withdrawing substituent was a wise choice. A reaction with vinyl butyl ether, in which the alkene has an electron-releasing substituent fails with the electrophilic radical from cyclohexyl iodide, but works with a nucleophilic radical from chloromalonic ester.

Second, concentrations can be adjusted. In this example the addition step was favored over reduction of the cyclohexyl radical by the hydride by using a large excess of the acrylonitrile. The rates of chain-breaking radical–radical reactions will automatically be low because the steady-state concentrations of the radicals are low.

Third, the reactivity of the metal hydride can be adjusted. If the alkene is too unreactive to compete with the hydrogen donor, a less reactive metal hydride can be used. The rate of hydrogen donation is greatest for alkylmercury hydrides, less with alkyltin hydrides, and still less with tributylgermanium hydride. In this example tributyltin hydride worked very well, but cyclohexylmercury hydride gave almost as good a yield. Alkenes such as styrene or vinylidene chloride, for which the radical addition step is too slow to compete, give lower yields with alkylmercury hydride than with the less reactive tributyltin hydride [1].

A particularly neat variant of the hydride trapping chain uses a suitable hydrogen donor R–H both as the source of hydrogen and the radical R· to be added. In the example shown in Scheme IV the donated H is the α hydrogen

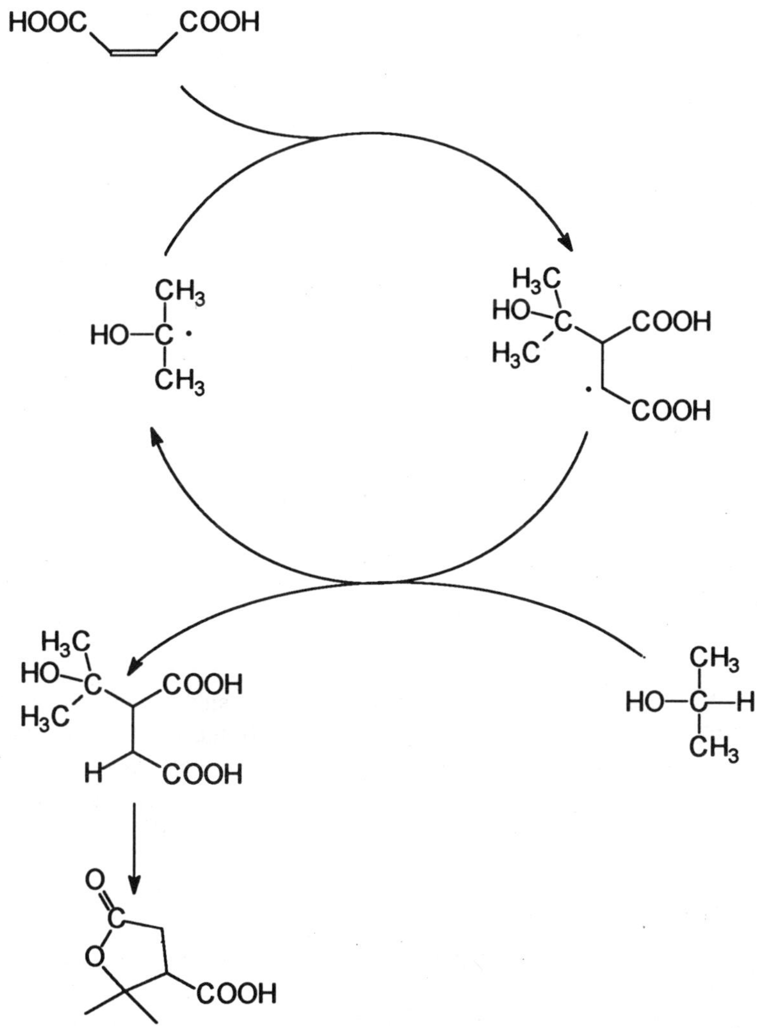

Scheme IV

of isopropyl alcohol and the radical is the 2-hydroxy-2-propyl radical which adds to maleic acid [15]. The reaction is initiated photochemically with benzophenone. The hydroxypropylsuccinic acid gives the lactone as the final product.

Another variation is shown in Scheme V [16]. In this reaction an intermolecular step is followed by a cyclization and then by an atom transfer to continue the chain.

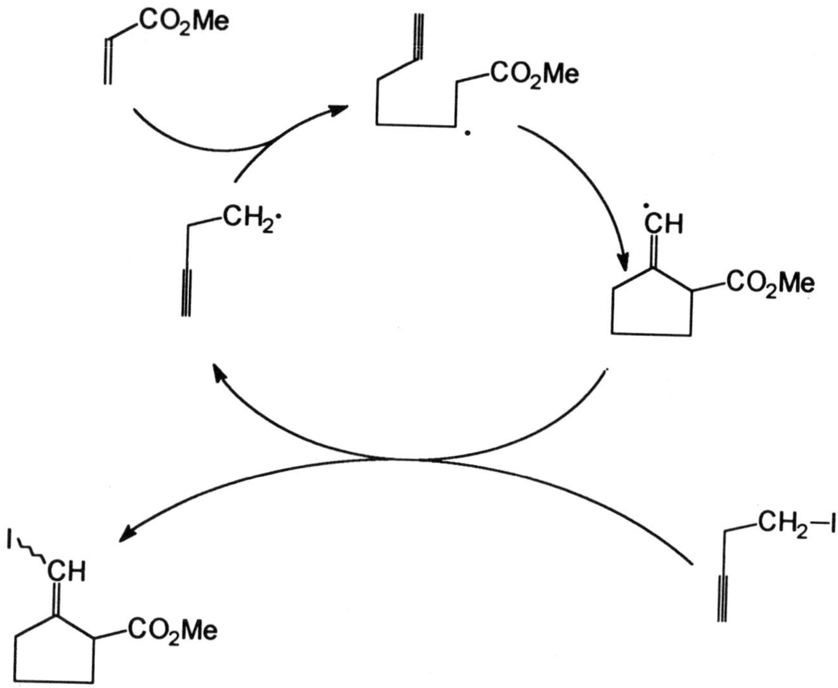

Scheme V

Scheme VI shows the use of an anisyltelluride substituent to attach a carbohydrate to maleimide [17]. The chains are initiated by the photolysis of *N*-acetoxy-2-thiopyridone, which produces methyl radical. After the attachment of the maleimide, conventional reactions are used to convert the product into the antibiotic showdomycin.

11.4.2 Intramolecular Reactions

Avoiding competing reactions is easiest when the addition is intramolecular, because then the ring-closing reaction has a less unfavorable entropy than competing bimolecular reactions. Another way of looking at it is that the C═C has a very high local concentration when it is part of the same molecule.

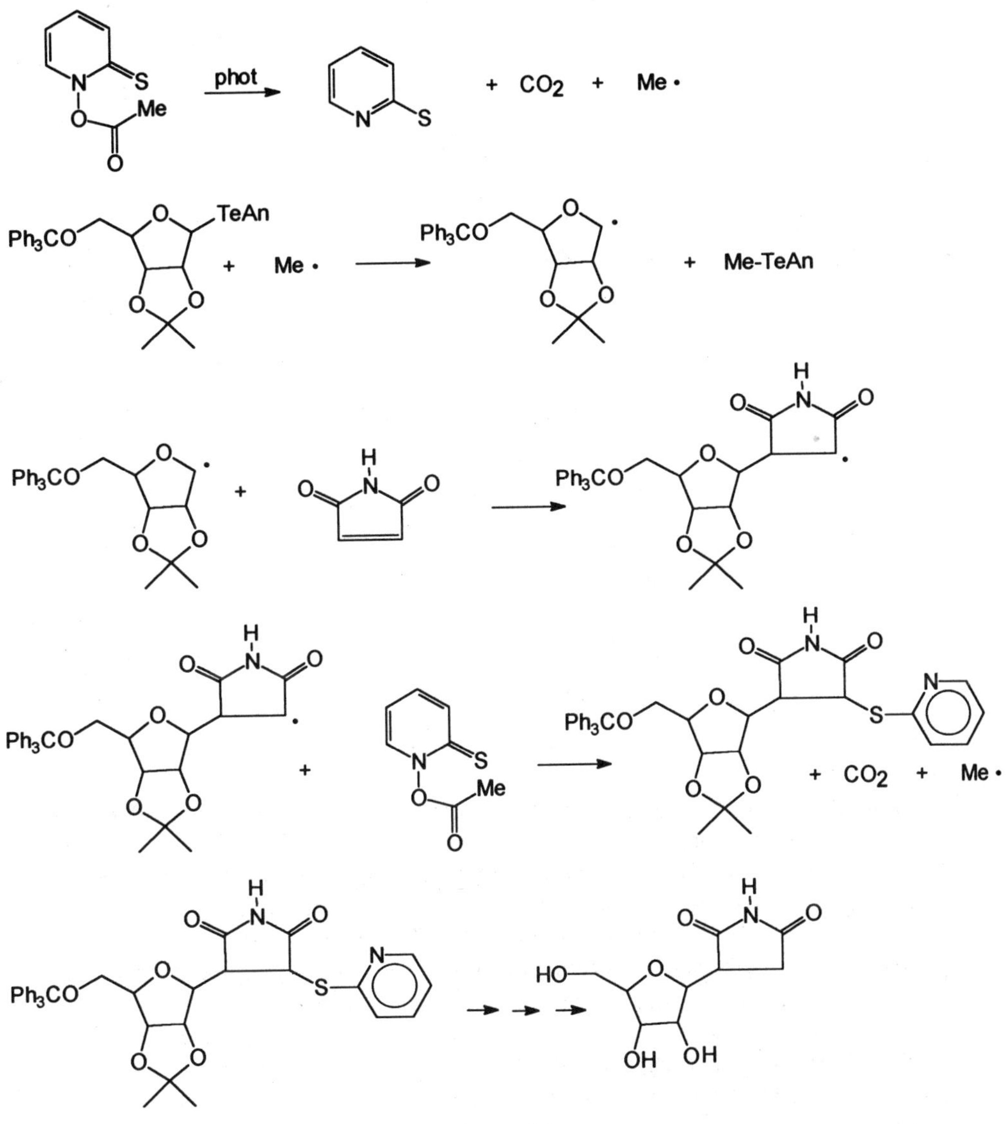

Scheme VI

Scheme VII shows an example of the cyclization of an activated alkene by Bu_3SnH [18].

5-Hexenyl Radicals. Although the free energy of the cyclohexyl radical is lower than that of cyclopentylmethyl [19], the cyclization of 5-hexenyl radicals in the presence of tributyltin hydride gives methylcyclopentane rather than cyclohexane. The formation of cyclopentylmethyl is about 50 times as fast (at 60°C) as the formation of cyclohexyl. Because the cyclopentylmethyl radical

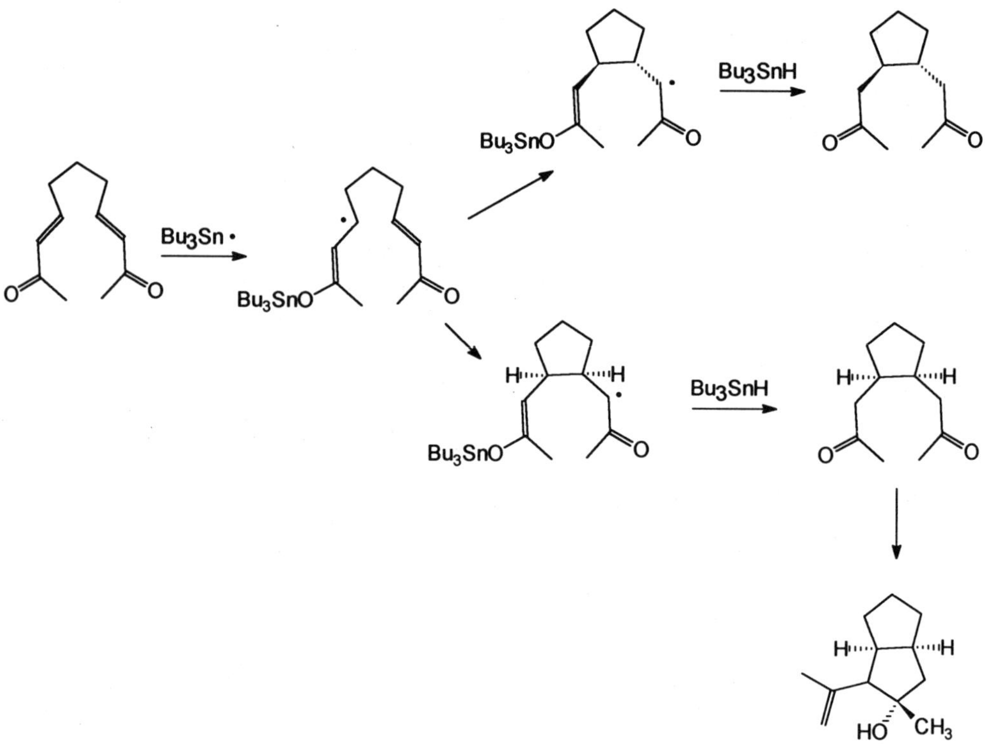

Scheme VII

is trapped so quickly, reversal of the reaction and eventual descent to the cyclohexyl radical does not occur.

The reactions of 5-hexenyl radicals also provide a sort of radical clock [9]. The rate constant for the cyclization to cyclopentylmethyl is given by $\log k = (10.42 \pm 0.32) - (6.85 \pm 0.42)/(2.303\ RT)$. Products will contain various proportions of cyclopentylmethyl and 5-hexenyl moieties depending on the relative rates of the cyclization and external reactions.

For the radical derived from 5-hexenylmalonic ester, because of adverse steric effects on the formation of the cyclopentyl radical, the rates of the alternative cyclization reactions are close enough so that a mixture of the two products is obtained (Eq. 11.12) [2,20,21].

A reaction involving the sequential closure of two six-membered rings is shown in Eq. 11.13 [22].

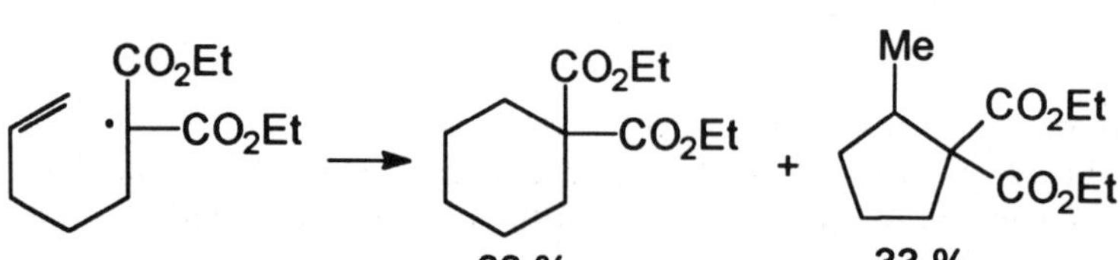

(11.12)

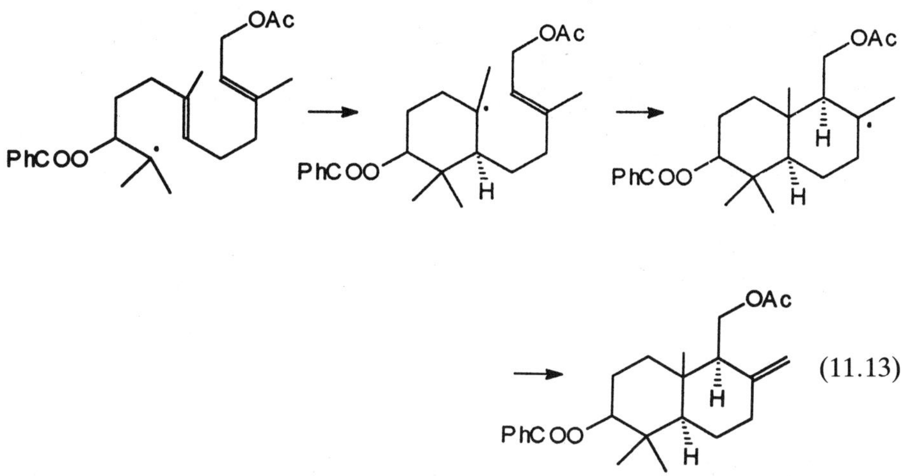

(11.13)

Hydrogen Translocation. A useful variation on the cycloaddition reaction changes the location of the radical site by H-atom transfer before the ring is closed [23,24]. An example is shown in Eq. 11.14 [23].

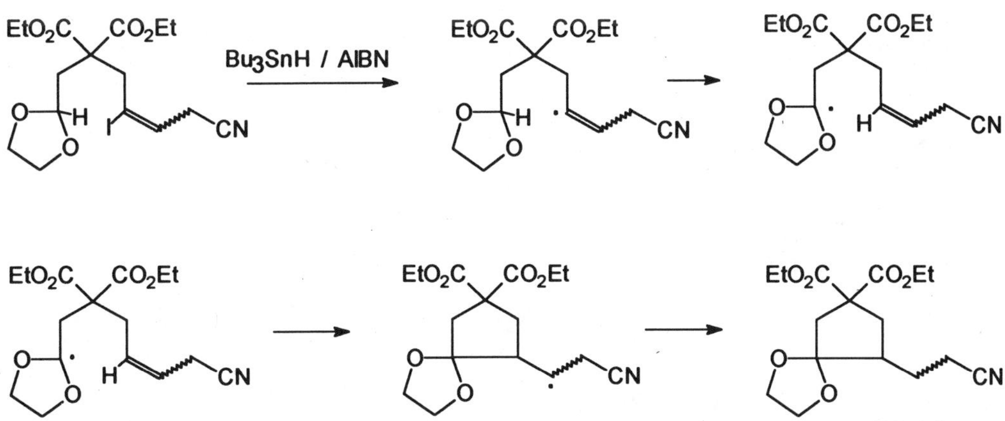

(11.14)

Halogen Transfer. Halogen atom transfer cyclizations have an advantage over the usual tin hydride cyclization in that the product retains a halogen substituent useful in further synthetic steps [25–28]. Irradiation of the alkyl iodide of Eq. 11.15 in the presence of hexabutylditin gave the cyclic vinyl iodide in 77% yield [25]. Initiation may be due either to photolysis of the ditin, in which case tributyltin radicals abstract iodine to give the alkyl radical, or by photolysis of the alkyl iodide. In any case the ditin is necessary, if only to scavenge iodine, which would otherwise intercept the alkyl radical.

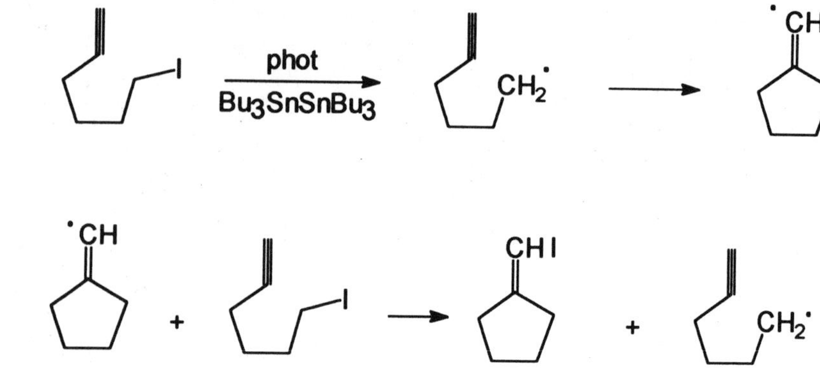

(11.15)

Equation 11.16 is a more complicated example, giving the spiro vinyl iodide in 53% yield with an E/Z isomer ratio of 83/17 [25].

A minor byproduct, 17% yield, is attributed to the 1,5 H transfer shown in Eq. 11.17.

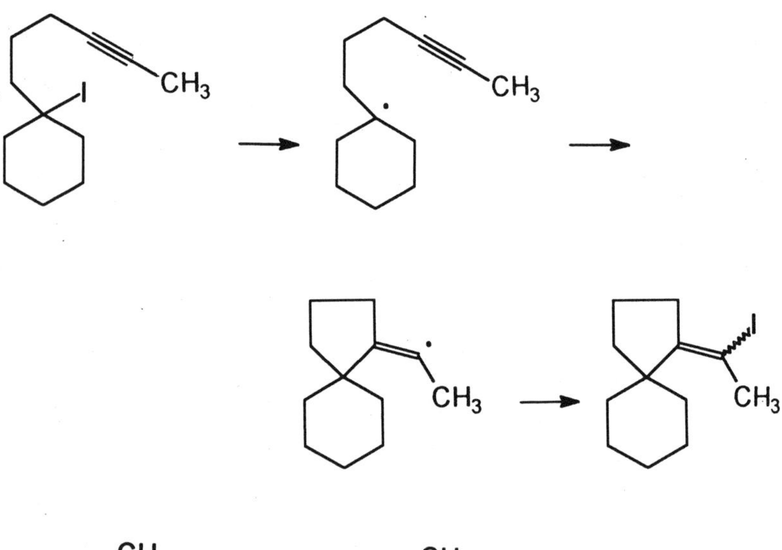

(11.16)

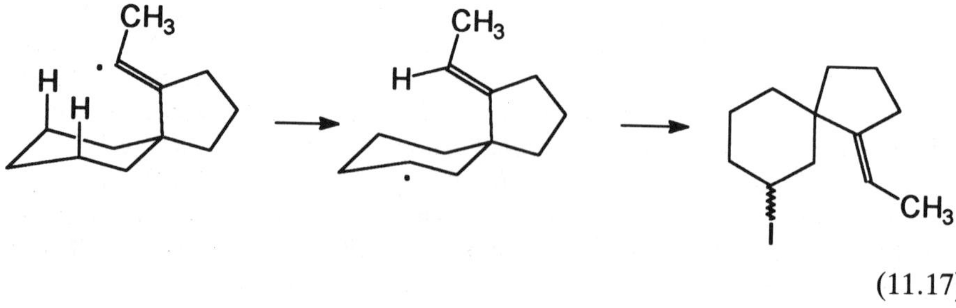

(11.17)

11.5 ADDITION TO C=X

11.5.1 Oxime Ethers

Conversion of the carbonyl function of a carbohydrate to an oxime ether function reverses the electronic character of the group and permits addition of a carbon radical [29–31]. An example is the cyclization of the benzyl-protected glucose hemiacetal in Scheme VIII [31].

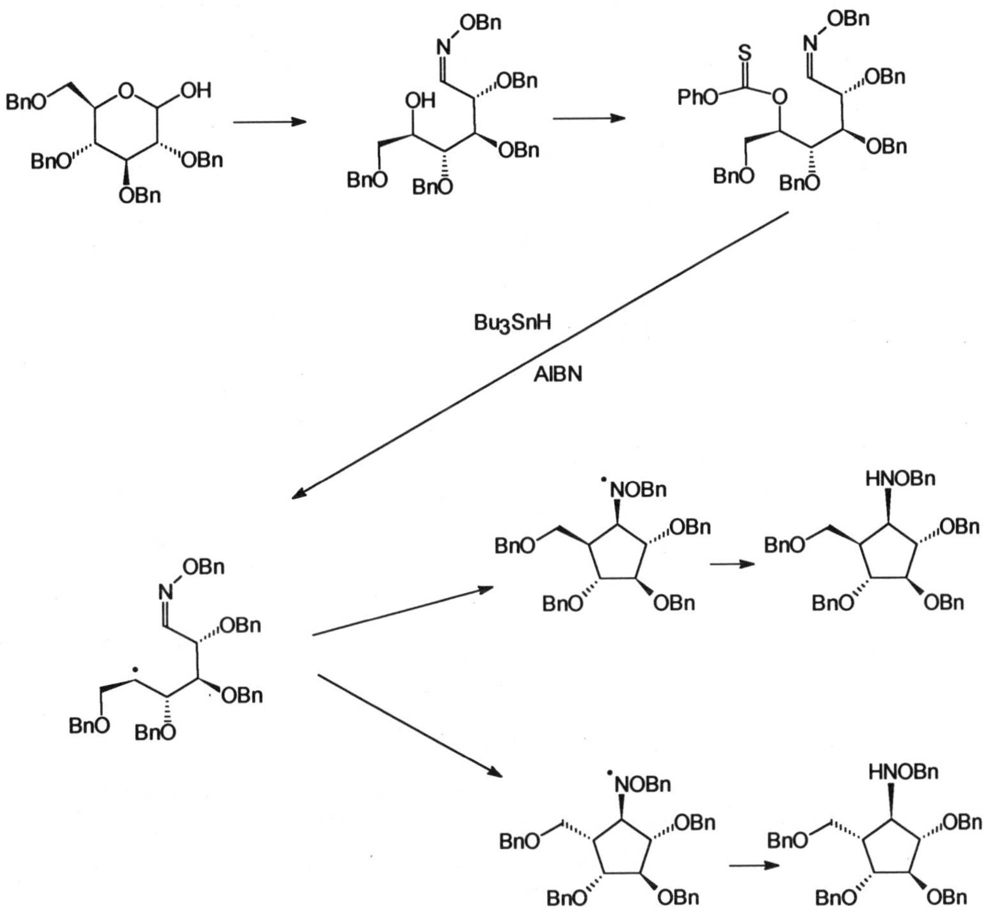

Scheme VIII

11.5.2 Isonitriles

Photolysis of 5-iodo-1-pentyne in the presence of hexabutylditin and aryl isocyanides gives 2,3-dihydrocyclopentaquinolines as in Eq. 11.18 [32].

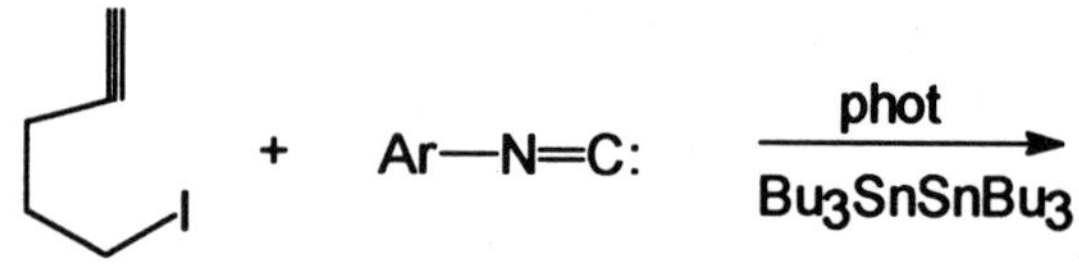

(11.18)

The mechanism involves two ring closures, one to give the 5-membered ring and a second involving attack on the aryl group either at an ortho position or an ipso addition. If the isocyanide is labeled with a para substituent, the aromatization of these intermediates gives two products, as shown in Scheme IX.

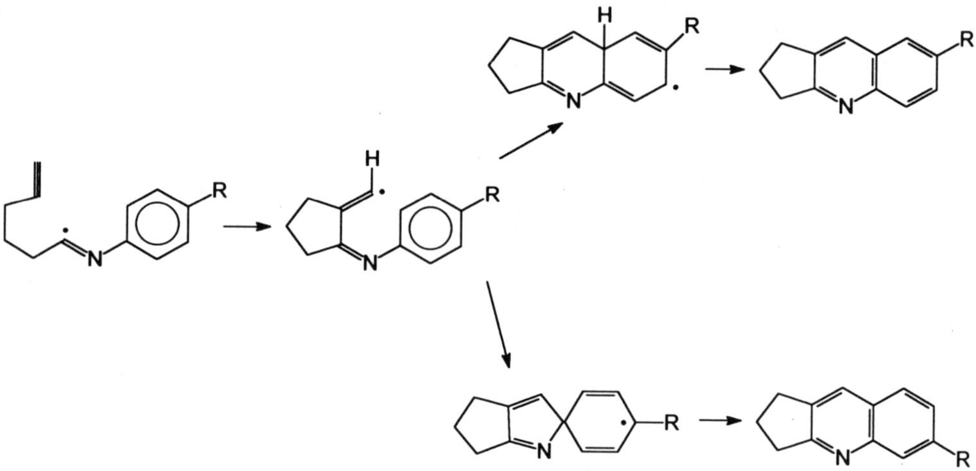

Scheme IX

11.5.3 Carbon Monoxide

The addition of carbon monoxide to a radical and subsequent oxidation of the acyl radical is a method of making carboxylic acids [33]. Carbon monoxide is bubbled through an aqueous mixture containing a precursor RH, hydrogen peroxide, and ferrous ion. Equation 11.19 is an example.

$$H_2NC(CH_3)_3 + CO \xrightarrow{H_2O_2/Fe^{2+}} H_2NC(CH_3)_2CH_2COOH \qquad (11.19)$$

Alkyl halides are converted to aldehydes in a chain reaction initiated by AIBN in the presence of tributyltin hydride and carbon monoxide (Eq. 11.20)

[34]. The reaction is successful with a variety of alkyl bromides and aryl iodides, but fails with benzyl bromide, which gives only toluene.

$$RX + Bu_3Sn\cdot \longrightarrow R\cdot + Bu_3SnX$$

$$R\cdot + CO \rightleftharpoons R\text{—}CO\cdot$$

$$R\text{—}CO\cdot + Bu_3SnH \longrightarrow R\text{—}CO\text{—}H + Bu_3Sn\cdot \qquad (11.20)$$

Using a tributylallylstannane rather than tributyltin hydride as the trap for the carbonylated radical gives allyl ketones [35].

$$R\text{—}CO\cdot + CH_2{=}CHR'CH_2SnBu_3 \longrightarrow R\text{—}CO\text{—}CH_2\text{—}CHR'{=}CH_2 \qquad (11.21)$$

11.6 DIMERIZATION

11.6.1 Mercury-Sensitized Dehydrodimerization

Mercury-photosensitized dehydrodimerization [36,37] in the vapor phase converts substrates R–H into dimers or pairs of substrates R–H and R′–H into cross dimers (Eq. 11.22–11.25).

$$R\text{—}H \longrightarrow R\cdot + H\cdot \qquad (11.22)$$

$$R\text{—}H + H\cdot \longrightarrow R\cdot + H_2 \qquad (11.23)$$

$$2\,R\cdot \longrightarrow R\text{—}R + R(-H) + RH \qquad (11.24)$$

$$R(-H) + H\cdot \longrightarrow R\cdot \qquad (11.25)$$

In the past this reaction has been restricted to small-scale operations and low extents of conversion, and therefore has not been useful. However, these limitations can be overcome by a simple experimental arrangement. The substrate or mixture of substrates, together with a small drop of mercury, is heated to provide the vapor to a cylindrical space irradiated by low-pressure 254-nm mercury lamps. At a steady rate of refluxing the vapor consists mostly of the substrate R–H and very little of the dimer. Disproportionation does not interfere, because one of the products is the starting material and the other reacts with hydrogen atoms to regenerate the radical.

The reaction is particularly good for making C–C bonds between highly substituted atoms. For example, the dimers from 2-methylbutane were 63% the tert-tert dimer, 34% the sec-tert dimer, and only 3% other isomers. Isobutane gave 2,2,3,3-tetramethylbutane as the sole product in spite of the 9:1 statistical advantage of the primary C–H bonds.

Alkanes can be functionalized by cross dimerization with alcohols or ethers using the same experimental technique.

11.6.2 Photolysis of Peroxides

Dimers R–R can be made by the photolysis of the neat peroxide $(RCOO)_2$ at $-78°C$ [38]. In two of the more favorable examples 1-heneicosene was prepared in 64% yield (Eq. 11.26), and 24-decyl-3α,7α,12α-triacetoxy-5β-cholane in 74% yield (Eq. 11.27).

$$CH_2{=}CH(CH_2)_8{-}COO{-}OOC{-}(CH_2)_{10}CH_3 \xrightarrow{h\nu} CH_2{=}CH(CH_2)_{18}CH_3 + 2\ CO_2 \quad (11.26)$$

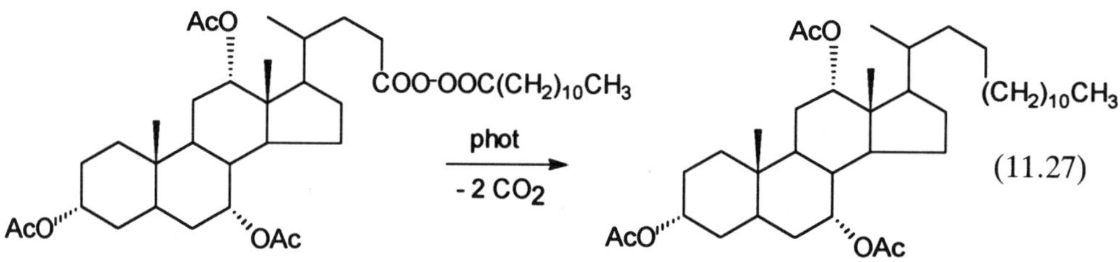

(11.27)

11.7 REGIO- AND STEREOCHEMISTRY IN CYCLIC SYSTEMS [19,39,40]

11.7.1 Cyclization of 5-Hexenyl Radicals

The formation of the 5-membered ring rather than the 6-membered ring from the 5-hexenyl radical not only has a less negative entropy of activation, but also a lower enthalpy of activation [41]. A stereoelectronic requirement may explain the lower enthalpy of activation [42,43]. According to this hypothesis, overlap of the semioccupied $2p$ orbital with one lobe of the vacant π^* orbital of the double bond can be accommodated better in forming the 5-membered ring than the 6-membered ring. In fact, the 5-membered ring usually predominates, but to an extent that varies with substituents on the radical.

When the 5-hexenyl radical has a substituent, the 5-membered ring will have two substituents, which can be either cis or trans. Cyclization of 1- or 3-substituted 5-hexenyl radicals leads mostly to the cis product; cyclization of 2- or 4-substituted 5-hexenyl radicals gives mostly the trans product [19,40]. For example the cis/trans ratio of the substituted cyclopentanes from the reaction of 1-bromo-3-methyl-5-hexene with tributyltin hydride at 60°C is 2.7; for 1-bromo-2-methyl-5-hexene it is 0.56.

These stereochemical results are understandable in terms of a model for the transition state (**III**) that resembles the chair form of cyclohexane [40]. The transition state for closing the ring is assumed to occur near the reagent end of the reaction coordinate. This means that the distance between the

III

reaction centers 1 and 5 is long, not very different from the distance between the nonbonded carbons labeled 1 and 5 in the cyclohexane. As in cyclohexane, bulky substituents in the transition state for the ring closure will prefer to occupy pseudoequatorial rather than axial positions. Thus for the 2-methyl-5-hexenyl radical the favored transition state is **IV**, giving the trans product.

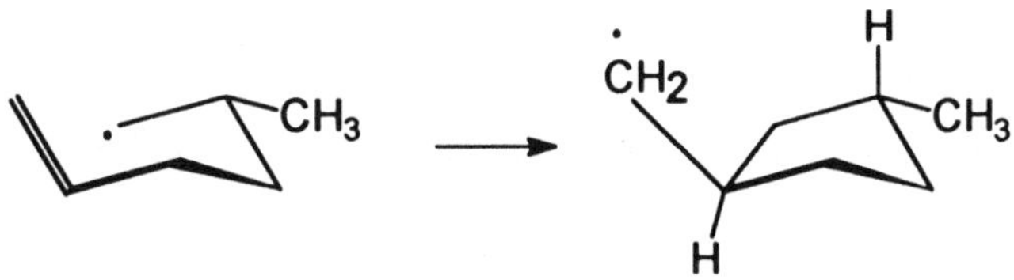

IV

Another example is the cyclization of the aldopyranose derivative shown in Scheme X [39]. The transition state shown is the one with the substituents at carbons 2, 3, and 4 all equatorial, favoring the first product shown (74%).

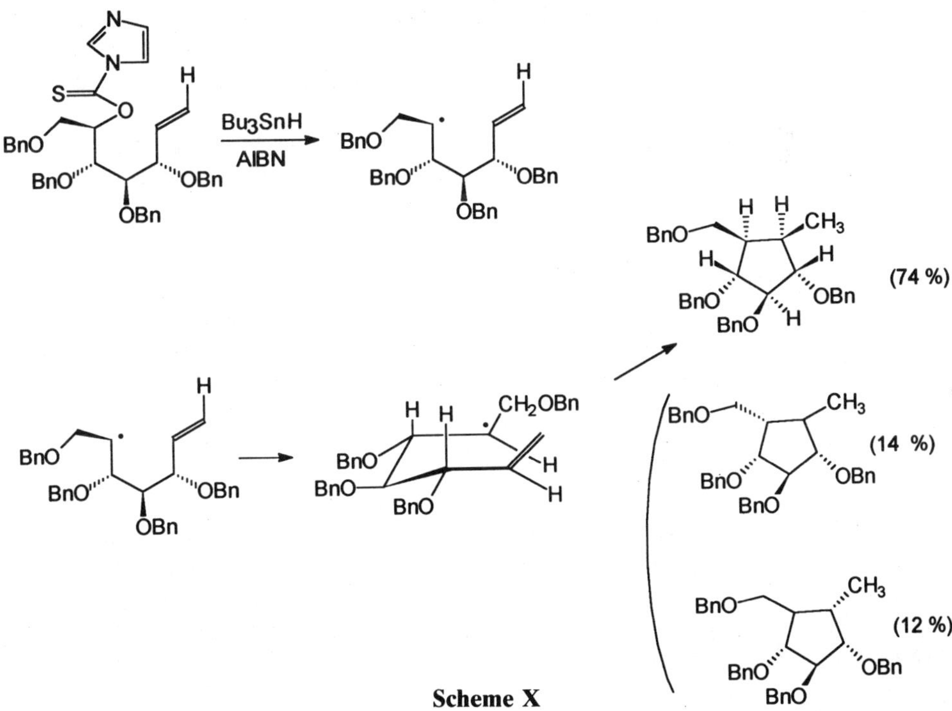

Scheme X

11.7.2 Reactions of 5-Membered Cyclic Radicals

When the attacking radical is a 5-membered ring there is a preference for reaction at the less-hindered face. For example, the reaction of acrylonitrile with the moderately hindered 2-ethoxycyclopentyl radical (Eq. 11.28) the product is 77% anti [44].

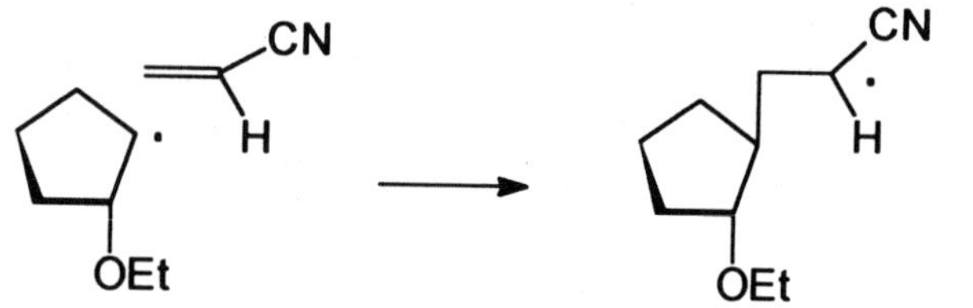

(11.28)

An acetamido group in the 2-position of the ring gives at least 98% anti product.

11.7.3 Reactions of 6-Membered Cyclic Radicals

With 6-membered cyclic radicals the factors affecting the stereochemical results are more complex [44].

In the 4-*t*-butylcyclohexyl radical, the 4-*t*-butyl group will be equatorial and the radical at the 1-position will have two nearly equivalent shallow pyramidal conformations. For reactions in which the radical abstracts OH, D, Cl, or Br the product is 69–80% axial, rather than largely equatorial as it would be if the product were equilibrium controlled rather than rate controlled.

A plausible explanation for this is shown in **V**a and b. If the transition state is "early," that is, with a long C-1 to X distance, steric repulsion between

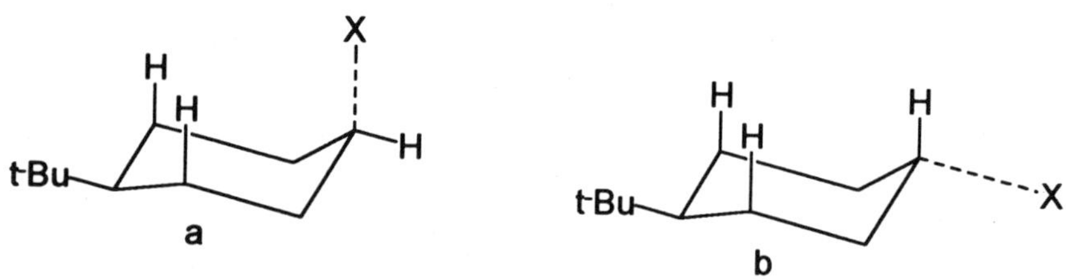

V

an axial incoming X and the axial hydrogens will be unimportant because they are too far away. On the other hand, with an incoming equatorial X, if the C-1 to H bond is rehybridized there will be repulsion between that nearly axial hydrogen and the other axial hydrogens. A similar explanation for preferential axial reaction of nucleophiles at a C-1 carbonyl has been supported by MM2 molecular mechanics calculations [45]. When the incoming X is much bulkier, equatorial reaction is favored, because 1,3 diaxial interactions become more important. For example, the addition of $C_3H_7CH{=}C(COOEt)_2$ gives 88% equatorial product.

Axial substituents on carbon 2 of the 4-*t*-butylcyclohexyl radical favor axial reaction at C-1, giving predominantly the anti product. Equatorial substituents at C-2 strongly favor equatorial reaction [44].

11.8 ACYCLIC STEREOCHEMICAL CONTROL

When a cyclohexyl radical reacts with a prochiral alkene, a chiral center is created at the point of attack and a new, stereogenic, radical center at the adjacent carbon [46]. Subsequent abstraction of hydrogen by the new radical could give two stereoisomers, but one of them will be favored by 1,2-asymmetric induction. The results can be opposite in cyclic and acyclic systems. For example, the addition of cyclohexyl radical to methylmaleic anhydride (Eq. 11.29) gives the erythro product. Addition of the cyclohexyl radical to methylfumaric diethyl ester gives a very similar intermediate radical, but the final product is the threo isomer.

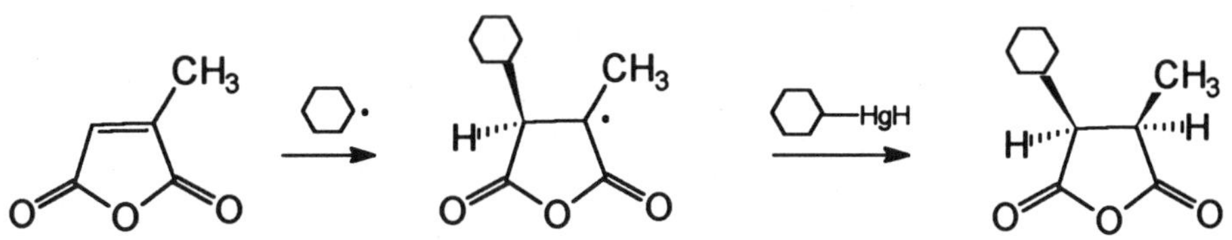

(11.29)

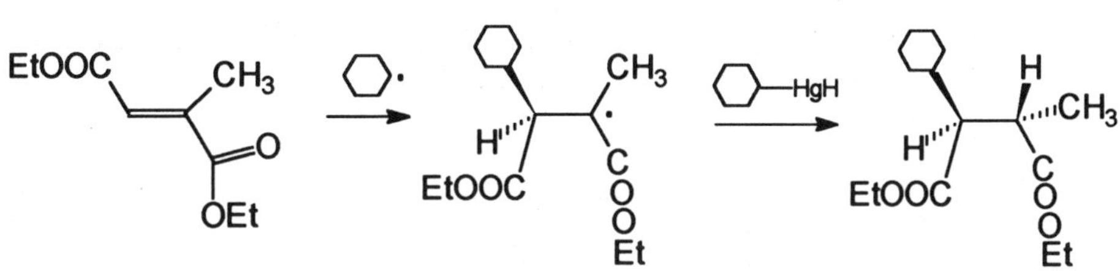

(11.30)

In Eq. 11.29 the hydrogen is added to the face of the prochiral radical that is trans to the cyclohexyl group. In Eq. 11.30, rotation about the C–C bond causes the other face of the prochiral radical to be less shielded. In general a radical **VI** will give the threo product whenever R is larger than X; R and X shield opposite faces of the radical center.

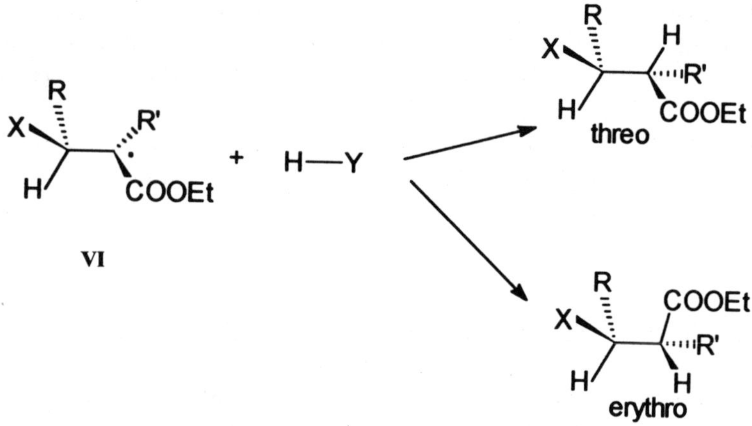

VI

When a prochiral radical reacts with a prochiral alkene as in Eq. 11.31, two new chiral centers are created.

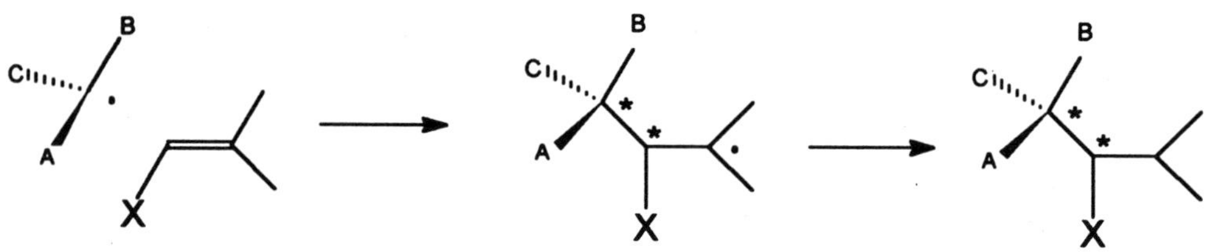

(11.31)

In order for a resident chiral group to exert stereocontrol on an acyclic reaction it must fulfill two conditions. First, it must be in a fixed position relative to the new stereogenic center. In the case of an amide or imide group on a radical or alkene the NR_2^* group is large relative to the carbonyl oxygen, and this fixes its position as shown in Scheme XI.

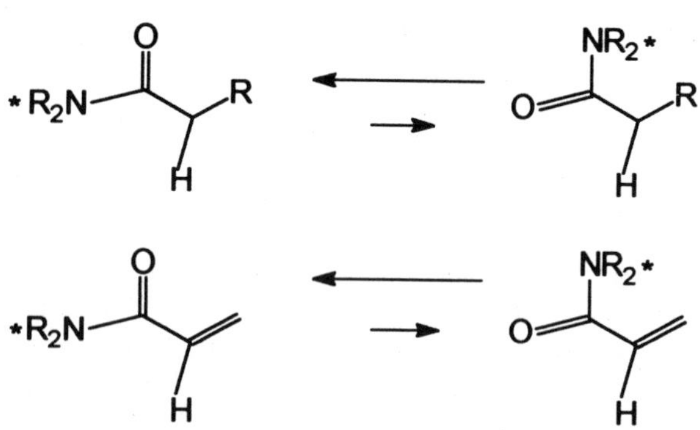

Scheme XI

The second requirement is that the resident chiral group must differentially shield one of the diastereotopic faces of the radical or the alkene. Equations 11.32 and 11.33 illustrate both points [47]. The chain reaction is initiated by photolysis (Eq. 11.32), producing a cyclohexyl radical which adds to the alkene to give a prochiral radical. Under the influence of the resident chiral oxazolidine group, which specifically shields one face of the prochiral radical, the next step in the chain gives the diastereoisomer shown with 98% selectivity.

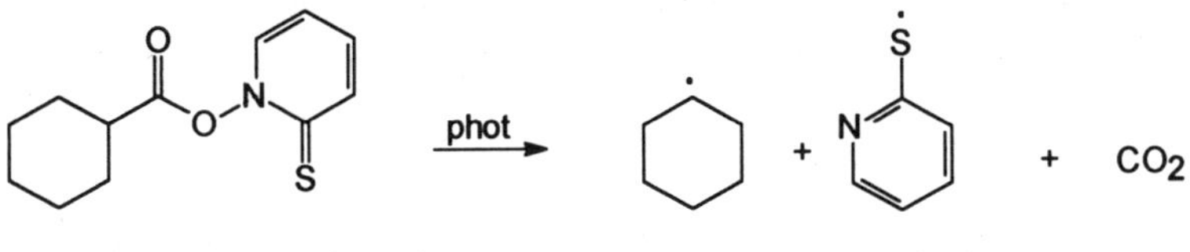

(11.32)

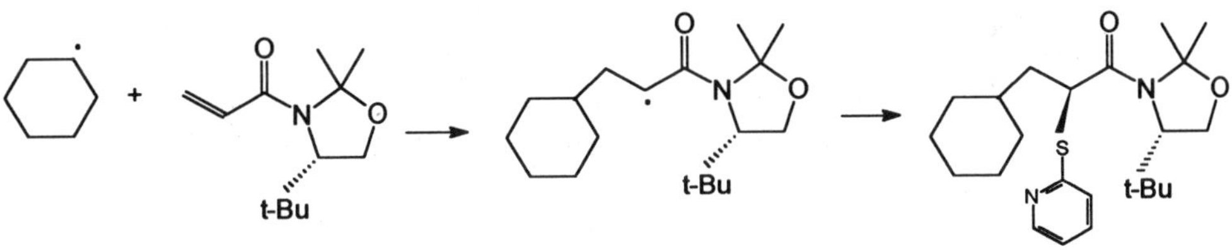

(11.33)

The chiral oxazolidine was even more effective in the alkene shown in Eq. 11.34, cyclohexyl radicals adding with 99% diastereoselectivity. Hydrolysis of the adduct gave R acid.

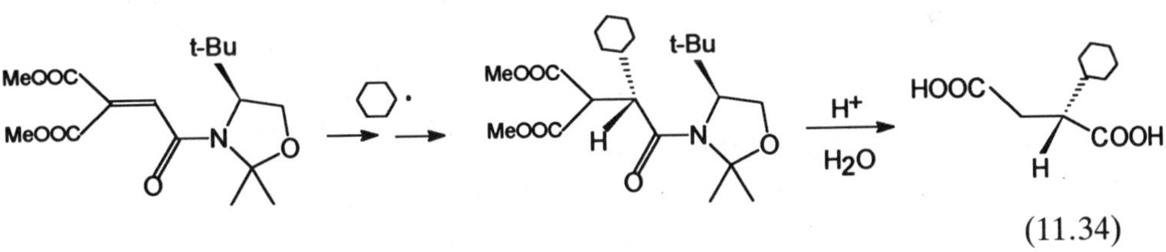

(11.34)

REFERENCES

1. Giese, B. *Radicals in Organic Synthesis: Formation of Carbon–Carbon Bonds,* Vol. 5, Pergamon: Oxford, 1986.
2. Hart, D. J. *Science* **1984,** *223,* 883.
3. Breslow, R.; Brandl, M.; Hunger, J.; Turro, N.; Cassidy, K.; Krogh-Jespersen, K.; Westbrook, J. D. *J. Am. Chem. Soc.* **1987,** *109,* 7204.
4. Abu-Raqabah, A.; Symons, M. C. R. *J. Am. Chem. Soc.* **1990,** *112,* 8614.

5. Breslow, R.; Brandl, M.; Hunger, J.; Adams, A. D. *J. Am. Chem. Soc.* **1987,** *109,* 3799.
6. Breslow, R. *Acc. Chem. Res.* **1980,** *13,* 170.
7. Breslow, R.; Heyer, D. *J. Am. Chem. Soc.* **1982,** *104,* 2045.
8. Breslow, R.; Wife, R. L.; Prezant, D. *Tetrahedron Lett.* **1976,** 1925.
9. Chatgilialoglu, C.; Ingold, K. U.; Scaiano, J. C. *J. Am. Chem. Soc.* **1981,** *103,* 7739.
10. Corey, E. J.; Suggs, J. W. *J. Org. Chem.* **1975,** *40,* 2554.
11. Barton, D. H. R.; Zard, S. Z. *Pure & Appl. Chem.* **1986,** *58,* 675.
12. Hayashi, T.; Iwaoka, T.; Takeda, W.; Ohki, E. *Chem. Pharm. Bull.* **1978,** *26,* 1786.
13. Barton, D. H. R.; Jaszberenyi, J. C.; Theodorakis, E. A. *J. Am. Chem. Soc.* **1992,** *114,* 5904.
14. Giese, B.; González-Gómez, J. A.; Witzel, T. *Angew. Chem. Int. Ed.* **1984,** *23,* 69.
15. Schlenk, G. O.; Koltzenburg, G.; Grossmann, H. *Angew. Chem.* **1957,** *69,* 177.
16. Curran, D. P.; Chen, M.-H. *J. Am. Chem. Soc.* **1987,** *109,* 6558.
17. Barton, D. H. R.; Ramesh, M. *J. Am. Chem. Soc.* **1990,** *112,* 891.
18. Enholm, E. J.; Kinter, K. S. *J. Am. Chem. Soc.* **1991,** *113,* 7784.
19. Beckwith, A. L. J. *Tetrahedron* **1981,** *37,* 3073.
20. Julia, M.; Maumy, M. *Bull. Soc. Chim. Fr.* **1969,** 2415.
21. Walling, C.; Pearson, M. S. *J. Am. Chem. Soc.* **1964,** *86,* 2262.
22. Breslow, R.; Olin, S. S.; Groves, T. T. *Tetrahedron Lett.* **1968,** 1837.
23. Curran, D. P.; Kim, D.; Liu, H. T.; Shen, W. *J. Am. Chem. Soc.* **1988,** *110,* 5900.
24. Snieckus, V.; Cuevas, J.-C.; Sloan, C. P.; Liu, H.; Curran, D. P. *J. Am. Chem. Soc.* **1990,** *112,* 896.
25. Curran, D. P.; Chen, M.-H.; Kim, D. *J. Am. Chem. Soc.* **1989,** *111,* 6265.
26. Curran, D. P.; Chen, M.-H.; Spletzer, E.; Seong, C. M.; Chang, C.-T. *J. Am. Chem. Soc.* **1989,** *111,* 8872.
27. Curran, D. P.; Morgan, T. M.; Schwartz, C. E.; Snider, B. B.; Dombroski, M. A. *J. Am. Chem. Soc.* **1991,** *113,* 6607.
28. Curran, D. P.; Seong, C. M. *J. Am. Chem. Soc.* **1990,** *112,* 9401.
29. Hart, D. J.; Seely, F. L. *J. Am. Chem. Soc.* **1988,** *110,* 1631.
30. Corey, E. J.; Pyne, S. G. *Tetrahedron Lett.* **1983,** 2821.
31. Bartlett, P. A.; McLaren, K. L.; Ting, P. C. *J. Am. Chem. Soc.* **1988,** *110,* 1633.
32. Curran, D. P.; Liu, H. *J. Am. Chem. Soc.* **1991,** *113,* 2127.
33. Coffman, D. D.; Cramer, R.; Mochel, W. E. *J. Am. Chem. Soc.* **1958,** *80,* 2882.
34. Ryu, I.; Kusano, K.; Ogawa, A.; Kambe, N.; Sonoda, N. *J. Am. Chem. Soc.* **1990,** *112,* 1295.
35. Ryu, I.; Yamazaki, H.; Kusano, K.; Ogawa, A.; Sonoda, N. *J. Am. Chem. Soc.* **1991,** *113,* 8558.
36. Brown, S. H.; Crabtree, R. H. *J. Am. Chem. Soc.* **1989,** *111,* 2946.
37. Brown, S. H.; Crabtree, R. H. *J. Am. Chem. Soc.* **1989,** *111,* 2935.

38. Feldhues, M.; Schäfer, H. J. *Tetrahedron* **1985,** *41,* 4213.
39. RajanBabu, T. V. *Acc. Chem. Res.* **1991,** *24,* 139.
40. Beckwith, A. L. J.; Eaton, C. J.; Lawrence, T.; Serelis, A. K. *Aust. J. Chem.* **1983,** *36,* 545.
41. Beckwith, A. L. J.; Moad, G. *J. Chem. Soc. Chem. Commun.* **1974,** 472.
42. Struble, D. L.; Beckwith, A. L. J.; Gream, G. E. *Tetrehedron Lett.* **1968,** 3701.
43. Beckwith, A. L. J.; Gream, G. E.; Struble, D. L. *Aust. J. Chem.* **1972,** *25,* 1081.
44. Giese, B. *Angew. Chem. Int. Ed.* **1989,** *28,* 969.
45. Wu, Y.-D.; Houk, K. N. *J. Am. Chem. Soc.* **1987,** *109,* 908.
46. Porter, N. A.; Giese, B.; Curran, D. P. *Acc. Chem. Res.* **1991,** *24,* 296.
47. Porter, N. A.; Bruhnke, J. D.; Wu, W.-X.; Rosenstein, I. J.; Breyer, R. A. *J. Am. Chem. Soc.* **1991,** *113,* 7788.

12

ELECTRON TRANSFER AND RELATED CONCEPTS

12.1 INTRODUCTION

Electron transfer from a donor molecule D to a separate acceptor molecule A involves a diffusive step to form [D A], which is called the precursor complex, the actual electron transfer to form the successor complex [$D^{+\cdot}$ $A^{-\cdot}$], and a second diffusive step to give the separated products.

$$D + A \rightleftharpoons [D\ A] \rightleftharpoons D^{+\cdot}\ A^{-\cdot}] \rightleftharpoons D^{+\cdot} + A^{-\cdot} \qquad (12.1)$$

Considering only the actual electron transfer step, the process can be represented as a change from a potential energy surface corresponding to the precursor complex [D A] to another corresponding to [$D^{+\cdot}$ $A^{-\cdot}$]. (When the electron transfer is intramolecular we will be talking about the reagent and product molecules rather than complexes.) Since the two complexes differ somewhat in the positions of their nuclei, including those of the surrounding solvent molecules, the positions of the two potential energy curves along the "reaction coordinate" are different (Fig. 12.1). The Franck–Condon principle requires that the nuclear and solvent molecules not move during the electron transfer. This means that the nuclear positions in the reagent must be changed to match those of the product before the electron transfer takes place. This adjustment involves thermal activation to a transition state near the intersection of the two potential energy curves, a point at which the two species have the same nuclear configuration as well as the same energy. Note that even when the reaction is isothermal, as it is in Fig. 12.1, a reorganization of the positions of the nuclei is required and this requires the input of some vibrational activation energy.

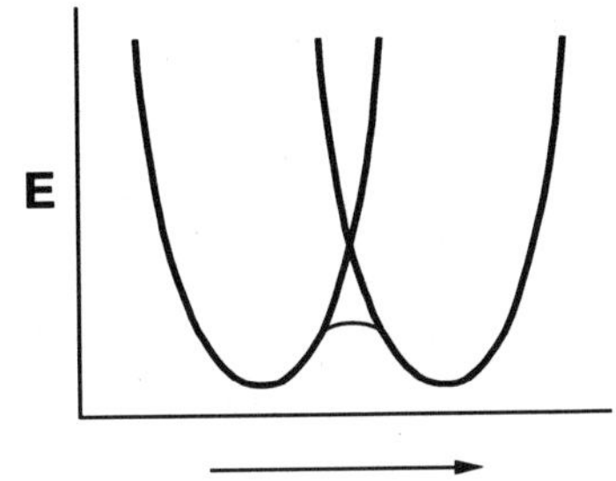

FIG. 12.1 Avoided crossing of two potential energy surfaces.

In a strictly classical model the electron-transfer would take place exactly at the intersection. A quantum mechanical treatment modifies this by allowing the transition, called an avoided crossing, to begin slightly before the intersection [1].

The activation energy for an electron transfer step will generally depend not only on a reorganization energy characteristic of the transition state but also on the overall free energy change $\Delta G°$. Of course, if the reaction is endothermic (uphill in enthalpy, strictly speaking uphill in free energy, $\Delta G° > 0$), $\Delta G^{\ddagger}$ cannot be less than $\Delta G°$.

The activation energy for exothermic electron-transfer reactions should also include a reorganization energy like that found for the isothermal reaction shown in Fig. 12.1, but the activation energy should also be affected in some way by any change in the exothermicity. Looking at Fig. 12.2 we see that as the reaction becomes more exothermic the intersection of the two curves might shift to lower energies for a while, but then might actually rise again. Of course this picture is greatly oversimplified and embodies various dubious assumptions.

Electron transfer can be treated theoretically as an activated radiationless transition. The simplest approach assumes that electronic and nuclear motions are separable, giving rise to independent electronic and vibrational factors in the expression for the rate constant (Eq. 12.2) [2,3].

$$k = 2\pi/\hbar|\mathbf{V}(r)|^2 \text{ FCWD} \tag{12.2}$$

The electronic factor is $|\mathbf{V}(r)|^2$, the square of the electronic coupling matrix element, and the nuclear factor is represented by FCWD, which stands for Franck–Condon weighted density of states. This factor corresponds to the need for vibrational activation, which can be appreciated qualitatively from Figs. 12.1 and 12.2. It turns out to depend not only on the exothermicity of the reaction, $\Delta G°$, as suggested by Fig. 12.2 but also on a quantity called the *reorganization energy*. The activation energies $\Delta G^{\ddagger}$ in the Eyring equation for a series of related reactions also depend both on $\Delta G°$ and on the reorganization energy.

Figure 12.1 was drawn for a reaction in which the overall free energy change

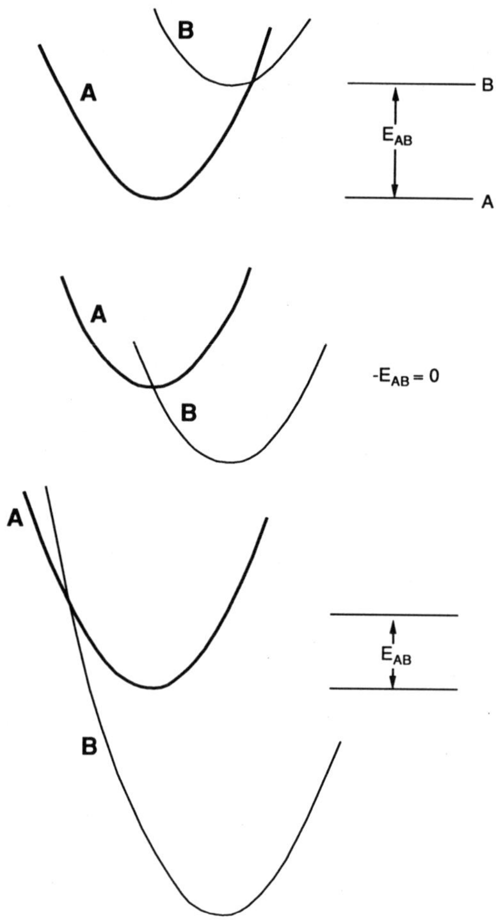

FIG. 12.2 Shift of the crossing point as the reaction becomes more exothermic.

$\Delta G°$ is zero, which is sometimes the case for a suitably selected pair of donor and acceptor species but is always the case for identity reactions. The reactions in Eq. 12.3, the rates of which were obtained by measuring the broadening of the ESR lines, are examples [4].

$$\text{Ar—H}^{+\cdot} + \text{Ar—H} \rightleftharpoons \text{Ar—H} + \text{Ar—H}^{+\cdot} \qquad (12.3)$$

The transition state for Eq. 12.3 is a resonance hybrid,

$$\text{Ar—H}^{+\cdot}\ \text{Ar—H} \longleftrightarrow \text{Ar—H}\ \text{Ar—H}^{+\cdot}$$

Although this must resemble the radical–cation plus hydrocarbon precursor complex, there are important differences. The bond lengths and angles in the reagents or products correspond to the lowest vibrational energy levels, but the bond lengths and angles of the transition state are modified and are more like those in vibrationally excited states. When the transition state relaxes to the product or reagent state, energy (which we will call reorganization energy) is released. Similarly, the solvent cage surround the transition state has a different structure from that surrounding the precursor or successor complex. When this solvation relaxes, it also releases reorganization energy.

For nonidentity reactions, like Eq. 12.1, the reorganization energy is defined as the mean of the activation energies of two corresponding identity reactions (Eqs. 12.4–12.5).

$$\mathrm{D\ D} \longrightarrow \mathrm{D^{+\cdot}\ {}^{-\cdot}D} \tag{12.4}$$

$$\mathrm{A\ A} \longrightarrow \mathrm{A^{+\cdot}\ {}^{-\cdot}A} \tag{12.5}$$

12.1.1 Inner- and Outer-Sphere Electron Transfers

The requirement for an electron transfer is some overlap, however, slight, between the initially occupied orbital and the eventually occupied orbital [5]. This overlap can be either direct or indirect via the orbitals of a bridging structure or group.

In a nomenclature derived from the chemistry of inorganic ions and their spheres of coordinated ligands or solvent molecules, electron transfers are divided somewhat arbitrarily into two categories. For molecules that are in contact either directly or through bridging ligands, the electronic interaction ($|\mathbf{V}(r)|^2$) between the molecules is strong, and the transfer process is called *inner-sphere* electron transfer.

More rarely, electron transfer can take place between molecules whose orbitals overlap so slightly that the energy of interaction is less than 1 kcal mole^{-1}. In this extreme case the transfer process is called *outer-sphere* electron transfer.

The term "outer-sphere" electron transfer refers to the mechanism, not just the overall result. The same overall result could be achieved by an inner-sphere electron transfer followed by dissociation of an intermediate molecule as in Eq. 12.6.

$$\mathrm{D + A} \longrightarrow \mathrm{[D\ A]} \longrightarrow \mathrm{D^{+} - A^{-}} \longrightarrow \mathrm{[D^{+\cdot}\ A^{-\cdot}]} \longrightarrow \mathrm{D^{+\cdot} + A^{-\cdot}} \tag{12.6}$$

True outer-sphere electron transfers are quite rare because competing processes like that of Eq. 12.6, in which the transition state is stabilized by a considerable overlap between the orbitals of the two reagents, will usually have a much lower energy of activation. Outer-sphere electron transfer sometimes becomes competitive when the energy of a bonded complex $\mathrm{D^{+} - A^{-}}$ is prohibitively high because of steric hindrance or low bond strength [6].

12.2 RATE–EQUILIBRIUM RELATIONSHIPS

In a series of closely related reactions, in which the only differences are minor changes in the structure of a reagent or the solvent, the rate and equilibrium constants are often related. This is also true of electron transfer reactions, both inner and outer sphere. On a qualitative level, the reactions with the more favorable equilibrium constants are usually faster, although there are exceptions.

12.2.1 Simple Linear Relationships

Sometimes the relationship is particularly simple, a linear free-energy relationship between the free energies of activation $\Delta G^{\ddagger}$ and the free energy changes ΔG°. Brønsted acid–base catalysis laws are examples of this.

As suggested by Fig. 12.1, the transition state has a lot in common with the reagents or the products, and usually with both. It is convenient to indicate the perturbations caused by the small changes in structure or solvent by an operator δ. For example, $\delta G^{\ddagger}$ is the effect on the free energy of the transition state, δG°_{R} is the effect on the free energy of the reagents, $\delta\Delta G^{\circ}$ $(= \delta G^{\circ}_{P} - \delta G^{\circ}_{R})$ is the effect on ΔG°, and so on.

We might expect the transition state to be intermediate in its sensitivity, so that [7]

$$\Delta G^{\ddagger} = \alpha\delta G^{\circ}_{P} + (1 - \alpha)\delta G^{\circ}_{R} \qquad (12.7)$$

The factors α and $(1 - \alpha)$ are used on the theory that if the transition state resembles the product more it must resemble the reagent less. From Eq. 12.7 it follows that

$$\delta\Delta G^{\ddagger} = \alpha\delta\Delta G^{\circ} \qquad (12.8a)$$

$$\Delta G^{\ddagger} = \text{constant} + \alpha\Delta G^{\circ} \qquad (12.8b)$$

$$\delta(\ln k) = \alpha\delta(\ln K) \qquad (12.8c)$$

For different reaction series the slope α is expected to vary in the range zero to plus one, being closer to zero for a series with "early" transition states and closer to unity for series with "late" or more product-like transition states. Plots of Eq. 12.8b or 12.8c are often linear over a moderate range, although α in a given series really should change somewhat as ΔG° changes in the series [8]. Values outside the 0–1 range can occur in special cases [9,10]. For example, in the reaction of substituted phenylnitromethanes with hydroxide ion, the Brønsted α value is 1.29. The negative charge is thought to be localized mainly on carbon in the transition state, but mainly on oxygen in the product nitronate ion [10]. A good way of putting it is that the delocalization of the electron into the nitro group lags behind the proton transfer [11]. This makes the product a poor model for the transition state. When the variable is a substituent in the base that abstracts the proton, a normal Brønsted β value of 0.56 is obtained.

Another failure of Eqs. 12.8a–c is the prediction of zero sensitivity of the rates to substituents when $\delta(\ln K)$ is zero. In Eq. 12.9, for which all the $\Delta G°$ values are necessarily zero, the electron transfer rate constants for various $XC_6H_4NO_2$ span a factor of 20 [4].

$$XC_6H_4NO_2^{-}\cdot + XC_6H_4NO_2 \rightleftharpoons XC_6H_4NO_2 + XC_6H_4NO_2^{-}\cdot \qquad (12.9)$$

Apparently effects on the reorganization energy of the transition state are not adequately modeled by effects on G_P° and G_R°. If groups X such as CN or a second nitro group are excluded, the range is reduced to a factor of 2.6. For a series of substituted benzonitriles, the range is also about a factor of 2.5.

12.2.2 The Marcus Equation

The Marcus treatment of electron transfer [5,12,13] (Eqs. 12.10–12.12), in which the parameters are partly justified by theory, includes the reorganization energy and differs from the simple linear free energy relationship of Eqs. 12.11 in that it includes a second-order term in $\Delta G^{\circ 2}$ in addition to the first order term in $\Delta G°$.

$$\Delta G^{\ddagger} = Z_1 Z_2 e^2 f / D r_{12} + (\tfrac{1}{4})\lambda(1 + \Delta G^{\circ\prime}/\lambda)^2 \qquad (12.10)$$

In Eqs. 12.10–12.13 Z_1 and Z_2 are the charges on the two reagents, f is an ionic strength factor, D is the dielectric constant of the medium, and r_{12} is the sum of the radii of the two reagents, assumed to be spherical. The parameter λ we will call the Marcus reorganization energy. The quantity $\Delta G^{\circ\prime}$ is given by Eq. 12.11.

$$\Delta G^{\circ\prime} = \Delta G^\circ + (Z_1 - Z_2 - 1)(e^2 f / D r_{12}) \qquad (12.11)$$

For the sake of simplicity we will restrict our discussion to reactions in which $Z_1 = +1$ and $Z_2 = 0$, so that $\Delta G^{\circ\prime}$ is just $\Delta G°$, and Eq. 12.10 becomes Eq. 12.12.

$$\Delta G^{\ddagger} = (\tfrac{1}{4})\lambda(1 + \Delta G^\circ/\lambda)^2 \qquad (12.12a)$$

$$\Delta G^{\ddagger} = \text{constant} + (\tfrac{1}{2})\Delta G^\circ + \Delta G^{\circ 2}/4\lambda \qquad (12.12b)$$

When the reaction is an identity reaction, $\Delta G°$ is zero. Like the simple linear free energy relationship, the Marcus equation fails to account for the variation of the rates of reaction 12.9 with changes in the substituent X. From

Eq. 12.12a we can see that the free energy of activation $\Delta G^{\ddagger}_{\text{o}}$, the value of $\Delta G^{\ddagger}$ when ΔG° is zero, should be constant at $\lambda/4$.

For a reaction like Eq. 12.1, which is not an identity reaction, λ is the *mean* of the values for the two identity reactions corresponding to electron transfer between D and $D^{-}\cdot$ and A and $A^{-}\cdot$. In terms of this mean "intrinsic barrier" $\Delta G^{\ddagger}_{\text{o}}$, the Marcus equation becomes Eq. 12.13.

$$\Delta G^{\ddagger} = \Delta G^{\ddagger}_{\text{o}}(1 + \Delta G^{\circ}/4\Delta G^{\ddagger}_{\text{o}})^2 \quad (12.13a)$$

$$= \Delta G^{\ddagger}_{\text{o}} + 0.5\Delta G^{\circ} + \Delta G^{\circ 2}/16\Delta G^{\ddagger}_{\text{o}} \quad (12.13b)$$

Equation 12.14 shows that the α of the simple linear rate-equilibrium model becomes $0.5 + \Delta G^{\circ}/16\Delta G^{\ddagger}_{\text{o}}$ in the Marcus equation:

$$\delta\Delta G^{\ddagger} = (0.5 + \Delta G^{\circ}/16\Delta G^{\ddagger}_{\text{o}})\Delta G^{\circ} \quad (12.14)$$

Despite the appearance of Eq. 12.14, $\Delta G^{\ddagger}$ values calculated by the Marcus equation frequently have long nearly linear portions when plotted against ΔG°.

As we shall see, the Marcus equation can be derived by a Taylor series expansion about an extent of reaction parameter set at 0.5. An implicit assumption is that $\Delta G^{\ddagger}_{\text{o}}$ is a constant in the series of reactions, although this was not a good approximation in the case of reaction 12.9.

12.2.3 Derivation of the Marcus Equation [13–16]

A derivation due to Ritchie [14], easier to follow than most, assumes that the free energy of the system (ΔG_{sys}) at any stage in the activation process can be represented by Eq. 12.15.

$$\Delta G_{\text{sys}} = n\Delta G^{\circ} + G_{\text{D}} + G_{\text{A}} \quad (12.15)$$

In Eq. 12.15 ΔG° is the standard free energy change between the reagent or precursor complex and the product or successor complex. The value of n, the extent of reaction parameter, is zero for the precursor complex, and 1 for the successor complex.

Both G_{D} and G_{A} are additional contributions to the energy associated with the donor and the acceptor at various points along the path of the reaction. They depend on the extent of reaction parameter n, as shown in Eq. 12.16a and b.

$$G_{\text{D}} = G'_{\text{D}}g_1(n) \quad (12.16a)$$

$$G_{\text{A}} = G'_{\text{A}}g_2(1 - n) \quad (12.16b)$$

The functions $g_1(n)$ and $g_2(1 - n)$ are zero at the reagent and product stages of the reaction. As the transfer of the electron begins to require an input of reorganization energy due to changes in bonding or solvation, both $g_1(n)$ and $g_2(1 - n)$ increase smoothly to their maximum value of 1 and then decrease as reorganization energy is released. Hence G'_D and G'_A are the maximum values of G_D and G_A, respectively. For identity reactions, symmetry forces n to be 0.5 at the transition state. Setting both G_D and G_A equal to G'_D in Eq. 12.16 shows that G'_D must be equal to one half the $\Delta G^\ddagger$ value for the $D^{-\cdot} + D$ identity reaction. Similarly, G'_A must be equal to one half the $\Delta G^\ddagger$ for the other identity reaction. In effect, we are viewing the activation process as an average of the curves for the two identity reactions, but with a vertical offset $n\Delta G°$ proportional to the extent of the reaction.

For nonidentity reactions the transition state will not generally occur exactly half way at $n = 0.5$. But if we make another assumption, that the $n^\ddagger$ is not very far from 0.5, we can represent ΔG_{sys} by means of a Taylor expansion of Eq. 12.15 about $n = 0.5$. This assumption means that the Marcus equation should be limited in its application to a range $|\Delta G°| \leq 4|\Delta G^\ddagger_o|$ [17].

$$\begin{aligned}\Delta G_{sys} = n\Delta G° &+ G'_D[1 + g'_1(n - 0.5) + g''_1(n - 0.5)^2/2] \\ &+ G'_A[1 - g'_2(n - 0.5) + g''_2(n - 0.5)^2/2] \end{aligned} \quad (12.17)$$

The functions g'_i and g''_i are the first and second derivatives of g_1 and g_2 evaluated at $n = 0.5$.

At the transition state, since ΔG_{sys} is a maximum, $d\Delta G_{sys}/dn$ will be zero (Eq. 12.18).

$$\begin{aligned}d\Delta G_{sys}/dn &= \Delta G° + (G'_D g'_1 - G'_A g'_2) + (G'_D g''_1 + G'_A g''_2)(n^\ddagger - 0.5) \\ &= 0\end{aligned} \quad (12.18)$$

Equation 12.18 can be rearranged to Eq. 12.19:

$$(n^\ddagger - 0.5) = (-\Delta G° - G'_D g'_1 + G'_A g'_2)/(G'_D g''_1 + G'_A g''_2) \quad (12.19)$$

Putting $n^\ddagger$ and $n^\ddagger - 0.5$ into Eq. 12.18, and noting that $\Delta G° = g'_2 G'_A - g_1 G'_D$ and that $G'_D + G'_A = \Delta G^\ddagger_o$, we obtain Eq. 12.20 for a first-order approximation to $\Delta G^\ddagger_{sys}$ and Eq. 12.21 for the second-order approximation

$$\Delta G^\ddagger_{sys} = 0.5\Delta G° + \Delta G^\ddagger_o \quad (12.20)$$

$$\Delta G^\ddagger_{sys} = 0.5\Delta G° + \Delta G^\ddagger_o - (\tfrac{1}{2})(n^\ddagger - 0.5)^2(G'_D g''_1 + G'_A g''_2) \quad (12.21)$$

To use the second-order approximation we have to make some assumption about the shapes of the functions g_1 and g_2. The precise choice of g_1 and g_2

is quite arbitrary except for two conditions: at $n = 0.5$ they have to be equal to 1 and have zero first derivatives.

A choice that leads to the Marcus equation is Eq. 12.22.

$$g_1(n) = g_2(n) = 4n(1 - n) \tag{12.22}$$

The second derivatives g_1'' and $g_2'' = -8$.

Putting these values into Eq. 12.21 gives the Marcus equation (Eq. 12.23).

$$\Delta G^{\ddagger} = 0.5\Delta G^{\circ} + \Delta G_{\circ}^{\ddagger} + (\tfrac{1}{16})(\Delta G^{\circ})^2/\Delta G_{\circ}^{\ddagger} \tag{12.23}$$

12.2.4 Deviations from the Marcus Equation

One of the predictions of the Marcus equation is that when the reaction is already *very* exothermic, making it still more exothermic should *decrease* rather than increase the rate. This is also suggested by the curves in Fig. 12.2. In terms of the quantum mechanical theory based on Eq. 12.2, the FCWD factor passes through a maximum. As can be seen by differentiating $\Delta G^{\ddagger}$ in Eq. 12.23 with respect to ΔG°, the Marcus equation predicts that the fall-off in rate should begin to appear when $-\Delta G^{\circ}$ exceeds $4\Delta G_{\circ}^{\ddagger}$.

The Marcus equation can almost always be made to fit the experimental data quite well in the "normal" region in which the rate and exothermicity are increasing together. When it fails, it is usually a failure of the inverted region to appear. Instead of the expected decrease at high exothermicities, the rate merely stops increasing and levels off.

One type of reaction in which this kind of failure is often seen is fluorescence quenching by the electron transfer mechanism [18–22]. Equations 12.24 and 12.25 represent the quenching, by electron transfer, of the fluorescence of an excited donor D* by an electron acceptor A or of an excited acceptor A* by a donor D. The ΔG° are calculated from the redox potentials, the zero–zero excitation energy, and the coulombic energy of the ion–radical pair being formed.

$$D^* + A \longrightarrow D^{+\cdot}\ A^{-\cdot} \tag{12.24}$$

$$A^* + D \longrightarrow A^{-\cdot}\ D^{-\cdot} \tag{12.25}$$

A specific example is a series of reactions in which the fluorescence of excited phenanthrenes or naphthalene is quenched by electron acceptors such as benzoquinone or tetracyanoethylene or by various amines acting as electron donors [20]. The rates are plotted against ΔG° in Figs. 12.3 and 12.4. The filled triangles represent reactions in which the excited phenanthrene or naphthalene is acting as the electron acceptor. The unfilled triangles represent

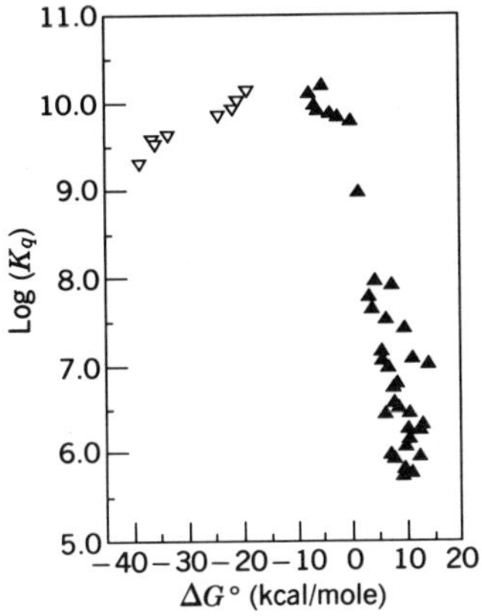

FIG. 12.3 Election transfers in cyclohexane. Reprinted with permission from J-M. Chen, T-H. Ho, and C-Y. Mou, *J. Phys. Chem.*, **1990,** *94*, 2889. Copyright 1969 American Chemical Society.

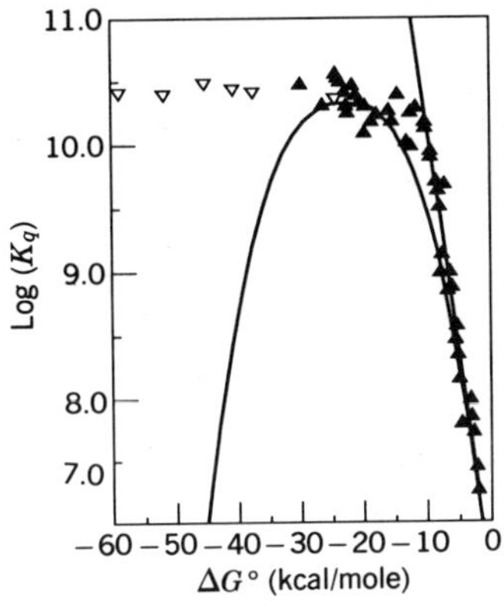

FIG. 12.4 Electron transfers in acetonitrile. Reprinted with permission from J-M. Chen, T-H. Ho, and C-Y. Mou, *J. Phys. Chem.*, **1990,** *94*, 2889. Copyright 1969 American Chemical Society.

reactions in which the phenanthrene or naphthalene is acting as the donor. Equations 12.26 and 12.27 are examples of both types of reaction.

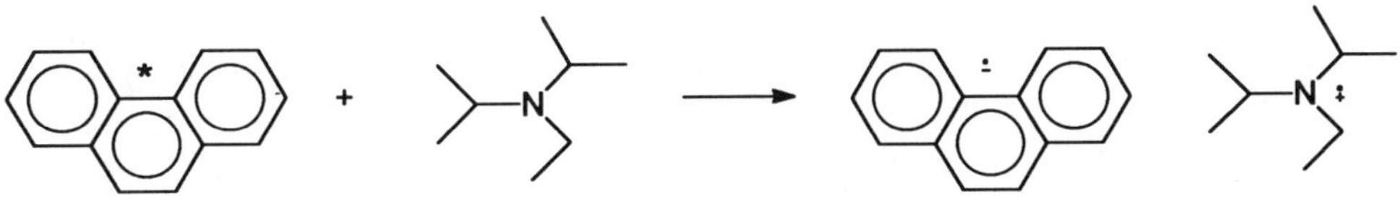

(12.26)

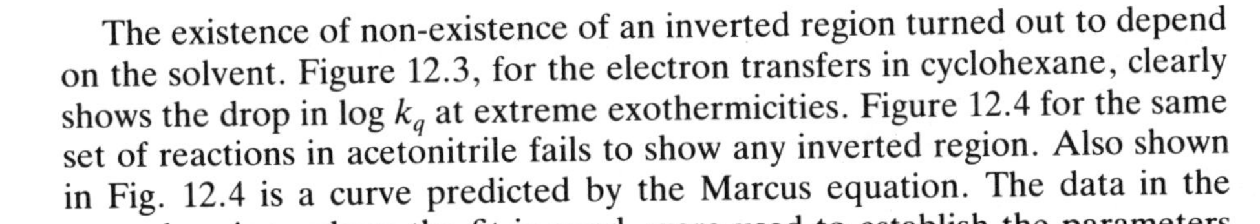

(12.27)

The existence of non-existence of an inverted region turned out to depend on the solvent. Figure 12.3, for the electron transfers in cyclohexane, clearly shows the drop in log k_q at extreme exothermicities. Figure 12.4 for the same set of reactions in acetonitrile fails to show any inverted region. Also shown in Fig. 12.4 is a curve predicted by the Marcus equation. The data in the normal region, where the fit is good, were used to establish the parameters for the rest of the curve. When ΔG° is more negative than -30 kcal mole^{-1}, the observed rates are still diffusion controlled and orders of magnitude higher than the predicted rates.

A similar failure of the Marcus equation in the inverted region has been reported by Rehm and Weller [18,19] for the fluorescence quenching of a different set of aromatic hydrocarbons, also in acetonitrile. They found that their rates, leveling off at the diffusion-controlled limit, fitted Eq. 12.28 [16,18,19].

$$\Delta G^{\ddagger} = 0.5\Delta G^{\circ} + [(\Delta G_{\circ}^{\ddagger})^{2} + (0.5\Delta G^{\circ})^{2}]^{1/2} \tag{12.28}$$

For reactions less exothermic than about -10 kcal mole^{-1} the fits of the Marcus equation, Eq. 12.28, and a simple linear relationship are about equally good.

The inverted region also failed to appear in the quenching of excited *trans*-stilbene or 4,4′-dimethyl-*trans*-stilbene by CN-substituted ethylenes in acetonitrile [21].

Mulliken charge-transfer complexes [Ar,NO^+] formed from aromatic hydrocarbons Ar and nitrosonium ion undergo a photochemical electron transfer to give a pair [$Ar^{+\cdot}$,NO·]. The charge transfer complex is reformed by a two-phase process [23]. Phase I is a first-order inner-sphere reaction represented by k_1 in Eq. 12.29.

$$[\mathrm{Ar,NO^{+}}] \underset{k_1}{\overset{h\nu_{CT}}{\rightleftharpoons}} [\mathrm{Ar^{+}\cdot,NO\cdot}] \tag{12.29}$$

This rate constant decreases as the driving force ΔG° (inner sphere) for different aromatic hydrocarbons decreases, and can be fitted to the Marcus equation. In phase II the rate of reformation of the Mulliken complex obeys second-order kinetics; these rate constants (k_2) are nearly constant at about 10^9 M^{-1} sec^{-1}, and much faster than the outer-sphere rate constants calculated by the Marcus equation. The explanation proposed [23] is that the actual path in phase II is not a simple outer-sphere bimolecular reaction, but a circuitous process in which the cation radical pair is reformed and then undergoes an inner-sphere electron transfer.

$$\mathrm{Ar^{+}\cdot} + \mathrm{NO\cdot} \xrightarrow{k_2} [\mathrm{Ar^{+}\cdot,NO\cdot}] \longrightarrow [\mathrm{Ar,NO^{+}}] \tag{12.30}$$

A trivial reason for the nonappearance of the inverted region, applicable only to bimolecular reactions, would be that the reaction at high exothermicities became diffusion controlled. At still higher exothermicities, perhaps not covered in the range of reactions studied, the rate might eventually have descended below the diffusion-controlled level. That is, a range of exothermicities in which the reaction is diffusion controlled may cut the top off of the Marcus inverted parabola, as in Fig. 12.5. The rate of the real electron transfer step may actually be decreasing, but unless it decreases to a value that brings it below the diffusion rate again, it will not be seen.

Various other reasons have been suggested for the lack of an inverted region. These include: formation of the product in an electronically or vibrationally excited state, specific solvation, quantum mechanical tunneling, and transfer through bridging solvent molecules [21,22,24–26]. For reactions

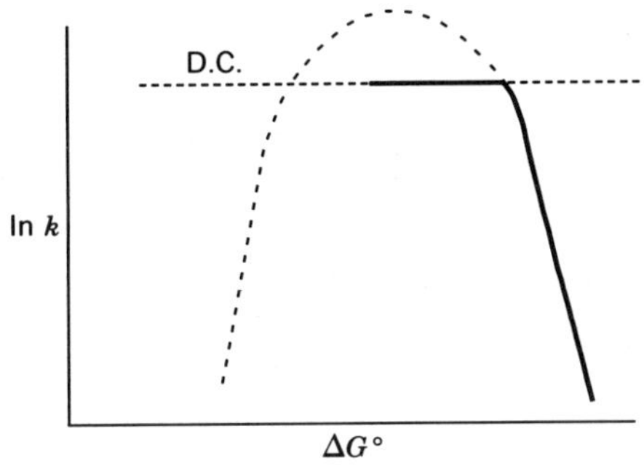

FIG. 12.5 Effect of diffusion control.

in acetonitrile in particular [20], the failure may occur because acetonitrile has a relatively low LUMO. This allows it to accept the electron and pass it on, acting as a bridge. If so, this adds a second electron transfer mechanism, one that acts when the donor and acceptor are at longer than contact distances. The rate enhancement due to this additional mechanism, which might reasonably be especially important in the exergonic region, may counteract the decrease in rate predicted by the Marcus equation.

It is also worth recalling that free-energy relationships typically arise from *small* perturbations corresponding to a family of closely related reactions. Too big a perturbation or the introduction of a substituent that exerts its effect on the reaction zone by a different interaction mechanism makes the reaction deviate from the correlation [27]. Sometimes the deviant point turns out to be a member of a family of its own correlated by a different line parallel to the first one or a different Marcus curve parallel to the first Marcus curve. It may be that cases where the inverted region fails to appear or the data show only a scatter diagram involve too drastic a change in structure. A set of reactions corresponding to Eq. 12.31 illustrate this point

$$[\mathrm{ArH}^{+}\cdot\ \mathrm{A}^{-}\cdot] \longrightarrow [\mathrm{ArH\ A}] \qquad (12.31)$$

The donors $A^{-}\cdot$ in this series were 9,10-dicyanoanthracene and 2,6,9,10-tetracyanoanthracene anion radicals and the acceptors $AH^{+}\cdot$ were a series of aromatic hydrocarbon cation radicals [28]. The geminate ion–radical pairs were created by quenching the excited state of the hydrocarbon by electron transfer, and what is being studied here is the back electron transfer to regenerate the ground state of the hydrocarbon. The solvent was acetonitrile, but it did not interfere with the inverted region, possibly because the geminate ion pairs were not solvent separated. Although the rates decreased smoothly with increasing exothermicity, as predicted for the Marcus inverted region, the plot (Fig. 12.6) shows a dispersion into three separate curves, one for each of three types of cation radical characterized by different numbers of aromatic rings. One curve fits benzene variously substituted with methyl groups. A second, with lower rates at a given exothermicity fits methyl-substituted

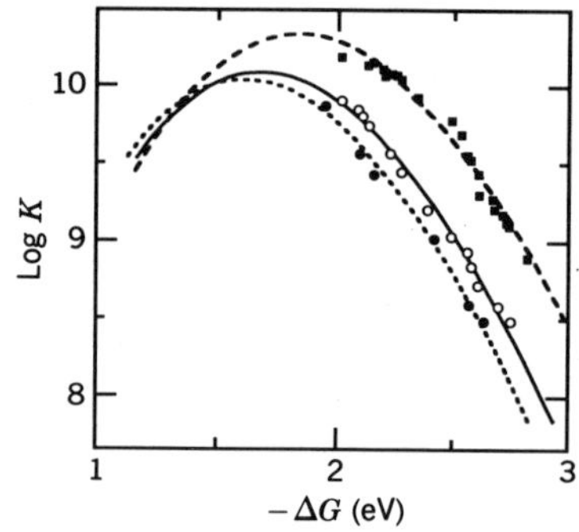

FIG. 12.6 Dispersion caused by excessive changes in structure. Reprinted with permission from I. A. Gould, D. Ege, J. E. Moser, and S. Farid, *J. Am. Chem. Soc.*, **1990,** *112,* 4290. Copyright 1990 American Chemical Society.

naphthalenes and biphenyls. A third, a little lower still fits methyl-substituted phenanthrenes.

A random selection of points from all three curves of Fig. 12.6 could easily have been mistaken for a scatter diagram. Similar effects probably account for the roughness found in other Marcus correlations such as those shown in Figs. 12.3 and 12.4

Another kind of failure of the Marcus equation has to do with the temperature dependence of the rates in the inverted region [26]. According to the theory in its classical form, the drop in rate in the inverted region has to do with the reappearance of an activation energy. Electron transfer from a 4-biphenylyl anion radical group across an androstane spacer to various quinone groups showed the inverted region, but the rates were nearly independent of temperature. A quantum mechanical treatment ascribes the decreased rates in the inverted region, not to an activation energy, but to decreased tunneling.

12.2.5 Examples of the Inverted Region

Because of the extreme exothermicity of the reactions that might show an inverted region, the reaction mixture at equilibrium is usually almost entirely product. Since the equilibrium constant cannot be measured directly, $\Delta G°$ is determined electrochemically from the redox potentials $E(\mathrm{D/D^{+\cdot}})$ and $E(\mathrm{A^{-\cdot}/A})$ of the reagents.

The Marcus equation always seems to work particularly well whenever the electron donor and acceptor are constrained in some way to make the reaction a unimolecular one. These constraints include building the donor and acceptor into the same molecule, creating them as pairs embedded in a rigid matrix, and creating them as geminate pairs in solution. In each case the separation of the reagents is fixed or nearly fixed, and bridging at variable distances by solvent molecules such as acetonitrile is excluded or less important.

We will discuss electron transfer in geminate ion radical pairs first [28–31] since these pairs are formed by fluorescence quenching reactions like the ones already discussed (e.g., Eqs. 12.26 and 12.27). The reactions, called back electron transfer, resemble the reverse of equations like 12.26 and 12.27 except that the product is not in an electronically excited state.

It is interesting that when both fluorescence quenching and the back reaction have been studied for the same D and A, the back reactions obeyed the Marcus equation even when the fluorescence quenching reactions did not [29].

A range in ΔG° from -0.55 to -2.83 eV was covered by using radical–ion pairs derived from various aromatic hydrocarbons and tetracyanoethylene for the less exothermic (normal) region and pairs derived from perylene and tetramethyl-*p*-phenylenediamine, pyromellitic dianhydride, or maleic anhydride for the more exothermic region [29]. The bell-shaped curve was clearly observed for the back reaction even though the solvent was acetonitrile. It will be recalled that this is the same solvent that failed to show any decrease in rate for highly exothermic fluorescence quenching, a process in which the pairs are nongeminate, since they are formed by diffusion [18,20].

In Rigid Matrices. Biphenyl anion radicals were made by radiolysis of biphenyl and various aromatic hydrocarbons in a 2-methyltetrahydrofuran glass [3]. Biphenyl captures a solvated electron to give the biphenyl anion radical $Bip^{-\cdot}$ which then transfers an electron to a hydrocarbon molecule A that happens to occupy an adjacent position in the matrix. The rates of the electron transfer were measured by the decrease in the $Bip^{-\cdot}$ absorbance.

$$Bip^{-\cdot}\, A \longrightarrow Bip\, A^{-\cdot} \tag{12.32}$$

If the rates were measured at 10^{-6} s after the irradiation, they showed a maximum at a ΔG° of about -0.8 eV, but if they were measured after 100 s, the maximum rate was at a ΔG° of -1.1 eV. This shift may reflect an increase in solvent reorganization energy when the reaction is carried out in a matrix that has had time to anneal after the irradiation.

Intramolecular Electron Transfer. Intramolecular electron transfer has been studied with the D and A attached to saturated hydrocarbon spacers [2,32,33], in cytochrome complexes [34] and in capped porphyrins [35].

A biphenyl anion radical and an electron-accepting aromatic group attached to a rigid 5-α-androstane steroid space (Fig. 12.7) give reactions that are the intramolecular analog of Eq. 12.32 [2,32]. The reactions were studied at 296 K in methyltetrahydrofuran solution. Changing the acceptor group A_X from a hydrocarbon (open circles), to a quinone (filled circles), and then to

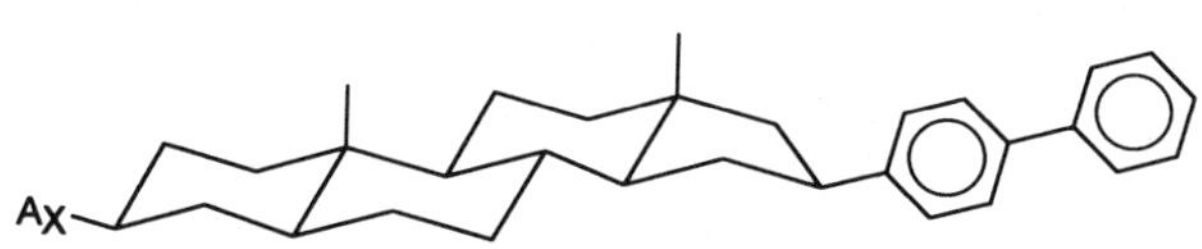

FIG. 12.7 The donor and acceptor are separated by an androstane spacer.

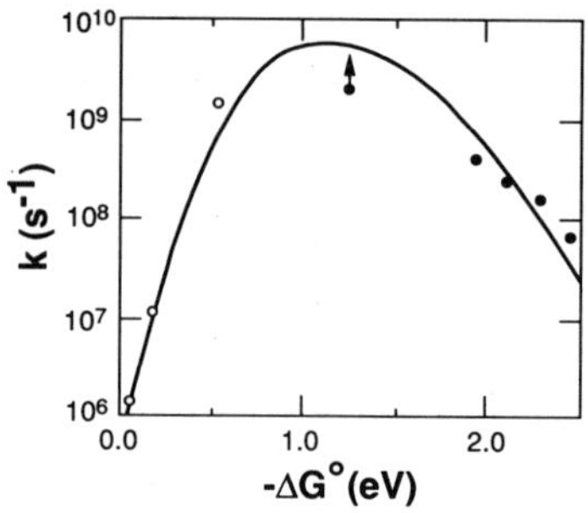

FIG. 12.8 Electron transfer through an androstane spacer. Reprinted with permission from G. L. Closs, L. T. Calcattera, N. J. Green, K. W. Penfield, and J. R. Miller, *J. Phys. Chem.*, **1986,** *90,* 3673. Copyright 1986 American Chemical Society.

quinones of higher oxidation potential revealed both the normal and inverted regions (Fig. 12.8).

12.3 STEREOELECTRONIC EFFECTS AND THROUGH-BOND INTERACTIONS

The effect of varying the spacer was studied in a series with A_X constant as β-naphthyl and the donor as 4-biphenyl anion radical [2]. The spacers (Fig. 12.9) varied in the distance of separation of the two groups, the number of intervening bonds, and in some instances equatorial versus axial stereochemistry. The spacers other than 5-α-androstane had the two groups separated by three, two, or one cyclohexane moieties, and finally by just a methylene group [2]. For weak through-space coupling the rate should decrease exponentially with distance (Eq. 12.33) as a consequence of the $|\mathbf{V}(r)|^2$ factor in Eq. 12.2.

$$\mathbf{V} = \mathbf{V}_0 \exp\{-[\beta(R_{DA} - R_0)/2]\} \tag{12.33}$$

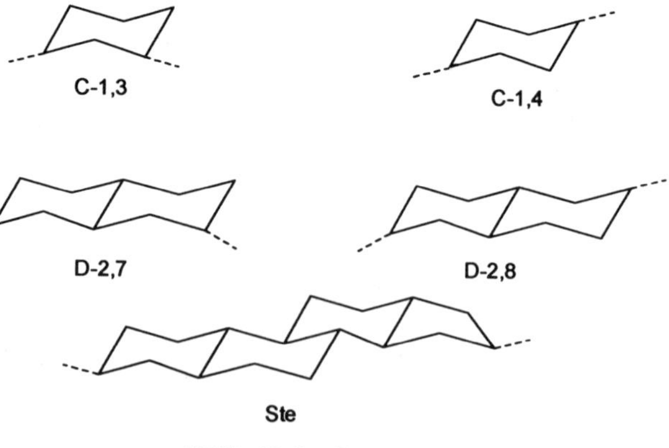

FIG. 12.9 Spacers.

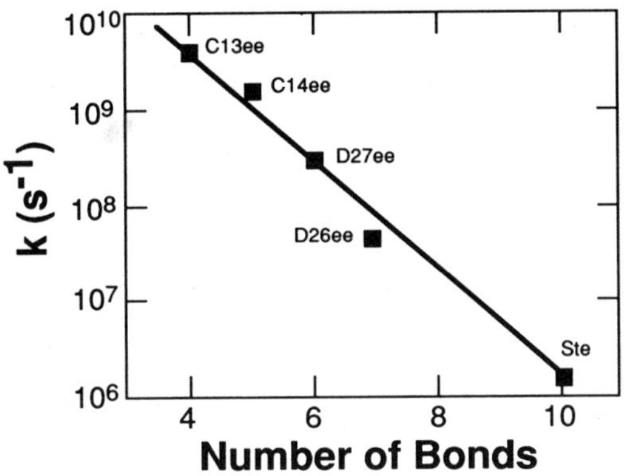

FIG. 12.10 Effects of changes in the spacer. Reprinted with permission from G. L. Closs, L. T. Calcaterra, N. J. Green, K. W. Penfield, and J. R. Miller, *J. Phys. Chem.*, **1986,** *90,* 3673. Copyright 1986 American Chemical Society.

In Eq. 12.33 R_0 is the van der Waals distance and β is the attenuation factor. As the distance between the groups decreased, the rate did in fact increase. However, when "distance" was the center-to-center distance, the trend in log k versus D was very rough, almost a scatter diagram. It was considerably better when D was the edge-to-edge distance, and best of all (Fig. 12.10) when D was simply the number of intervening sigma bonds along the shortest path.

The line in Fig. 12.10 corresponds to Eq. 12.34, in which β' is 1.15 bond^{-1} and $\mathbf{V}_0$ is 1.1×10^3 cm^{-1}.

$$2 \ln \mathbf{V} = 2 \ln \mathbf{V}_0 - \beta'(\text{no. of bonds} - 1) \qquad (12.34)$$

It was especially striking that the rate depended not only on the number of intervening bonds, but also on the equatorial versus axial stereochemistry. Figure 12.10 includes only the equatorial isomers. The axial–axial (aa) and ae or ea isomers fell below the line.

Similar results were obtained for fluorescence quenching by the electron transfer mechanism [36]. The donor, excited dimethoxynaphthalene, was separated from the acceptor, dicyanoethylene, by rigid hydrocarbon spacers of various lengths. The number of bonds in the minimum path varied from 4 to 12.

12.4 TRIPLET ENERGY TRANSFER

One model for triplet–triplet energy transfer explains it as simultaneous transfer of two electrons (Fig. 12.11) [37]. According to this model an electron transfers from the singly occupied HOMO of the triplet energy donor to the

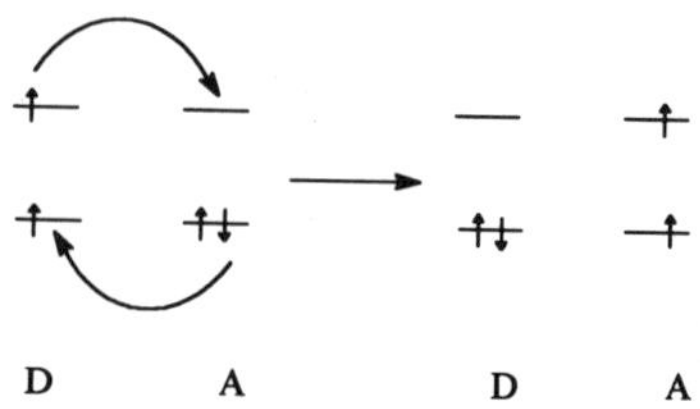

FIG. 12.11 A model for triplet energy transfer.

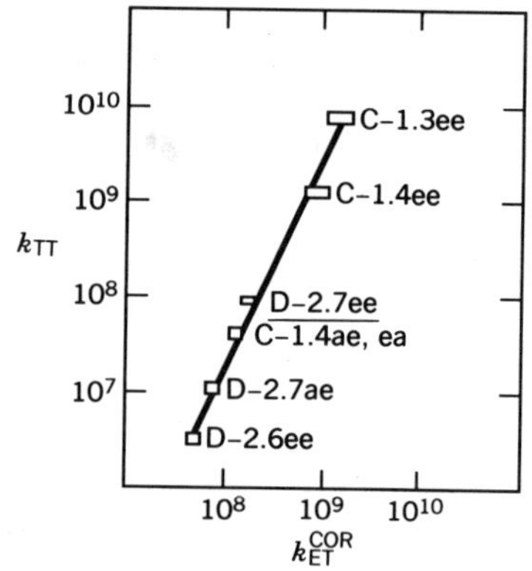

FIG. 12.12 Correlation of triplet energy transfer and electron transfer. Reprinted with permission from G. L. Closs, P. Piotrowiak, J. M. MacInnis, and G. R. Fleming, *J. Am. Chem. Soc.* **1988,** *110,* 2652. Copyright 1988 American Chemical Society.

LUMO of the acceptor, while another electron moves in the other direction from the doubly occupied HOMO of the acceptor to the lower singly occupied orbital of the triplet energy donor.

If this model is valid, a correlation might be expected between the rate of triplet energy transfer and the rate of *single* electron transfer in a related system. Consider for example single electron transfer (k_{ET}) in a system already discussed, 2-naphthyl acting as the acceptor and 4-biphenylyl as the donor, with cyclohexane and decalin spacers [37]. Triplet energy transfer (k_{TT}) in a closely related system uses the same set of spacers, 2-naphthyl as the triplet donor and 4-benzophenoneyl as the acceptor. There is an excellent linear relationship (Fig. 12.12) between ln k_{TT} and ln k_{ET}, with a slope of about 2. Even compounds with axial stereochemistry (ae or ea) fall on the line.

12.5 THE S_{RN} MECHANISMS

In this section we will try to sort out three reaction mechanisms that can follow second-order kinetics and give similar products. The three mechanisms are the $S_{RN}1$ and $S_{RN}2$ processes, which involve electron transfer, and the familiar S_N2 displacement reaction. To simplify the discussion somewhat we will assume that the nucleophile is an anion, Nu^-, although this is not necessary.

The $S_{RN}1$ reaction is a chain reaction carried by noncaged radicals. It is subject to inhibition by free-radical traps and acceleration by radical initiators or light. The intermediate radicals have been detected and identified by ESR spectroscopy [38,39].

The $S_{RN}2$ reaction is a nonchain mechanism involving electron transfer between Nu^- and the substrate R-X to form a radical-pair intermediate which collapses to give the product.

A reaction such as Eq. 12.35 that appears overall to be a nucleophilic substitution (S_N2) may involve one of the S_{RN} processes instead of, or in addition to, the S_N2 mechanism.

$$Nu:^- + R\text{-}X \longrightarrow Nu\text{—}R + :X^- \qquad (12.35)$$

Three factors favor S_{RN} mechanisms over the competing S_N2 reaction: (1) a low reduction potential for RX, (2) a low oxidation potential for the nucleophile, and (3) steric hindrance to bond formation in either reagent.

The $S_{RN}1$ reaction is favored over the $S_{RN}2$ reaction when the radical $RX^{-\cdot}$ is stable enough to escape from the solvent cage before decomposition to $R\cdot$ and X^- occurs.

To simplify the discussion we will *define* an S_N2 reaction as a reaction in which a two-electron bond is formed between the nucleophile and a group R at the same time or nearly at the same time that an R–X bond is broken to form X^-. By this definition, the S_N2 reaction has a transition state but no intermediates, radical or otherwise. This allows the S_N2 reaction to have a *transition state* that resembles a radical pair, but distinguishes it from the $S_{RN}2$ mechanism, which has a radical-pair *intermediate*.

12.5.1 The $S_{RN}1$ Mechanism

A general mechanism for the $S_{RN}1$ reaction is shown in Scheme I [39–41].

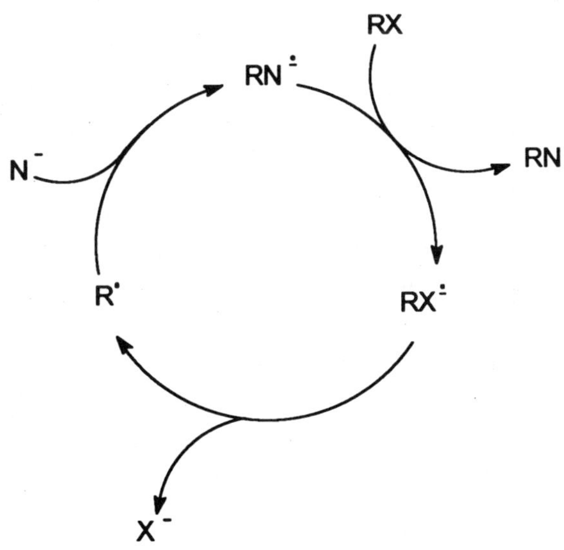

Scheme I

Bunnett [41] invented the name $S_{RN}1$ for these reactions because he found the mechanism hard to remember until he noticed the analogy with the classic S_N1 nucleophilic substitution mechanism. The dissociation of $RX^{-}\cdot$ into $R\cdot$ and X^- corresponds to the ionization of RX, and the association of $R\cdot$ and N^- corresponds to the bonding of R^+ and a nucleophile.

An example of an $S_{RN}1$ reaction is the reaction shown in Eq. 12.36 between *p*-nitrocumyl chloride and the carbanion derived from diethyl malonate [42].

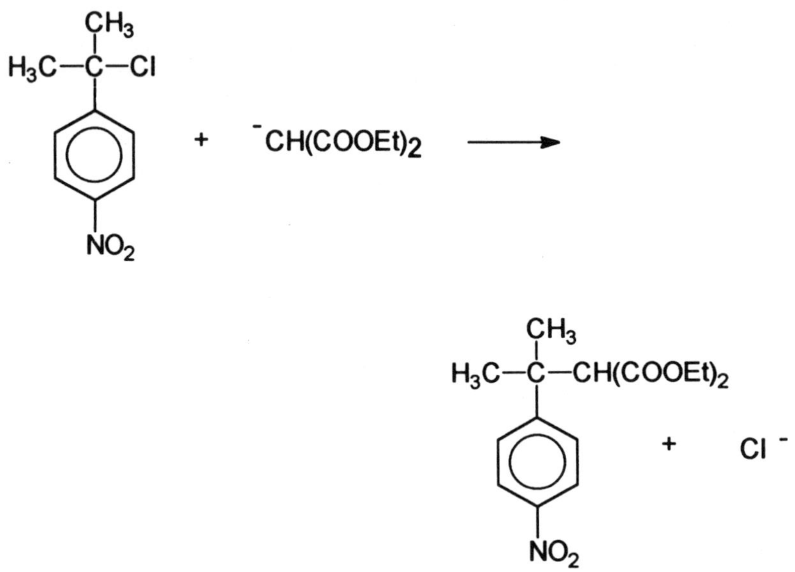

(12.36)

The $S_{RN}1$ chains are subject to inhibition by radical traps such as *p*-dinitrobenzene, di-*t*-butylnitroxyl radical, oxygen, or cupric ion [39,40,42]. In reaction 12.36 the yield of alkylation product is 90% in the absence of oxygen, but in the presence of oxygen the *p*-nitrocumyl radicals are diverted to give an 85% yield of *p*-nitrocumyl alcohol. Strong electron acceptors such as *p*-dinitrobenzene break the $S_{RN}1$ reaction chains by removing the electron from the chain-carrying $RN^{-}\cdot$ [42].

Some reagents give different displacement products, O- versus C-alkylation, by the $S_{RN}1$ and $S_{RN}2$ or S_N2 mechanisms, a difference which shows up when the $S_{RN}1$ process is prevented by an added trap for the chain-carrying radicals. For example, the sodium salt of 2-carbomethoxycoumaran-3-one reacts with *p*-nitrobenzyl chloride to give either the O-alkylation product or a mixture of C- and O-alkylation products, depending on the presence of a trace of cupric ion as inhibitor (Eq. 12.37) [40]. In the absence of cupric ion the reaction was fast ($k = 300 \times 10^5\ M^{-1}\ sec^{-1}$) and 90% of the product was the C-alkylate. In the presence of a trace of cupric ion to inhibit the chains, the remaining reaction was slow ($k = 2 \times 10^5$), and the product was a 40/60 mixture of the C- and O-alkylates.

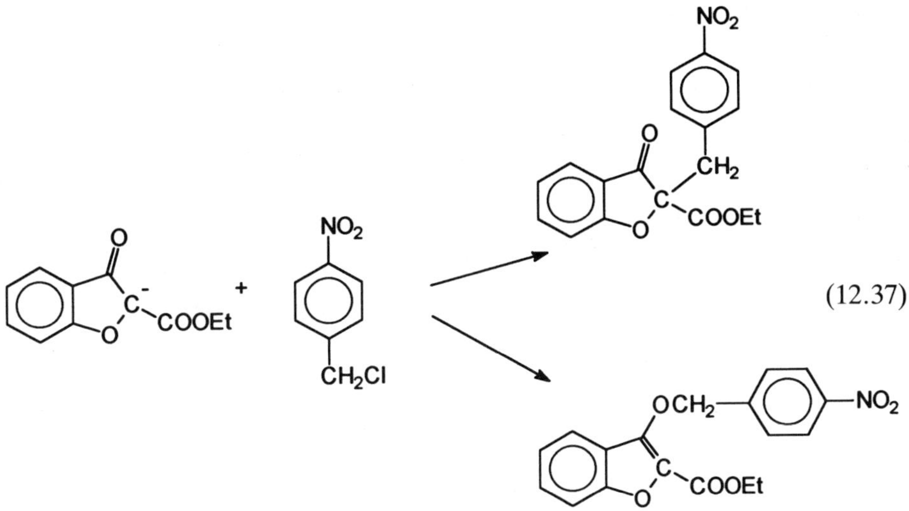

(12.37)

The $S_{RN}1$ reactions can also be accelerated by adding small amounts of another nucleophile (Nu'^-) whose lower oxidation potential allows it to act as an initiator.

$$Nu'^- + RX \longrightarrow Nu'\cdot + RX^-\cdot \qquad (12.38)$$

Most of the radicals are trapped by the original nucleophile rather than Nu'^- because of the difference in concentrations, so the product is still almost entirely R–Nu. An example is the reaction of *p*-nitrocumyl chloride with nitrite ion (Eq. 12.39).

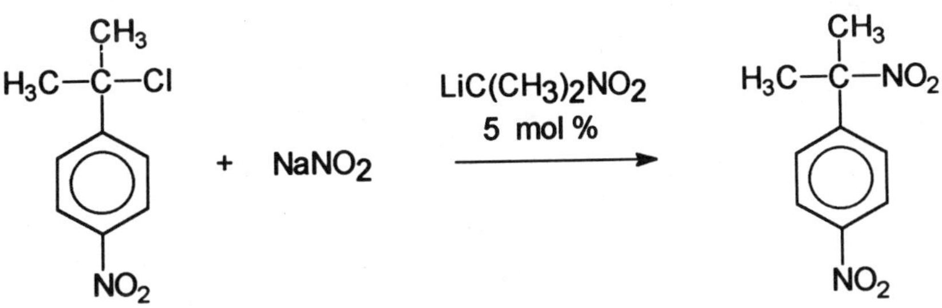

(12.39)

This reaction fails to proceed in the dark unless 5 mol% of the lithium salt of 2-nitropropane is added, but then gives the α-nitro derivative in 93% yield [42].

An aromatic substitution reaction by the $S_{RN}1$ mechanism is initiated by potassium in liquid ammonia, essentially solvated electrons, which add to ArX to give $ArX^-\cdot$ [41]. Because the crucial step is electron transfer, aromatic $S_{RN}1$ reactions are not sensitive to steric hindrance. Equation 12.40 is an example.

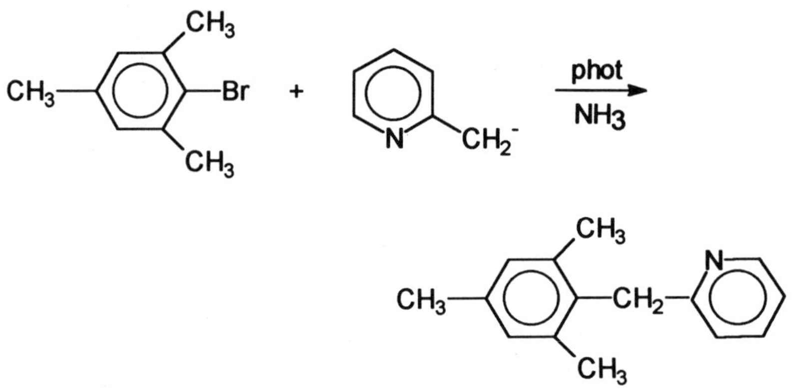

(12.40)

12.5.2 The $S_{RN}2$ Mechanism

An $S_{RN}2$ reaction is a nonchain, bimolecular process in which the nucleophile transfers an electron to R–X, followed by the formation of R–N *within the solvent cage.*

For alkyl halides the paths indicated by the dashed arrows (Scheme II) are less likely because stable aliphatic $RX^{-\cdot}$ intermediates have never been detected in electron capture experiments [43]. Radicals escaped from the $S_{RN}2$ cage may initiate an $S_{RN}1$ chain, as we have already seen. The $S_{RN}2$ reaction becomes more important when loss of X^- from $RX^{-\cdot}$ occurs before cage escape.

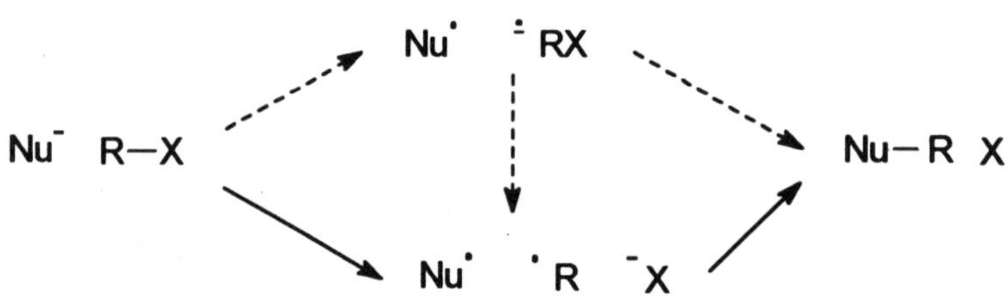

Scheme II

12.5.3 $S_{RN}2$ versus S_N2 Reactions [44]

The S_N2 process is often represented with curved arrows, implying that the electrons move in pairs. However, as pointed out by Pross [6,45] and Shaik [46] such a pairwise movement is rather unlikely and the same ultimate result could be obtained by movements of electrons one at a time. Thus the S_N2 reaction could equally well be represented (Eq. 12.41) as the result of an inner-sphere single electron transfer with simultaneous transfer of a bridging ligand [6,44–46].

$$Nu^- \; R{\cdot\cdot}X \longrightarrow \underset{\textbf{I}}{Nu\cdot \; ^-{\cdot}R{-}X} \longrightarrow Nu{:}R \; X^- \qquad (12.41)$$

Because of our narrow definition of the S_N2 process, **I** would have to be just a transition state or a point along the reaction coordinate; if it is an intermediate we would classify the reaction as $S_{RN}2$. However, if the transition state of an S_N2 reaction resembles **I** there could be substituent effects on the rate not easy to distinguish from those for an $S_{RN}2$ reaction [44].

Steric Effects on the Reaction Rate. Because a covalent bond is not being formed, electron-transfer reactions, in contrast to S_N2 reactions, are expected to be insensitive to steric hindrance. In typical S_N2 reactions of *n*-butyl and *sec*-butyl derivatives the effect of steric hindrance is expected to make the primary compound react considerably faster than the secondary compound with a given nucleophile. With thiophenoxide as the nucleophile and the leaving groups X listed in Table 12.1, the ratio decreases markedly as the half-wave potential for the reduction of CH_3–X becomes less negative [47]. This suggests a contribution of an ion radical structure in the transition state, but does not by itself prove the formation of a radical intermediate. Unfortunately, the sequence of relative rates is in the same as that of increasing electronegativity of the departing group, and could be ascribed to increasing $Nu^- \; R^+ \; X^-$ character in the transition state of the S_N2 reaction [47,48]. Although the styrene trapping test showed at least some radical formation from the *p*-nitrobenzenesulfonate, it was not applied to the other substrates. The stereochemical test, using optically active 2-butyl *p*-nitrobenzenesulfonate showed complete inversion.

Substituted fluorenide carbanions 9-G-Fl^- react slower with benzyl chloride (Eq. 12.42) as the bulk of the substituent G is increased in the series Me, MeO, and Me_2N, the result expected for an S_N2 reaction on the basis of steric effects [49].

TABLE 12.1 Correlation of *n*-Butyl/*sec*-Butyl Rates and Redox Potentials

Leaving Group	Relative Rates	MeX Reduction V
Bromide	25.7	−1.96
Iodide	12.8	−1.63
Tosylate	5.3	−1.25
Mesylate	5.9	
p-Nitrobenzene-sulfonate	2.5	−0.85

Source: Reprinted with permission from S. Bank and D. A. Noyd, *J. Am. Chem. Soc.*, **1973,** *95*, 8203. Copyright 1973 American Chemical Society.

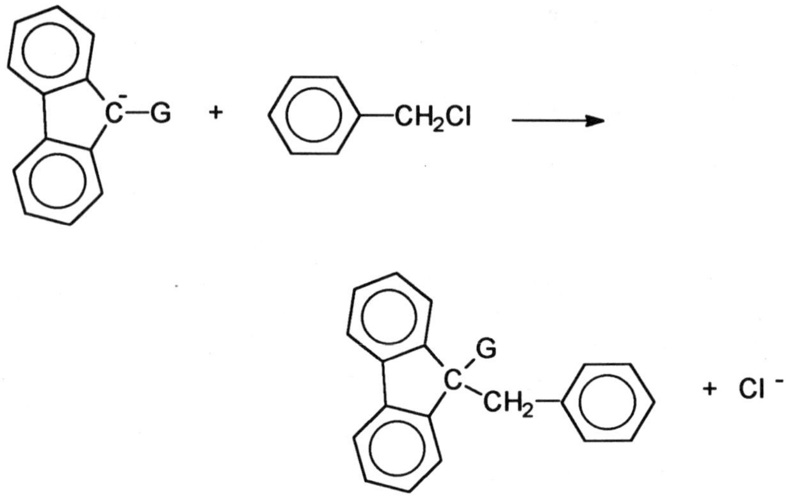

(12.42)

The same progressive change to a slower reaction is seen when the alkyl halide is isobutyl bromide. With 1-nitrocyclohexyl tosylate, a much better electron acceptor, the sequence of rates is reversed; ($Me_2N > MeO > Me$). This is the sequence of decreasing ease of oxidation.

Detection of Escaped Radicals. In extreme cases, the radical R· is stable and does not form a substitution product at all. Examples are the formation of tris-(*p*-nitrophenyl)methyl radical when the bromide is treated with any of several nucleophiles [50] and the formation of trityl radical by the reaction of trityl bromide with $LiAlD_4$ [51].

Both ESR [52] and CIDNP [53,54] and spin trapping of reactive cage-escaped radicals have also been observed [47]. Of course, these results, like those from other procedures for detecting extremely small amounts of escaped radicals, mean only that the $S_{RN}2$ process has some share in the mechanism and do not exclude S_N2 as the major mechanism. When an intermediate radical can be seen to decay at the same rate that the ultimate product is formed, the evidence is much stronger. An example is the reduction of dimesityl ketone by AlH_3 in tetrahydrofuran (Eq. 12.43) [52].

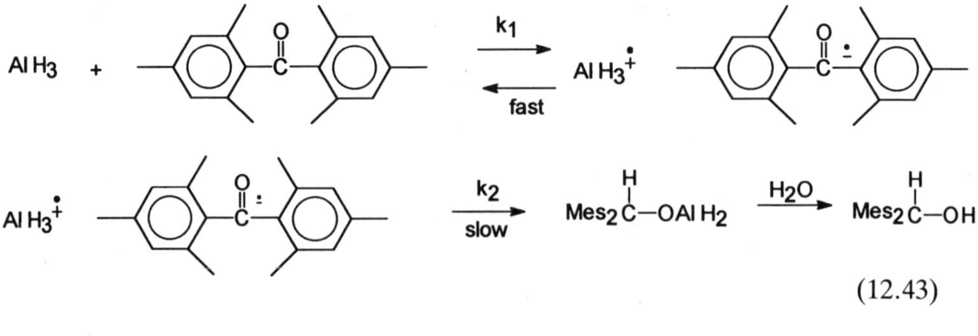

(12.43)

The ketyl intermediate is detected both by its violet color and its ESR signal. The disappearance of the original ketone can be followed by its UV absorption.

When sodium thiophenoxide was heated with *sec*-butyl *p*-nitrobenzenesulfonate in the presence of styrene, a styrene oligomer containing phenylthiyl groups was formed [47]. Similar experiments with various halides gave significant amounts of polystyrene in some cases and in others did not [55]. Benzyl bromide and benzyl chloride appeared on the basis of these experiments to be reacting at least in part by the $S_{RN}2$ mechanism, and benzyl iodide to a lesser extent. Methyl iodide and other aliphatic halides did not give significant amounts of polymer, apparently reacting by the nonradical S_N2 or S_N1 mechanisms exclusively.

Chiral and Radical Clock Probes. One way of testing for a radical pair intermediate is to use a reagent R-X that has the leaving group at a chiral center. The $S_{RN}2$ mechanism should give a product that is at least partly racemized. On the other hand, complete racemization cannot be expected even if none of the reaction goes by the S_N2 mechanism. The existence of stereoelectronic effects in intramolecular electron transfer suggests that the electron transfer step in the $S_{RN}2$ mechanism might have a strongly preferred orientation of the reagents and that orientation might be retained at least in part if cage recombination is fast enough.

Another way of testing for radical pair intermediates is to use a radical that is known to undergo a fast unimolecular reaction rearrangement such as the cyclization of 5-hexenyl radicals [51]. The proportion of rearranged to unrearranged product can also be used as a radical clock to estimate the lifetime of a radical when the rearrangement rate happens to be known [56]. Equation 12.44 is an example of the use of a chiral probe and Eq. 12.45 is an example of the use of a radical clock, chosen for its close structural resemblance to the chiral probe [51].

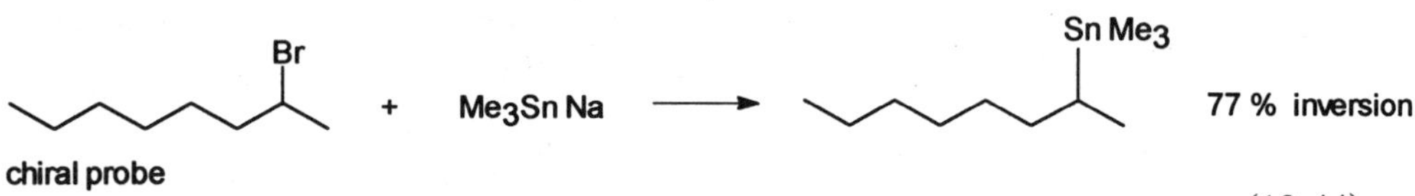

(12.44)

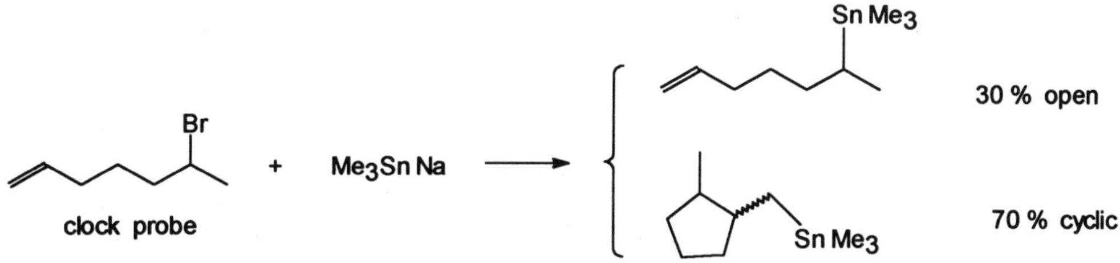

(12.45)

TABLE 12.2 2-Octyl Compounds in DMF at Room Temperature

Nucleophile	RX	% Inversion	Nucleophile	RX	% Inversion
LiSPh	OTs	100.0	$LiPPh_2$	OTs	100.0
	Cl	99.5		Cl	100.0
	Br	99.6		Br	93.8
	I	88.6		I	94.0
$LiSPr^i$	OTs	100.0	$NaSnMe_3$	OTs	100.0
	Cl	99.8		Cl	88.0
	Br	85.7		Br	77.2
	I	85.1		I	81.1
LiCN	OTs	100.0	$LiAlD_4$	OTs	100.0
	Br	83.6		Cl	100.0
	I	82.9		Br	100.0
				I	75.8

Source: Reprinted with permission from E. C. Ashby, *Acc. Chem. Res.*, **1988,** *21,* 414.

In the displacement reactions on 2-octyl compounds (Eq. 12.44 for example) the extent of inversion depends both on the nucleophile and on the leaving group, as shown in Table 12.2.

For a given leaving group, the nucleophiles that are better reducing agents give the most racemization. For example, with Cl as the leaving group the extent of racemization follows the sequence $Me_3Sn^- > i\text{-}PrS^- > Me_2P^-$. For a given nucleophile, the R-X that is most easily reduced gives the most racemization, and the sequence is I > Br > Cl > OTs.

In the reduction of 5,5-dimethyl-6-X-1-hexenes by $LiAlH_4$ or $LiAlD_4$ both an open chain (5,5-dimethyl-1-hexene) and a cyclic (1,1,3-trimethylcyclopentane) product are formed and the proportions also vary with the leaving group X, as shown in Table 12.3. The formation of these products can be accounted for by Scheme III, but the interpretation is somewhat obscured by the fact that some of the cyclic hydrocarbon is formed by reduction of the cyclic iodide [51,57,58]. The results both of the experiments with optically active reagents

TABLE 12.3 Reduction of 5,5-Dimethyl-6-X-1-hexenes by $LiAlH_4$ in THF at RoomTemperature

	% Yields of RH (RD)	
R-X	Open	Cyclic
ROTs	1.8	0
RCl	0	0
RBr	9.8	7.5
RI	2.5	96
RI ($LiAlD_4$)	5.5	89
	(69% d_1)	(59% d_1)

Source: Reprinted with permission from E. C. Ashby, *Acc. Chem. Res.*, **1988,** *21,* 414.

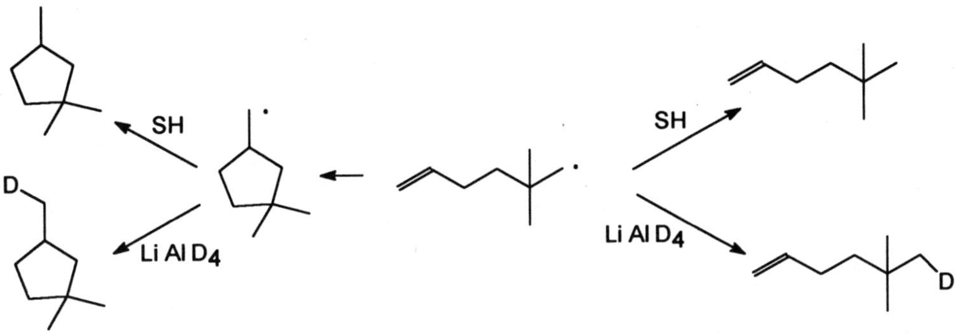

Scheme III

and those with cyclizable reagents strongly support varying amounts of participation by the $S_{RN}2$ mechanism.

12.5.4 Related Dehalogenation Reactions

Aryl halides are dehalogenated (Eq. 12.46) by a radical-initiated chain reaction between the aryl halide and an alkoxide ion having at least one α hydrogen atom [59].

$$ArX + R_2CH{-}O^- \longrightarrow ArH + R_2C{=}O + X^- \qquad (12.46)$$

Chains are initiated by typical radical sources: a radical abstracts a hydrogen atom from the alkoxide ion to give the anion radical. The other steps are like those shown in Scheme IV [59] for aryl iodides and methoxide ion.

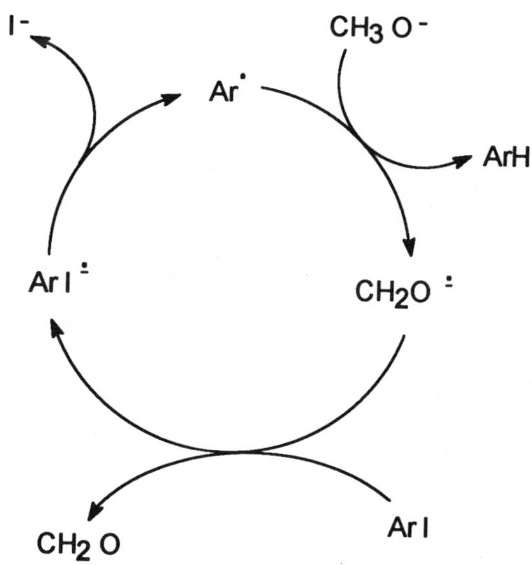

Scheme IV

For aryl iodides, insensitivity of the rate to the structure of the aryl group suggests that the electron added to ArI to give $ArI^{-}\cdot$ resides in the Ar-I σ^* antibonding orbital rather than in a π^* LUMO of the aromatic ring. For aryl bromides involvement of the π^* LUMO is more probable.

12.6 RADICAL PAIRS VERSUS ION PAIRS

Many dissociation reactions, both thermal and photochemical, ultimately give both radical and ion pairs and radical- and cation-derived products. Early examples of such "hybrid" thermal dissociations were found in the thermolysis of diacyl peroxides or peresters. Many of these are discussed in Chapter 7.

Singlet or triplet excited states can have similar capabilities, as we have seen from the example of the photolysis of α-naphthylmethyl esters, discussed in Chapter 10 [60,61].

A number of mechanisms for hybrid dissociation reactions are conceivable, differing in the point at which the nature of the product is decided. This might be at the reagent (Mechanism I), at a hybrid intermediate (Mechanism II), or at an electron transfer step in a radical or ion pair (Mechanisms III and IV). The problem of deciding among these possibilities is a difficult one, much like the problem of deciding between $S_{RN}2$ and S_N2 mechanisms.

Mechanism V, which is much less plausible than the others, has a single transition state supplying two channels, one leading to radicals and the other leading to ions. One of the postulates of transition state theory is that a

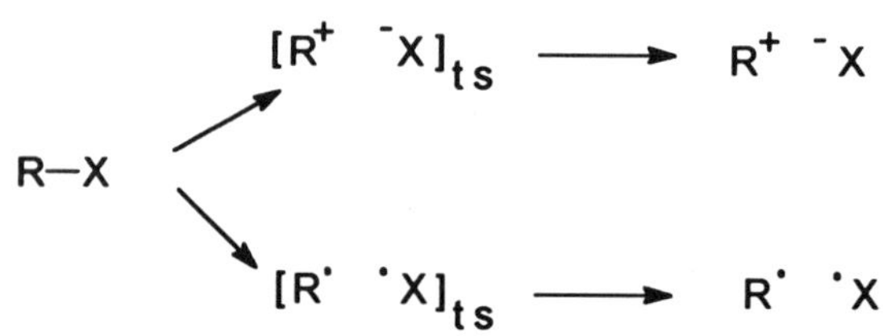

Mech. I

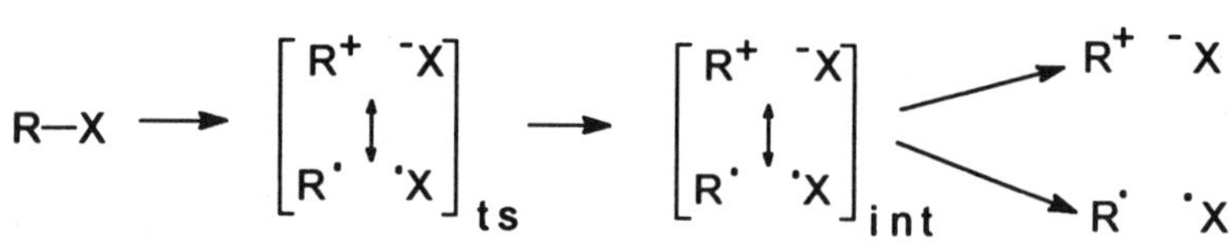

Mech. II

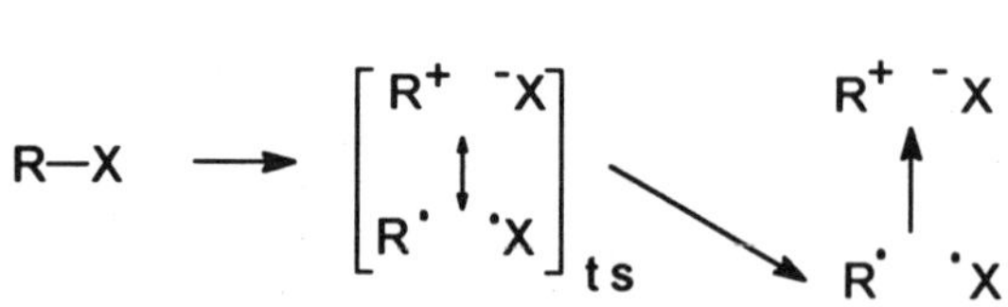

Mech. III

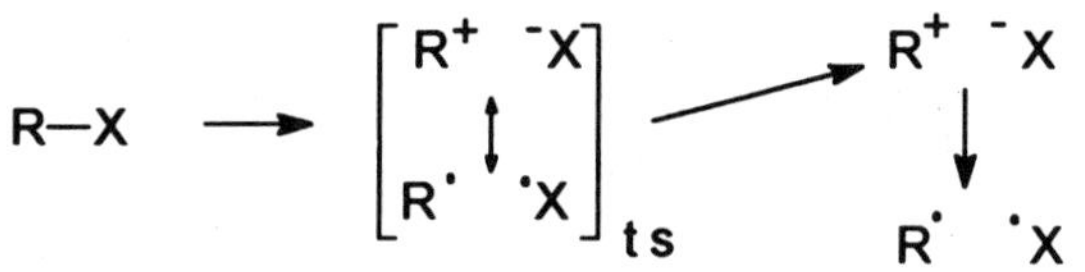

Mech. IV

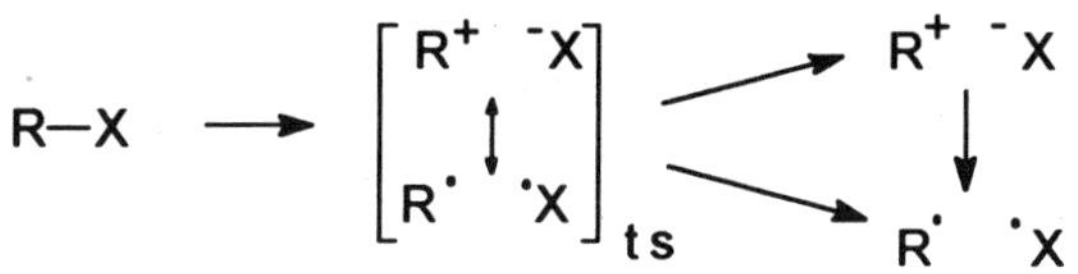

Mech. V

reacting system, once over the transition state, is rapidly deactivated all the way down to the first product or intermediate: it acts as though the system has momentum in the direction of the product. Given a choice of two directions, what would determine the choice of channel? If it is the conformation of the reagent or its solvation shell, we again have two transition states rather than one. For quantum mechanical uncertainty to be the deciding factor, the two channels would have to be nearly identical in the arrangement of the reagents and solvent molecules.

Mechanism II, which features a hybrid intermediate, was originally proposed for diacyl peroxide thermolysis [62,63] and later for the photolysis of alkyl halides [64]. Mechanism II was preferred over Mechanism I for diacyl peroxide thermolysis because polar products accompanied the radical products in a series of peroxides despite very large differences in the decomposition rates [62]. It was also noted that peroxides that could be expected to give radicals of comparable stability seemed to show widely different yields of polar products. A single common rate-determining step and multiple product-determining steps seemed to be an economical way of decoupling the rates from the products. Equation 12.47 is an elaboration of this idea to accommodate the need for loss and rebonding of CO_2.

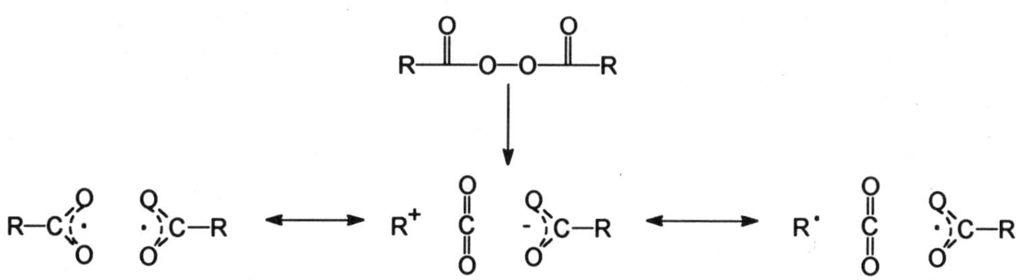

(12.47)

The mixing of the singlet radical pair wave function with that of the ion pair depends on a close approach of the covalent and ionic potential energy surfaces [65]. The relative energies of these surfaces depend on the differences in electronegativity of the two moieties, the dielectric constant, and the distance apart of the two radicals or of the two ionic charges.

In defense of mechanisms other than II, it should be noted that both the unimolecular part of peroxide thermolysis rates and initial product ratios are subject to considerable uncertainty. As discussed in Chapter 7, the observation of CIDNP in rearranged carbenium ion products showed that an electron transfer step like that in Mechanism III occurs in at least some diacyl peroxide thermolyses. A common intermediate for radical and ionic products was also postulated at one time for the decomposition of arenediazonium compounds in acidic methanol. However, the hybrid intermediate was abandoned as unnecessary after a reinvestigation showed that the rates of production of Ar–H (the radical product) and Ar–OCH_3 (the ionic product) were in fact quite different. The product and kinetic data are well accommodated by independent radical and ionic decomposition reactions [66].

12.6.1 Photolysis of Diacyl Peroxides

Photolysis of diacyl peroxides, also discussed in Chapter 7, seems to give only radicals. This result probably means either that the photolysis, which gives almost two moles of CO_2, does not give an intermediate R· ·OOCR pair at all, or gives a pair in which decarboxylation is much faster than electron transfer. Electron transfer may be slowed by one intervening molecule of CO_2.

12.6.2 Photolysis of Esters

The photolysis of 3-methoxybenzyl acetate in aqueous dioxane gives a mixture of 3-methoxybenzyl alcohol and radical products; 3,5-dimethoxybenzyl acetate gives only the alcohol; 4-methoxybenzyl acetate gives only radical-derived

TABLE 12.4[a] Products and Decarboxylation Rates in the Photolysis of α-Naphthylmethyl Esters of RCOOH

R	% Ether	% N-CH_2R	$k_{decarb} \times 10^{-9}$
CH_3	98	—	<1.3
CH_3CH_2	93	6	2.0
$(CH_3)_2CH$	80	6	6.5
$(CH_3)_3C$	71	9	11.0
$PhCH_2$	84	16	5.0
$PhCH_2CH_2$	92	7	2.3

[a]Rates estimated relative to that of a known acyloxy radical clock. Reprinted with permission from J. W. Hilborn and J. A. Pincock, *J. Am. Chem. Soc.*, **1991,** *113,* 2683. Copyright 1991 American Chemical Society.

products such as the corresponding bibenzyl [67]. The explanation advanced by Zimmerman for this substituent effect is that a methoxy substituent in the π,π^* first excited state of benzyl acetate releases electrons only to the meta position, in contrast to its behavior in the ground state. This explanation, if correct, favors Mechanism I in which benzyl cations ions and benzyl radicals are produced in independent concurrent steps.

In the photolysis of α-naphthylmethyl esters in methanol the effects of substituents suggest a mechanism like III with the yield of ion-derived product dependent on the rate of electron transfer. For α-naphthyl esters of acids RCOOH the proportions of ionic and radical products varies with the nature of R as shown in Eq. 12.48 and Table 12.4 [60]. As the rate of decarboxylation of RCOO· increases, electron transfer to form the carbenium ion pair is less able to compete and the yield of ether declines.

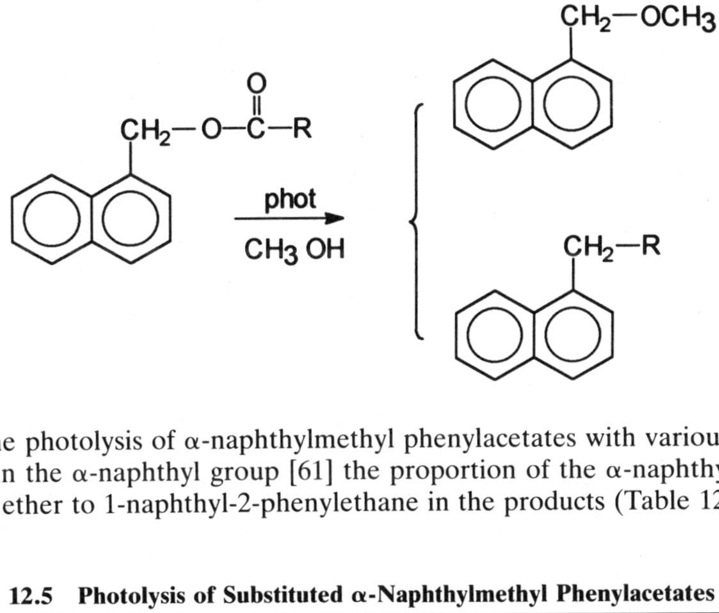

(12.48)

In the photolysis of α-naphthylmethyl phenylacetates with various substituents in the α-naphthyl group [61] the proportion of the α-naphthylmethyl methyl ether to 1-naphthyl-2-phenylethane in the products (Table 12.5) may

TABLE 12.5 Photolysis of Substituted α-Naphthylmethyl Phenylacetates

Substituent	% Ether	% N-CH_2Ph	E^0
None	84	16	0.47
4-CN	12	80	0.72
3-OCH_3	31	52	0.50
4-CH_3	93	7	0.35
4-OCH_3	74	24	0.04
4-OEt	80	22	0.04
4,8-$(OCH_3)_2$	47	51	−0.05
4,7-$(OCH_3)_2$	35	48	−0.06

Source: Reprinted with permission from D. P. DeCosta and J. A. Pincock, *J. Am. Chem. Soc.*, **1989,** *111,* 8948. Copyright 1989 American Chemical Society.

show a Marcus parabolic dependence on the oxidation potential of the α-naphthylmethyl radical. In any case it does not depend on the position of the methoxy groups in the same way as it did in the photolysis of benzyl acetates, since the yield of ether from the 4-methoxy compound is greater than that from the 3-methoxy compound. This suggests Mechanism III as the major process.

Photolysis of benzyl *p*-methylphenylacetate or *p*-methylbenzyl phenylacetate in MeOH gave only radical-derived products [68].

12.6.3 Photolysis of Halides

Photolysis of halomethylarenes was discussed in Chapter 10. In nonpolar solvents only radical products are produced, but photolysis in methanol gives both ion- and radical-derived products. The photolysis of 1-(iodomethyl)naphthalene appears to be a special case: iodine atoms complex with the RI starting material to give an $RI_2\cdot$ intermediate that is susceptible to a nucleophilic displacement by the methanol solvent [69].

The photolysis of aliphatic halides often shows evidence of carbenium ion intermediates, either by giving a product with a rearranged carbon skeleton or by giving a solvolysis product [70]. For example, photolysis of neat *n*-propyl iodide gives isopropyl iodide [71]. In methanol the photolysis of 1-iodonorbornane, a compound extremely resistant to solvolysis under thermal conditions, gives a mixture of norbornane and 1-methoxynorbornane (Eq. 12.49) [70].

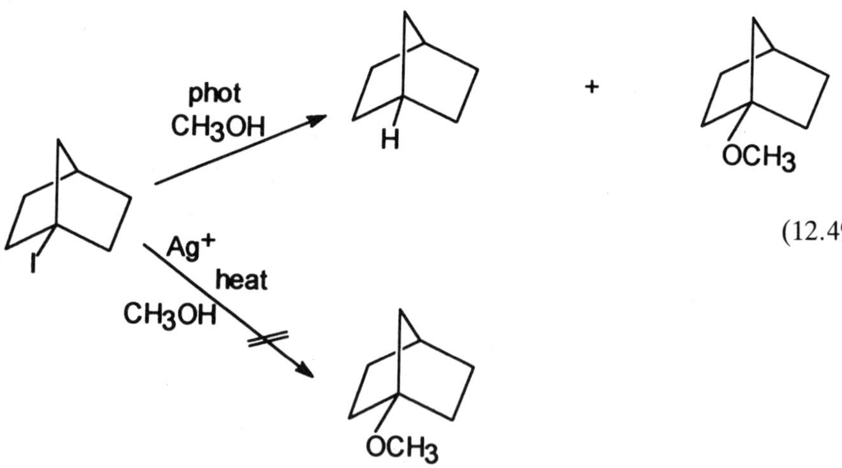

(12.49)

Running the photolysis of the norbornyl iodide in methanol gives 89% methyl ether and 11% norbornane. In the presence of oxygen as a radical trap, norbornane is not formed at all and the yield of methyl ether is lowered. In ethylene glycol, a much more viscous solvent, the yield of solvolysis product is increased to 99% and only a trace of norbornane is formed. Both of these observations support Mechanism III for the photolysis.

Using Mechanism I for photolysis reactions would most probably involve different excited states for the ionic and radical paths. If so, the product composition should vary with the wavelength of the light source. The product composition in reaction 12.48 was found to be wavelength independent [70].

Alkyl bromides on photolysis in methanol give a higher proportion of alkane and less of the alkyl methyl ether than do the corresponding iodides [72]. This has been attributed to a greater degree of escape of the bromine atom from the cage. The energies of the absorbed photon and the bond energies are such that the R··Br pair is formed with about 30 KJ per mole more energy than the R··I pair.

Photosolvolysis by Intramolecular Electron Transfer. Certain benzobicyclooctane derivatives can undergo photosolvolysis with loss of a nucleofugic group X [73–78]. The solvolysis is probably associated with electron transfer from a π,π^* orbital to the back lobe of the C–X σ^* orbital (Eq. 12.50).

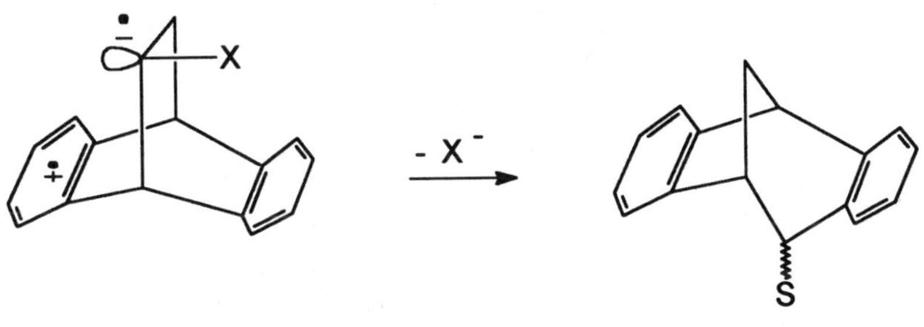

(12.50)

Methoxy substituents in the ring transferring the electron decrease the π,π^* fluorescence quantum yield and increase the solvolysis quantum yield.

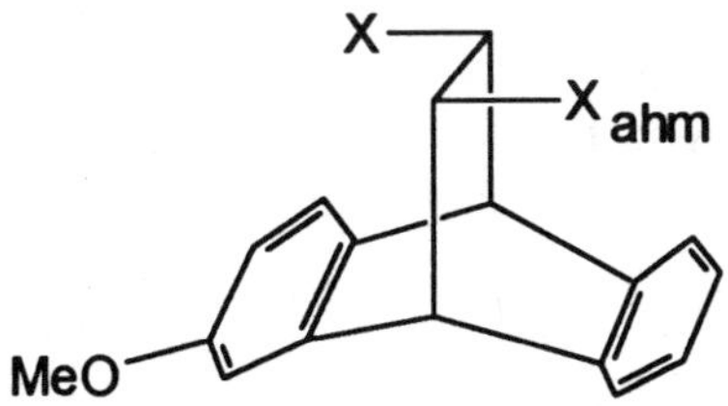

II

In the trans dichloro compound **II** [78] it is the anti-homo-meta chlorine, labeled X_{ahm}, that is solvolyzed. This result is another example of the Zimmerman "meta effect" in which an electron-donating substituent in the excited state increases electron density in the meta position but decreases it in the para position.

12.6.4 Other Photolysis Reactions

Fluorenol. Photolysis of 9-fluorenol in methanol or aqueous methanol gives a mixture of radical and ionic products via the excited singlet (Eq. 12.51) [79].

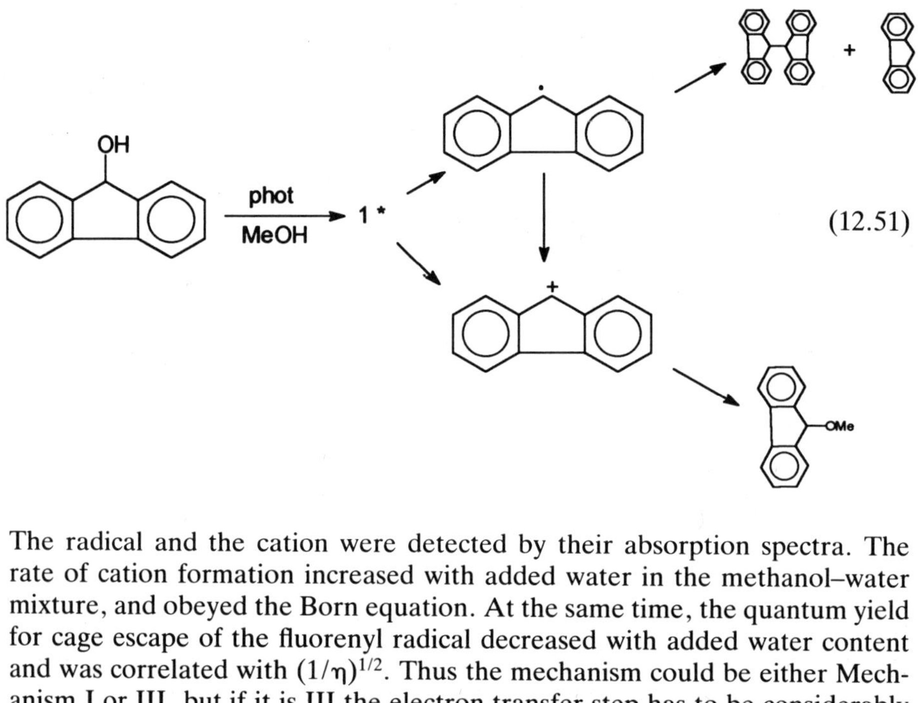

(12.51)

The radical and the cation were detected by their absorption spectra. The rate of cation formation increased with added water in the methanol–water mixture, and obeyed the Born equation. At the same time, the quantum yield for cage escape of the fluorenyl radical decreased with added water content and was correlated with $(1/\eta)^{1/2}$. Thus the mechanism could be either Mechanism I or III, but if it is III the electron transfer step has to be considerably faster than the decay of the excited singlet fluorenol.

Onium Salts. Photolysis of the sulfonium salt in Scheme V involves electron transfer from a π,π^* orbital of the anthracene chromophore to the σ^* orbital of the S–CH_2 bond [80]. As in the case of Eq. 12.50, this weakens the bond. Cleavage is probably concerted with the electron transfer. Rebonding of the cyanobenzyl radical at various anthracene ring positions gives five regioisomeric products. Electron transfer in the radical pair gives the sulfide and a cyanobenzyl cation which is trapped by the acetonitrile.

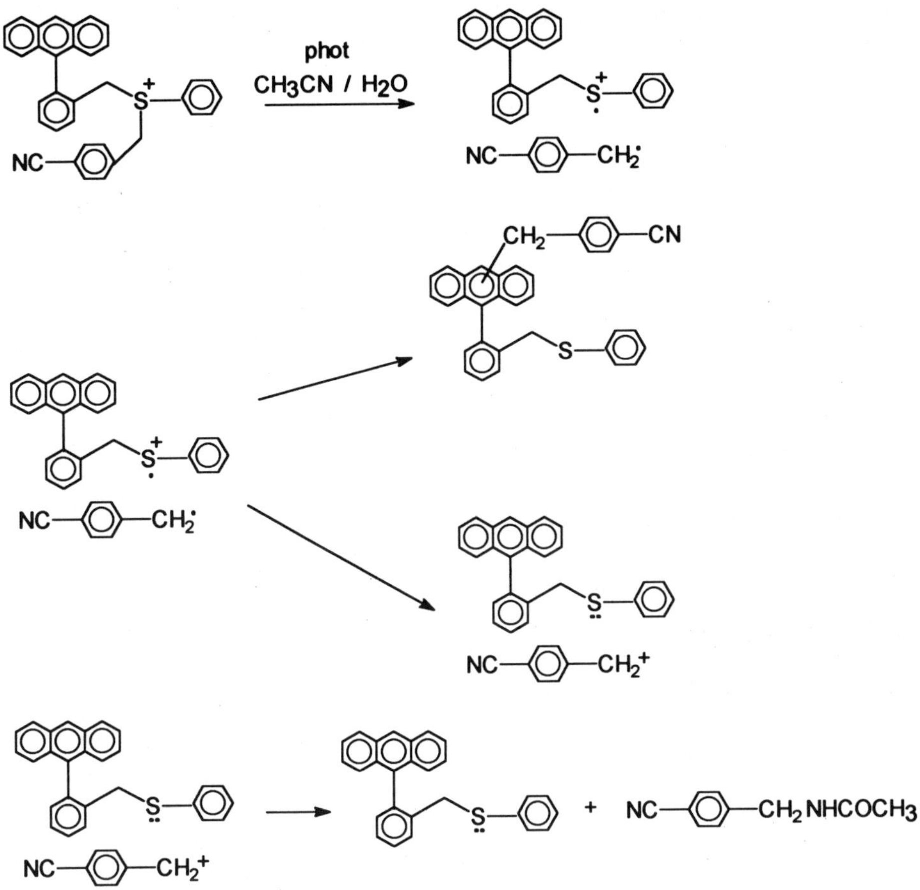

Scheme V

12.7 ZWITTERIONS VERSUS DIRADICALS

12.7.1 The Distinction

In terms of valence bond structures the difference between diradicals and zwitterions is fairly simple [65]. For orbitals ϕ_A and ϕ_B the diradical is represented as the singlet,

$$(2 + 2S^2_{AB})^{-1/2}\,[\phi_A(1)\phi_B(2) + \phi_B(1)\phi_A(2)],$$

or as the triplet,

$$(2 - 2S^2_{AB})^{-1/2}\,[\phi_A(1)\phi_B(2) - \phi_B(1)\phi_A(2)]$$

where S_{AB} is the overlap integral. There are also two zwitterionic states, both singlets, represented by

$$(2 + 2S^2_{AB})^{-1/2} [\phi_A(1)\phi_A(2) + \Phi_B(1)\phi_B(2)]$$

and

$$(2 - 2S^2_{AB})^{-1/2} [\phi_A(1)\phi_A(2) - \phi_B(1)\phi_B(2)]$$

As in the case of radical pairs, mixing of configurations can cause a diradical to have partial polar character, leading to reaction products resembling those from zwitterions.

12.7.2 Interconversion

Partial zwitterionic or diradical character, helped by changes in conformation or solvation, can lead to interconversion of the two forms. An example of an equilibrium between a stable zwitterion and a diradical is shown in Eq. 12.52 [81]. The diradicals or zwitterions were prepared by oxidation of the corresponding bisphenols.

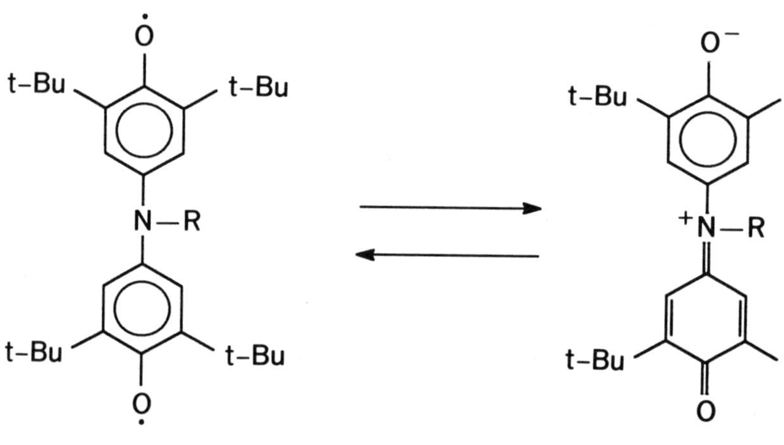

(12.52)

The rates of the conversion from the diradical to the zwitterion were obtained from NMR linewidths. For R = CH_3CO the rate of the diyl to zwitterion reaction in deuterochloroform at 240 K was 43 sec^{-1} and the rate in the other direction was 1.3×10^6.

12.7.3 Trimethylenemethanes

Trimethylenemethanes can be prepared by the thermolysis of azo compounds as illustrated by Eq. 12.53.

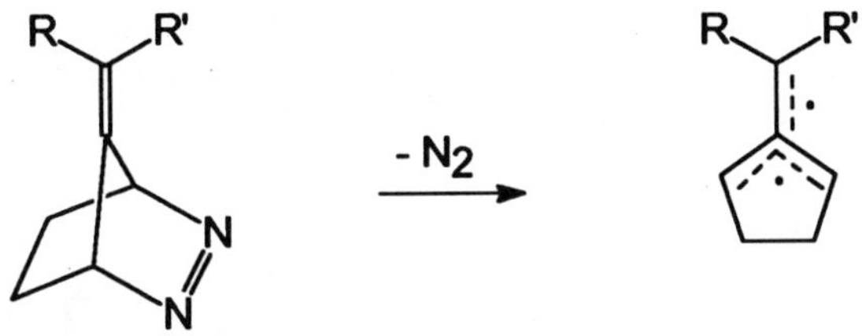

(12.53)

The diyls can exist in triplet spin states, for which there is evidence from ESR spectra, and in singlet states. The singlet states necessarily have a considerable dipolar character [82]. The most probable conformation, when R or R′ (of Eq. 12.53) is anything but hydrogen, is bisected (**IV**) rather than the planar conformation **III** [1].

III **IV**

Thermolysis of a diazene in water (Eq. 12.54) can give products that correspond to the addition of water to a zwitterion as well as products suggesting a diradical precursor. The latter were dimers and products resulting from the trapping of the diyl by the double bond in the side chain.

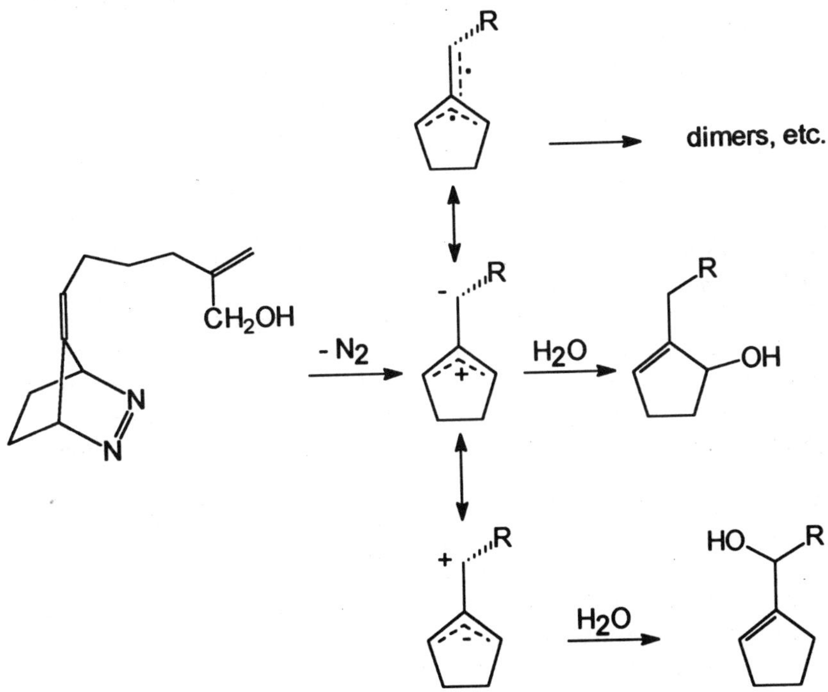

(12.54)

The acetal in Eq. 12.55 is in equilibrium with the trimethylenemethane [83]. In CD_3OD the zwitterion is trapped as the ether. In benzene oxygen traps the diradical as a cyclic peroxide.

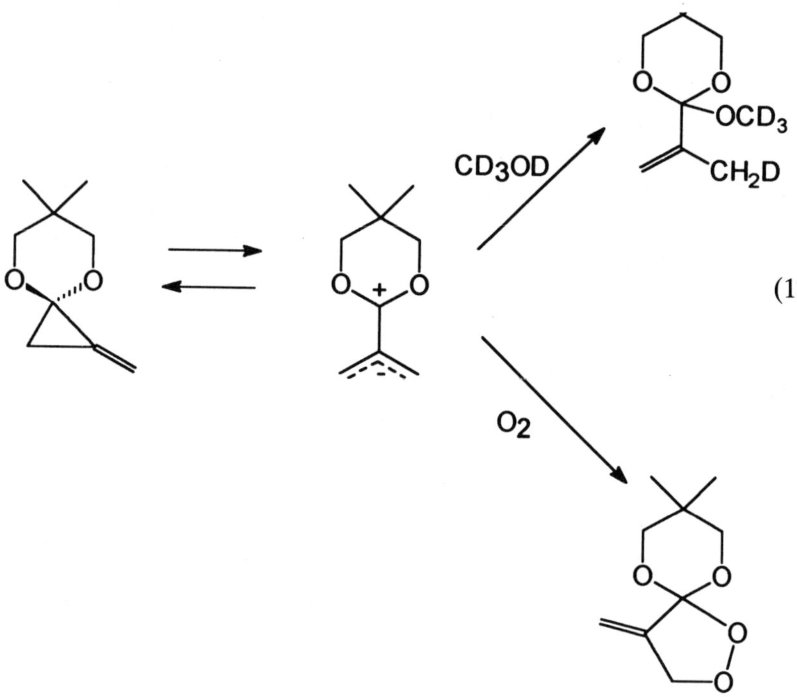

(12.55)

12.7.4 Tetramethylenes

The reaction of an alkene with another alkene can give both a [2 + 2] cycloadduct and polymer. Since the concerted mechanism is excluded by the Woodward–Hoffmann rules, the formation of the cycloadduct requires an intermediate. The intermediate can be either a tetramethylene diradical or a tetramethylene zwitterion [84–86]. The same intermediate is responsible for the polymerization, and the kind of polymer formed is diagnostic for the diradical or zwitterionic nature of the intermediate.

The polymerization in Scheme VI can be initiated either by adding the electrophilic olefin to the excess of *N*-vinylcarbazole or by adding the cyclobutane [84]. A highly colored electron donor–acceptor (EDA) complex is seen when the electrophilic olefin is used as the initiator, but not when the polymerization is initiated by the cyclobutane. The fact that the polymer is a homopolymer of the *N*-vinylcarbazole means that the process is cationic polymerization: the intermediate is the zwitterion rather than the diradical. The zwitterion can also be trapped by methanol.

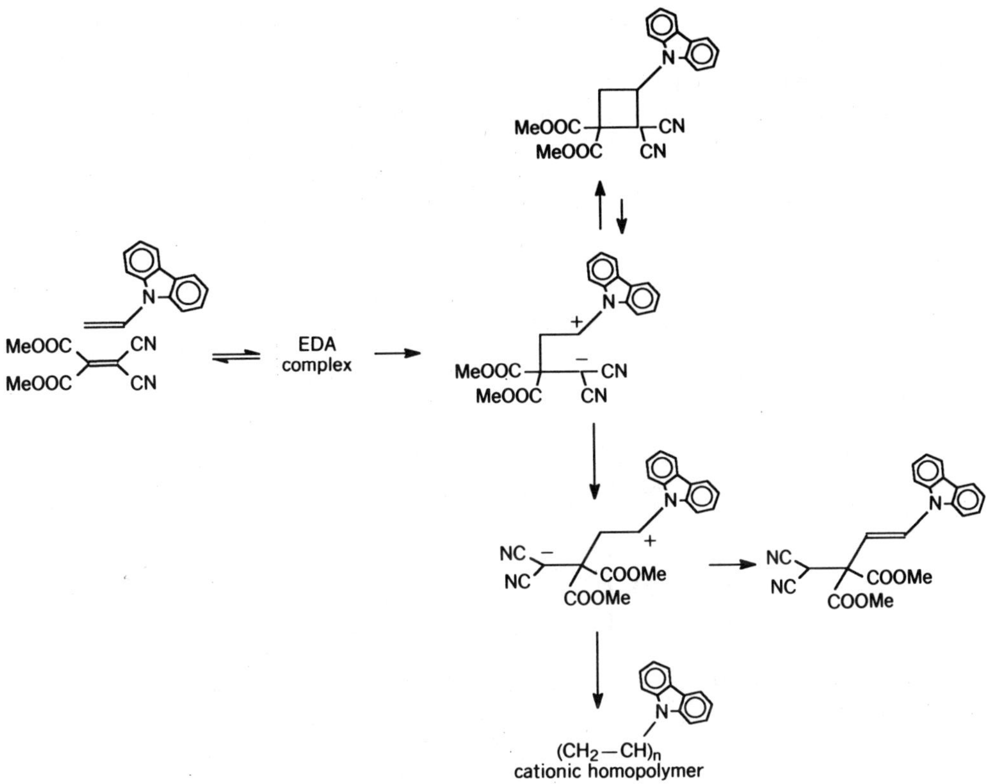

Scheme VI

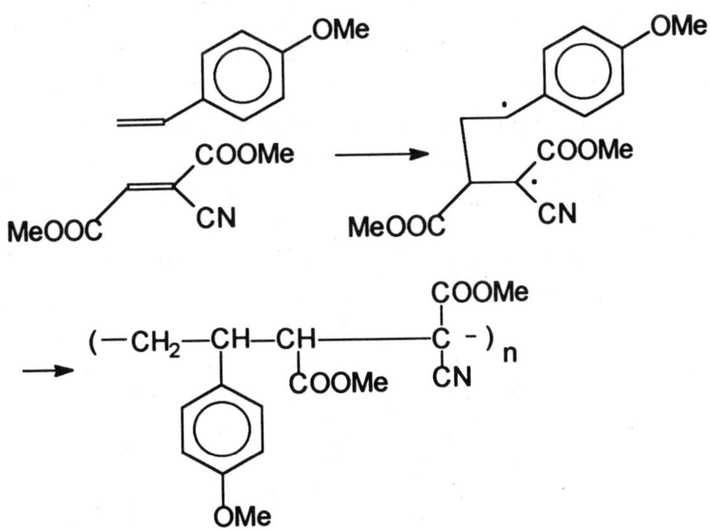

Scheme VII

When the substituents on the olefin are such as not to favor the zwitterion so strongly (Scheme VII), polymerization is initiated by the diradical, giving alternating copolymer instead of homopolymer.

Still other combinations of olefins give both types of polymerization, indicating the presence of both tetramethylene diradicals and zwitterions.

REFERENCES

1. Sutin, N. *Acc. Chem. Res.* **1982,** *15,* 275.
2. Closs, G. L.; Calcaterra, L. T.; Green, N. J.; Penfield, K. W.; Miller, J. R. *J. Phys. Chem.* **1986,** *90,* 3673.
3. Miller, J. R.; Beitz, J. V.; Huddleston, R. K. *J. Am. Chem. Soc.* **1984,** *106,* 5057.
4. Kowert, B. A.; Marcoux, L.; Bard, A. J. *J. Am. Chem. Soc.* **1972,** *94,* 5538.
5. Eberson, L. *Electron Transfer Reactions in Organic Chemistry,* Springer-Verlag: Berlin, 1987.
6. Pross, A. *Acc. Chem. Res.* **1985,** *18,* 212.
7. Leffler, J. E. *Science* **1953,** *117,* 340.
8. Rosenberg, S.; Silver, S. M.; Sayer, J. M.; Jencks, W. P. *J. Am. Chem. Soc.* **1974,** *96,* 7986.
9. Bordwell, F. G.; Boyle, W. J. J. *J. Am. Chem. Soc.* **1972,** *94,* 3907.
10. Pross, A.; Shaik, S. S. *J. Am. Chem. Soc.* **1982,** *104,* 1129.
11. Bernasconi, C. F. *Acc. Chem. Res.* **1992,** *25,* 9.
12. Marcus, R. A. *Faraday Disc.* **1964,** *29,* 21.
13. Marcus, R. A. *J. Phys. Chem.* **1968,** *72,* 891.
14. Ritchie, C. D. *Physical Organic Chemistry,* Marcel Dekker: New York, 1990.
15. Murdoch, J. R.; Magnoli, D. E. *J. Am. Chem. Soc.* **1982,** *104,* 3792.
16. Murdoch, J. R. *J. Am. Chem. Soc.* **1983,** *105,* 2159.
17. Magnoli, D. E.; Murdoch, J. R. *J. Am. Chem. Soc.* **1981,** *103,* 7465.
18. Rehm, D.; Weller, A. *Ber. Bunsenges. phys. Chem.* **1969,** *73,* 834.
19. Rehm, D.; Weller, A. *Israel J. Chem.* **1970,** *8,* 259.
20. Chen, J.-M.; Ho, T.-I.; Mou, C.-Y. *J. Phys. Chem.* **1990,** *94,* 2889.
21. Angel, S. A.; Peters, K. S. *J. Phys. Chem.* **1991,** *95,* 3606.
22. Zhou, J. S.; Rodgers, M. A. J. *J. Am. Chem. Soc.* **1991,** *113,* 7728.
23. Bockman, T. M.; Karpinski, Z. J.; Sankararaman, S.; Kochi, J. K. *J. Am. Chem. Soc.* **1992,** *114,* 1970.
24. Bixon, M.; Jortner, J. *J. Phys. Chem.* **1991,** *95,* 1941.
25. Saveant, J.-M. *Adv. Phys. Org. Chem.* **1990,** *26,* 1.
26. Liang, N.; Miller, J. R.; Closs, G. L. *J. Am. Chem. Soc.* **1990,** *112,* 5353.
27. Leffler, J. E.; Grunwald, E. *Rates and Equilibria of Organic Reactions,* Wiley: New York, 1963.
28. Gould, I. R.; Ege, D.; Moser, J. E.; Farid, S. *J. Am. Chem. Soc.* **1990,** *112,* 4290.

29. Mataga, N.; Asahi, T.; Kanda, Y.; Okada, T.; Kakitani, T. *Chem. Phys.* **1988,** *127,* 249.
30. Ohno, T.; Yoshimura, A.; Mataga, N. *J. Phys. Chem.* **1986,** *90,* 3295.
31. Ohno, T.; Yoshimura, A.; Mataga, N. *J. Phys. Chem.* **1990,** *94,* 4871.
32. Miller, J. R.; Calcaterra, L. T.; Closs, G. L. *J. Am. Chem. Soc.* **1984,** *106,* 3047.
33. Wasielewski, M. R.; Niemczyk, M. P.; Svec, W. A.; Pewitt, E. B. *J. Am. Chem. Soc.* **1985,** *107,* 1080.
34. McLendon, G.; Miller, J. R. *J. Am. Chem. Soc.* **1985,** *107,* 7811.
35. Irvine, M. P.; Harrison, R. J.; Beddard, G. S.; Leighton, P.; Sanders, J. K. M. *Chem. Phys.* **1986,** *104,* 315.
36. Verhoeven, J. W.; Paddon-Row, M. N.; Hush, N. S.; Oevering, H.; Heppener, M. *Pure & Appl. Chem.* **1986,** *58,* 1285.
37. Closs, G. L.; Johnson, M. D.; Miller, J. R.; Piotrowiak, P. *J. Am. Chem. Soc.* **1989,** *111,* 3751.
38. Bowman, W. R.; Symons, M. C. R. *J. Chem. Soc. Perkin II* **1983,** 25.
39. Russell, G. A.; Danen, W. C. *J. Am. Chem. Soc.* **1966,** *88,* 5663.
40. Kornblum, N.; Michel, R. E.; Kerber, R. C. *J. Am. Chem. Soc.* **1966,** *88,* 5662.
41. Bunnett, J. F. *Acc. Chem. Res.* **1978,** *11,* 413.
42. Kornblum, N. *Angew. Chem. Int. Ed.* **1975,** *14,* 734.
43. Knighton, W. P.; Grimsrud, E. P. *J. Am. Chem. Soc.* **1992,** *114,* 2336.
44. Lewis, E. S. *J. Am. Chem. Soc.* **1989,** *111,* 7576.
45. Pross, A. *Adv. Phys. Org. Chem.* **1985,** *21,* 99.
46. Shaik, S. S. *Prog. Phys. Org. Chem.* **1985,** *15,* 197.
47. Bank, S.; Noyd, D. A. *J. Am. Chem. Soc.* **1973,** *95,* 8203.
48. Eliel, E. L.; Haber, R. G. *J. Am. Chem. Soc.* **1959,** *81,* 1249.
49. Bordwell, F. G.; Wilson, C. A. *J. Am. Chem. Soc.* **1987,** *109,* 5470.
50. Leffler, J. E. *J. Am. Chem. Soc.* **1953,** *75,* 3598.
51. Ashby, E. C. *Acc. Chem. Res.* **1988,** *21,* 414.
52. Ashby, E. C.; Goel, A. B.; DePriest, R. N. *J. Am. Chem. Soc.* **1980,** *102,* 7779.
53. Russell, G. A.; Metcalfe, A. R. *J. Am. Chem. Soc.* **1979,** *101,* 2359.
54. Ward, H. R.; Lawler, R. G.; Cooper, R. A. *J. Am. Chem. Soc.* **1969,** *91,* 746.
55. Flesia, E.; Crozet, M. P.; Surzur, J.-M.; Jauffred, R.; Ghiglione, C. *Tetrahedron* **1978,** *34* 1699.
56. Bowry, V. W.; Lusztyk, J.; Ingold, K. U. *J. Am. Chem. Soc.* **1991,** *113,* 5687.
57. Park, S.-U.; Chung, S.-K.; Newcomb, M. *J. Org. Chem.***1987,** *52,* 3275.
58. Curran, D. P.; Kim, D. *Tetrahedron Lett.* **1986,** *27,* 5821.
59. Bunnett, J. F. *Acc. Chem. Res.* **1992,** *25,* 2.
60. Hilborn, J. W.; Pincock, J. A. *J. Am. Chem. Soc.* **1991,** *113,* 2683.
61. DeCosta, D. P.; Pincock, J. A. *J. Am. Chem. Soc.* **1989,** *111,* 8948.
62. Walling, C.; Waits, H. P.; Milovanovic, J.; Pappiaonnou, C. G. *J. Am. Chem. Soc.* **1970,** *92,* 4927.
63. Leffler, J. E.; More, A. A. *J. Am. Chem. Soc.* **1972,** *94,* 2483.
64. Cristol, S. J.; Stull, D. P.; Daussin, R. D. *J. Am. Chem. Soc.* **1978,** *100,* 6674.

65. Salem, L.; Rowland, C. *Angew. Chem. Int. Ed.* **1972,** *11,* 92.
66. Bunnett, J. F.; Yijima, C. *J. Org. Chem.* **1977,** *42,* 639.
67. Zimmerman, H. E.; Sandel, V. R. *J. Am. Chem. Soc.* **1963,** *85,* 915.
68. Meiggs, T. O.; Grossweiner, L. I.; Miller, S. I. *J. Am. Chem. Soc.* **1972,** *94,* 7986.
69. Slocum, G. H.; Kaufmann, K.; Schuster, G. B. *J. Am. Chem. Soc.* **1981,** *103,* 4625.
70. Kropp, P. J. *Acc. Chem. Res.* **1984,** *17,* 131.
71. McCauley, C. E.; Hamill, W. H.; Williams, R. R. J. *J. Am. Chem. Soc.* **1954,** *76,* 6263.
72. Kropp, P. J.; Adkins, R. L. *J. Am. Chem. Soc.* **1991,** *113,* 2709.
73. Cristol, S. J.; Bindel, T.H. *J. Org. Chem.* **1980,** *45,* 951.
74. Cristol, S. J.; Optiz, R. J.; Bindel, T. H.; Dickenson, W. A. *J. Am. Chem. Soc.* **1980,** *102,* 7977.
75. Cristol, S. J.; Ali, M. Z. *Tetrahedron Lett.* **1983,** *24,* 5839.
76. Cristol, S. J.; Seapy, D. G.; Aeling, E. O. *J. Am. Chem. Soc.* **1983,** *105,* 7337.
77. Cristol, S. J.; Aeling, E. O.; Strickler, S. J.; Ito, R. D. *J. Am. Chem. Soc.* **1987,** *109,* 7101.
78. Cristol, S. J.; Aeling, E. O.; Heng, R. *J. Am. Chem. Soc.* **1987,** *109,* 830.
79. Gaillard, E.; Fox, M. A.; Wan, P. *J. Am. Chem. Soc.* **1989,** *111,* 2180.
80. Saeva, F. D.; Breslin, D. T.; Luss, H. R. *J. Am. Chem. Soc.* **1991,** *113,* 5333.
81. Espersen, W. G.; Kreilick, R. *J. Am. Chem. Soc.* **1970,** *92,* 3894.
82. Masjedizadeh, M. R.; Fite, C.; Little, R. D. *Tetrahedron Lett.* **1990,** *31,* 1229.
83. Nakamura, E.; Yamago, S.; Ejiri, S.; Dorigo, A. E.; Morokuma, K. *J. Am. Chem. Soc.* **1991,** *113,* 3183.
84. Hall, H. K., Jr.; Padias, A. B. *Acc. Chem. Res.* **1990,** *23,* 3.
85. Huisgen, R. *Acc. Chem. Res.* **1977,** *10,* 199.
86. Bartlett, P. D. *Q. Rev., Chem. Soc.* **1970,** *24,* 473.

Index